IET CONTROL, ROBOTICS AND SENSORS SERIES 125

Short-Range Micro-Motion Sensing with Radar Technology

IET International book series on sensing – call for authors

The use of sensors has increased dramatically in all industries. They are fundamental in a wide range of applications from communication to monitoring, remote operation, process control, precision and safety, and robotics and automation. These developments have brought new challenges such as demands for robustness and reliability in networks, security in the communications interface, and close management of energy consumption. This Book Series covers the research and applications of sensor technologies in the fields of ICTs, security, tracking, detection, monitoring, control and automation, robotics, machine learning, smart technologies, production and manufacturing, photonics, environment, energy, and transport.

Proposals for coherently integrated International multi-authored edited or co-authored handbooks and research monographs will be considered for this Book Series. Each proposal will be reviewed by the IET Book Series Editorial Board members with additional external reviews from independent reviewers. Please email your book proposal to: vmoliere@theiet.org or author_support@theiet.org.

Short-Range Micro-Motion Sensing with Radar Technology

Edited by
Changzhan Gu and Jaime Lien

The Institution of Engineering and Technology

Published by The Institution of Engineering and Technology, London, United Kingdom

The Institution of Engineering and Technology is registered as a Charity in England & Wales (no. 211014) and Scotland (no. SC038698).

First published 2019

The Institution of Engineering and Technology
Michael Faraday House
Six Hills Way, Stevenage
Herts, SG1 2AY, United Kingdom

www.theiet.org

British Library Cataloguing in Publication Data
A catalogue record for this product is available from the British Library

ISBN 978-1-78561-760-7 (hardback)
ISBN 978-1-78561-761-4 (PDF)

Typeset in India by MPS Limited
Printed in the UK by CPI Group (UK) Ltd, Croydon

Contents

Chapter 1

Introduction

Changzhan Gu[1] *and Jaime Lien*[2]

In the past 10 years, we are in the so-called "mobile era" where iOS, Android, and other smart devices have become the mobile computing platform that is deeply integrated into our lives. In recent years, owing to the advancement of computing, it is believed that the trend is shifting to a new era where artificial intelligence (AI) will become ubiquitous and unlock things that were unthinkable before [1–4]. In the coming AI era where computing is more universally available and accessible, human–computer interaction (HCI) would need to be much more natural and intuitive, and above all more intelligent [5].

Since the debut of the modern computer, various techniques have been invented or developed as input technologies for our human beings to interact or control the functionality of a computing device. For decades, keyboard and mouse have probably been the most commonly used input devices. In gaming consoles like Sony's PlayStation or Microsoft's Xbox, a remote controller might be hand-held to perform hand gestures. For smartphones that we use frequently on daily basis, the interaction happens by physically touching the screen. As technology advances, new input devices and approaches provide additional freedom and flexibility. A good example may be the wireless mouse that is free of the wire constrain. It captures hand actions like button clicks and movement positions, which is then wirelessly transmitted to the receiving computing device. The wireless nature gives the user more freedom with mobility and placement of the interacting device. Nevertheless, all the above-mentioned input devices require that the users must physically hold, touch, click, or move to enter inputs to the computing devices. What if the input devices get lost or misplaced? In that case, the users won't be able to enter inputs with that mechanism. The ubiquitous computing calls for additional freedom by removing the need for a physical device as the input mechanism. In the past few years, owing largely to the advancement of cloud computing, big data, and improved accuracy of speech recognition, digital voice assistants such as Apple's Siri, Google Assistant, and Amazon's Alexa are coming from fiction to our daily lives transforming the way we interact with the physical world around us. Devices integrated with voice assistant, e.g. Google Home or Amazon Eco, are believed to

[1]Department of Electronic Engineering, Shanghai Jiao Tong University, China
[2]Google LLC, USA

be the gateway that would control the vast array of "Internet of things" smart devices such as smart refrigerator, thermostats, light bulb, air conditioner, indoor/outdoor camera, TV, and game console.

No doubt that voice control is a natural and more advanced way to interact with the computing devices. Gesture control, on the other hand, is also very common. We human beings probably have been using hand gestures to interact with the physical world for thousands of years. It is deeply integrated in our lives. For example, while having a conversation at home, we might say "can you pull the curtains it's really dark" while pointing at a window, or "can you turn up the volume" while pointing at a speaker; while answering people for directions, we might say "go straight a mile and then turn right" while pointing to that direction and performing the right turn gesture. Studies show that to interact with computing systems, people prefer a combination of speech and gesture over speech or gesture alone [6–8]. The multimodal interface combining both gesture and speech allows a more natural, intuitive, and intelligent way for HCI. The user may control the computing devices with their own choice of interaction that is the most natural and efficient way.

We humans have been using our hands as the natural tool to perform action gestures to interact with the physical world. The hand gesture is a simple and readily available mechanism for the user to input commands to the computing device. The small muscle groups in the twist and fingers can result in intuitive interaction by performing highly precise and controlled motions that allow for fluid, effective, and rapid manipulation. To capture the fine movement of the hand and fingers in the free space has not been an easy task. Attaching a movement sensing device to the hand doesn't remove the dependency on a peripheral device so it is not desirable. Alternatively, in the field of computer science, computer vision techniques have been developed to capture and identify hand gestures [8,9]. However, they still suffer from latency and technical constraints such as resolving the overlapped fingers and requiring instrumentation of the hands [10]. Moreover, it is not power-efficient and computing resource-efficient to process the frame-to-frame images to extract motion information, which is not desirable for wearable and mobile devices [10]. Moreover, the computer vision is often times sensitive to light and cannot capture motions behind obstacles like clothing, wood, and plastics, which limits the implementation of this sensing approach. In recent years, the advancement of other optical techniques including multi-camera [11], structured light [12,13], and time of flight [14] allows them to be integrated into consumer hardware devices such as gaming console [15] and smartphone [16]. These new optical techniques provide depth information that can greatly improve precision. However, they have their inherent constraints. For example, they may not scale well to the small form factor or extra low power required by wearable and mobile devices. It should be noted that there is no input technology that is universally optimal for all computing devices and use cases. The choice of input technologies is not only dependent on the specific computing device but also always application driven, which involves difficult trade-offs between size, power consumption, sensing precision, sensor system bill of material cost, ease of integration to the computing device, sensitivity to light and environmental noise, impact as aggressor to neighboring sensors, and many other considerations.

Due to the technical advancements in semiconductor, signal processing, and system integration, radar has emerged as a promising technique for short-range motion sensing—detecting the object's relative displacement motions with very high accuracy. It is an alternative solution for noncontact gesture sensing as it transmits radio frequency signals to the target, and the target's movement can be extracted by analyzing the backscattering reflections [17–19]. It is a promising and attractive form of HCI for emerging applications in wearable, mobile, and ubiquitous computing, where the traditional touch interaction is not desired or difficult if not impossible. During the past several years, breakthroughs have been made in this area by using radar to wirelessly and precisely detect hand gestures [20–25]. A good example is Google's Project Soli that uses millimeter wave 60 GHz radar sensor to capture micro gestures for HCI [26]. Soli is more than a new radar sensor chip or a new signal processing algorithm. Instead, it is a complete end-to-end radar sensing system specifically designed for tracking and recognizing fine hand gestures, as illustrated in Figure 1.1. The custom designed miniature millimeter wave radar chip, along with the developed digital signal processing techniques, machine learning algorithms, user experience (UX) and user interface (UI), and software integration, makes it possible for ubiquitous and intuitive gesture sensing [10]. Detecting the hand's action gestures is a subset and a specific application of radar motion sensing. This book will cover the broad design phases of radar motion sensing, which includes radar sensor hardware, digital signal processing, and machine learning. The book aims to not only give the general audiences an introduction to short-range radar motion sensing but also provide researchers and professionals a glimpse of the latest advancements in the area.

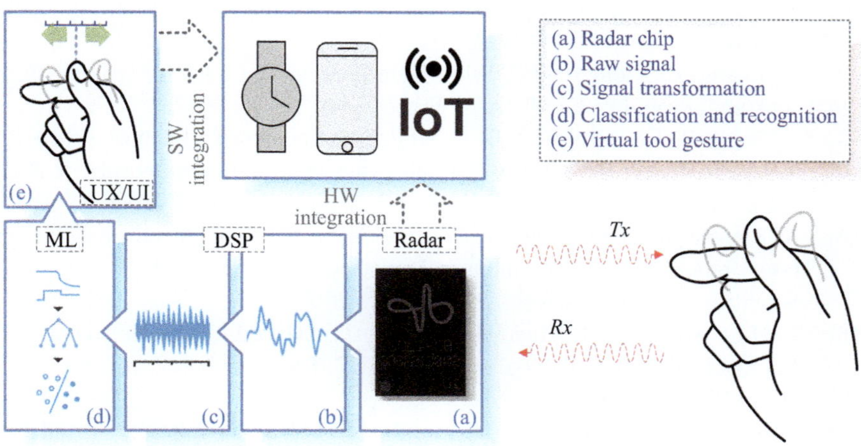

Figure 1.1 Google Soli: a complete end-to-end radar sensing system specifically designed for ubiquitous and intuitive gesture sensing [10]. ML – machine learning; SDP – digital signal processing; SW – software; HW – hardware

References

[1] D.M. West, and J.R. Allen "How artificial intelligence is transforming the world," Available at: https://www.merantix.com/news-posts/how-artificial-intelligence-is-transforming-the-world/ (accessed on April 24, 2019).

[2] L. Dormehl, Thinking Machines: The Quest for Artificial Intelligence—and Where It's Taking Us Next (New York: Penguin–TarcherPerigee, 2017).

[3] R. Rothe, "Applying deep learning to real-world problems," Medium, May 23, 2017.

[4] M. Purdy, and P. Daugherty, "Why artificial intelligence is the future of growth," Accenture, 2016.

[5] https://www.blog.google/products/assistant/personal-google-just-you/

[6] A.G. Hauptmann, "Speech and gestures for graphic image manipulation," ACM SIGCHI, Bulletin, vol. 20, pp. 241–245, 1989.

[7] J. Liu, and M. Kavakli, "A survey of speech-hand gesture recognition for the development of multimodal interfaces in computer games," ICME, pp. 1564–1569, 2010.

[8] S.S. Rautaray, and A. Agrawal, "Vision-based hand gesture recognition for human computer interaction: a survey," Artificial Intelligence Review, vol. 43, no. 1, pp. 1–54, 2015.

[9] T.B. Moeslund, A. Hilton, and V. Kruger, "A survey of advances in vision-based human motion capture and analysis," Computer Vision and Image Understanding, vol. 104, no. 2, pp. 90–126, 2006.

[10] J. Lien, N. Gillian, M.E. Karagozler, *et al.*, "Soli: Ubiquitous gesture sensing with millimeter wave radar," ACM Trans. Graph., vol. 35, no. 4, 2016, Art. no. 142.

[11] H. Lu, Y. Zhang, Y. Li, *et al.*, "Depth map reconstruction for underwater Kinect camera using inpainting and local image mode filtering," IEEE Access, vol. 5, pp. 7115–7122, 2017.

[12] Y. Zhang, Z. Xiong, and F. Wu, "Hybrid structured light for scalable depth sensing," Proc. Int. Conf. Image Process, pp. 17–20, 2012.

[13] Z. Song, R. Chung, and X.-T. Zhang, "An accurate and robust strip-edge-based structured light means for shiny surface micromeasurement in 3-D," IEEE Trans. Ind. Electron., vol. 60, no. 3, pp. 1023–1032, 2013.

[14] A. Kilpelä, R. Pennala, and J. Kostamovaara, "Precise pulsed time-of-flight laser range finder for industrial distance measurements," Review of Scientific Instruments, vol. 72, no. 2197, 2001.

[15] J. H.L. Shao, D. Xu, and J. Shotton, "Enhanced computer vision with Microsoft Kinect sensor: a review," IEEE Trans. Cybern., vol. 43, no. 5, pp. 1318–1334, 2013.

[16] https://www.extremetech.com/mobile/255771-apple-iphone-x-truedepth-camera-works (accessed on December 30, 2018).

[17] C. Li, and J. Lin, Microwave Motion Sensing and Analysis, Hoboken, NJ: John Wiley & Sons, ISBN: 978-0470642146, 2013.

[18] J. Lin, C. Li, C.-C. Chang, T.-H. Tsai, D. Zito, and S.-F. Chang, "Editors' choice—Review—Semiconductor integrated radar for sensing applications," ECS Journal of Solid State Science and Technology, vol. 7, no. 7, Q3126–Q3142, 2018.

[19] M.A. Richards, J.A. Scheer, and W.A. Holm, Principles of Modern Radar: Basic Principles, 2nd ed. (Edison, NJ: SciTech Publishing, 2010).

[20] Y. Kim, B. Toomajian, "Hand gesture recognition using micro-Doppler signatures with convolutional neural network," IEEE Access vol. 4 pp. 7125–7130, 2016.

[21] J. Song, S. Wang, J. Lien, I. Poupyrev, and O. Hilliges, "Interacting with Soli: exploring fine-grained dynamic gesture recognition in the radio-frequency spectrum," Proceedings of the 29th Annual ACM Symposium on User Interface Software and Technology UIST '16, 2016.

[22] T. Fan, C. Ma, Z. Gu, *et al.*, "Wireless hand gesture recognition based on continuous-wave Doppler radar sensors," IEEE Trans. Microw. Theory Tech., vol. 64, no. 11, pp. 4012–4020, 2016.

[23] F.K. Wang, M.C. Tang, Y.C. Chiu, *et al.*, "Gesture sensing using retransmitted wireless communication signals based on Doppler radar technology," IEEE Trans. Microw. Theory Tech., vol. 63, no. 12, pp. 4592–4602, 2015.

[24] C. Gu, and J. Lien, "A two-tone radar sensor for concurrent detection of absolute distance and relative movement for gesture sensing," IEEE Sensors Lett., vol. 1, no. 3, 2017.

[25] C. Li, Z. Peng, T.Y. Huang, *et al.*, "A review on recent progress of portable short-range noncontact microwave radar systems," IEEE Trans. Microw. Theory Techn., vol. 65, no. 5, pp. 1692–1706, 2017.

[26] https://atap.google.com/soli/ (accessed on December 30, 2018).

Chapter 2

Proximity RF/microwave biosensor techniques for vital sign detection

Hyunjae Lee[1], Byung-Hyun Kim[1], Sang-Gyu Kim[2], Yunseog Hong[3], Gi-Ho Yun[4], Hee-Jo Lee[5], Yong-Jun An[6], and Jong-Gwan Yook[1]

2.1 Introduction

With the growing interest in personal healthcare, many researchers have contributed to the development of a biosensor system concerning heart rate, respiration, blood pressure, glucose, and so on [1–3]. In particular, since tracking of the heart rate provides valuable information on arterial occlusions, arrhythmia, and autonomous nervous system pathologies, it can be used for the prevention or diagnosis of cardiovascular diseases, which accounts for one-third of the deaths worldwide. Typically, the well-known sensor systems are photoplethysmogram (PPG) and electrocardiogram (ECG) sensors [4–10]. However, as these schemes should be contacted on a human body, it is uncomfortable to monitor vital signs for a prolonged period. In view of the limitation or problem, the radio frequency (RF) and microwave sensor systems have become alternative method due to the several merits, such as noncontact, nondestructive, and nonobtrusive schemes.

There are mainly two methods in the vital sign RF/microwave sensor system. The first is the Doppler radar system, which is the phase shift detection caused by physiological movements including breathing and heartbeat [11–17]. Although Doppler radar has high sensitivity and long detection range, it has difficulty in detecting the vital signs when there are multiple subjects and motion artifacts. To resolve these problems, dual-radar systems, phased array antenna, and frequency-modulated continuous-wave radar have been proposed. Nevertheless, it is obvious that motion artifacts and multiple subjects are critical issues in radar-type vital sign sensor systems. Another method is the near-field proximity vital sign sensor. The

[1]Department of Electrical and Electronic Engineering, Yonsei University, South Korea
[2]Future Device R&D, LG Electronics, South Korea
[3]Radar and PGM Center, Hanwha Thales, South Korea
[4]Information and Communications Engineering Department, Sungkyul University, South Korea
[5]Department of Physics Education, Daegu University, South Korea
[6]Samsung Electronics, South Korea

variation of near field due to periodically vibrating body alters the oscillation frequency of the oscillator when the sensor is near the body such as a wrist and a chest. The main idea of the proximity vital sign sensor is to utilize the oscillation frequency deviation owing to the input impedance variation. Since proximity coupling RF sensors detect vital signs at a very close distance to a chest or a wrist, motion artifacts and multiple subject issues are insignificant problems. However, the frequency deviation of the oscillator caused by the physiological signal is very small owing to the tiny movement of the body, especially the heartbeat. For the solutions to the small signal deviation, various methods have been developed to enhance the sensitivity of the vital sign sensor.

This chapter will describe the methods of a proximity coupling RF vital sensor. The methods are mainly divided into two categories: the chest wall sensor and the wrist pulse sensor. For each part, the theoretical considerations, simulation results, and experimental results on methods for efficiently enhancing the sensitivity of the sensors are described in detail. In addition, we will introduce an autoregressive (AR) model which is one of the parametric spectral analyses suitable for vital sign detection [18].

2.2 Chest wall sensor

It is needless to mention that the heart rate can be detected most effectively is near a chest where the heart is located. Critical physiological signals have been detected in various ways by utilizing RF circuits, which enhance the sensitivity of the sensor system. Each detailed operation principle will be described in the following.

2.2.1 SAW filter system

Figure 2.1 illustrates the configuration of the vital sign sensor system based on surface acoustic wave (SAW) filter. It consists of a two-port planar resonator, SAW filter, envelope detector, an operational amplifier, and a digital signal processing (DSP) stage. As mentioned earlier in this chapter, the planar resonator detects the variation of the near field generated from the periodically vibrating human body part. The changes in the input impedance of the planar resonator provide the vital sign information. This operation can be mathematically expressed using simplified circuit

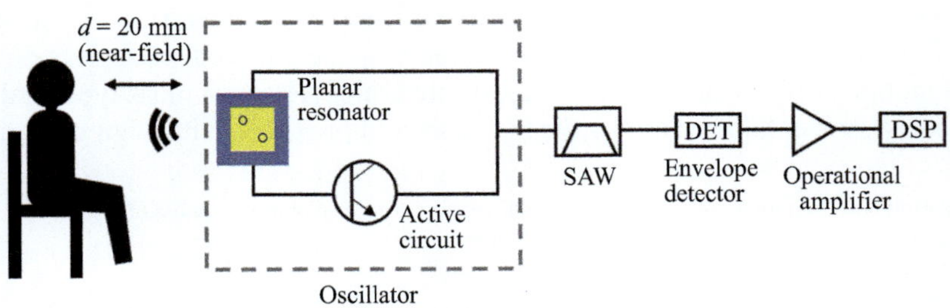

Figure 2.1 Block diagram of a vital sign sensor using a SAW filter

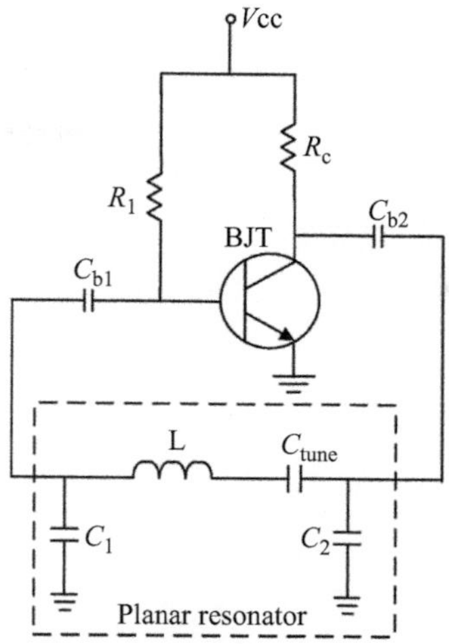

Figure 2.2 Schematic of the oscillator with a planar resonator. © 2012 IEEE, Reprinted with permission from [19]

schematic. Figure 2.2 shows the schematic of the oscillator with planar resonator and bipolar junction transistor (BJT). The oscillation frequency can be written as

$$\omega_0 = \sqrt{\frac{1}{C_T \cdot L}} \tag{2.1}$$

where C_T is the parallel combination of C_1, C_2, and C_{tune}. Then the changes of oscillation frequency can be written as

$$\Delta\omega = -\frac{\Delta C_T \cdot \omega_0}{2(\Delta C_T + C_T)} \tag{2.2}$$

In (2.2), the oscillation frequency is dependent on the overall capacitance and inductance of the resonator. When the sensor approaches the chest, the change in near field induces a change in the capacitance. Therefore, the deviation of the oscillation frequency is a function of the distance between the human body and planar resonator. However, since the movement of the heartbeat is only around 1–2 mm, the detection capability is not sufficiently enough to the medical application due to a tiny frequency deviation with conventional circuit topologies. To improve the sensitivity of the sensor, SAW filter is used at the output of the oscillator to amplify the small frequency shift. Figure 2.3(b) is the output spectrum of the typical SAW filter. The frequency characteristics are rapidly changed at the skirt region of the SAW filter. Therefore, if the range of frequency deviation of the oscillator is in the skirt range, small frequency change caused by physiological signal can be transformed into a large magnitude variation at the output voltage.

For the verification on the performance of the sensor system, the planar resonator with a simplified human body model is simulated in the three-dimensional

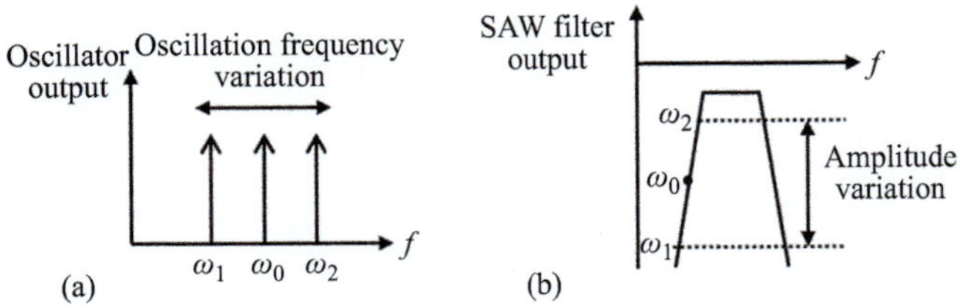

Figure 2.3 Output spectrums. (a) Oscillator. (b) SAW filter

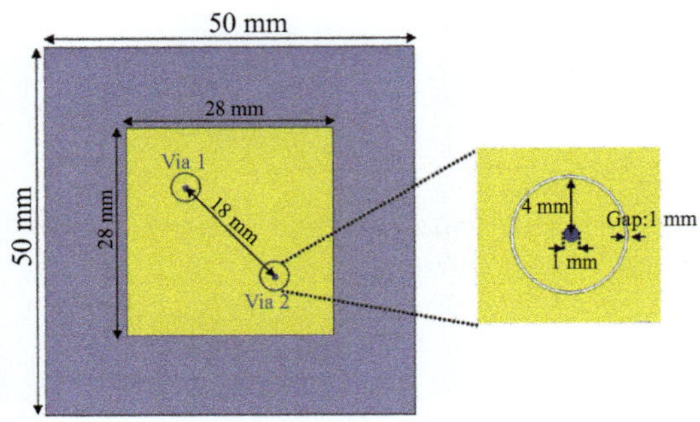

Figure 2.4 Dimensions and shape of the two-port planar resonator. © 2012 IEEE, Reprinted with permission from [19]

(3D) full-wave electromagnetic solver. Figure 2.4 shows the configuration of a two-port planar resonator. The size of the resonator is 28 mm × 28 mm and overall size including ground is 50 mm × 50 mm. Two ports are separated by 18 mm. The size of the human body is modeled as a cuboid of 30 cm × 30 cm × 30 cm. The permittivity and the conductivity of the human body model are 52.8 and 1.7 S/m, respectively [20]. The distance between the oscillator and the human body model changes from 1 to 60 mm. Figure 2.5 shows the simulated and measured deviation of the oscillation frequency as a function of the distance between the resonator and the human body model. It verifies that the input impedance of the resonator is dependent on the distance. In experimental results, when the distance between the resonator and the human body is in the range of 7 to 20 mm, the input impedance of the resonator shows sensitivity of 600 kHz/mm. When the distance is larger than 20 mm, the interaction between the oscillator and the human body becomes quite small. Thus, the maximum detection range of the resonator for the use of vital sign detection is estimated about 20 mm.

A small frequency deviation of a few hundred kilohertzs is challenging to detect at 2.4 GHz. Therefore, to detect the minute physiological signal, a small frequency change is transformed into large amplitude variation at the output voltage using a SAW filter. The performance of the SAW filter is presented in Figure 2.6. Note that the expected frequency deviation of the oscillator due to the

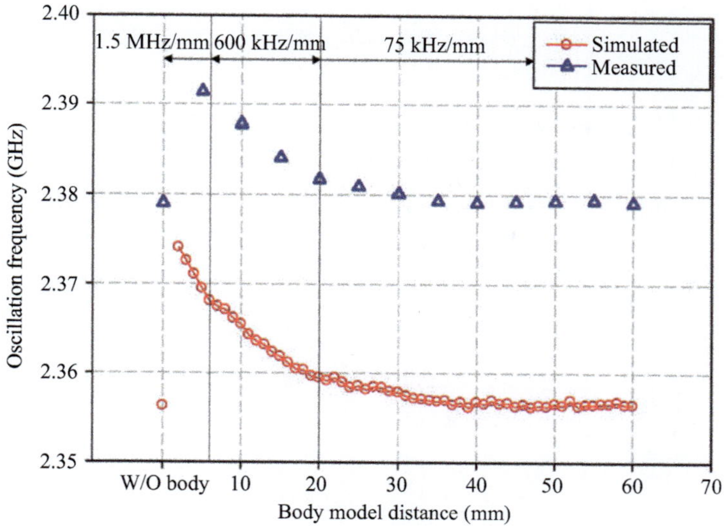

Figure 2.5 Oscillation frequency deviation in relation to the distance between the resonator and the human body model. © 2012 IEEE, Reprinted with permission from [19]

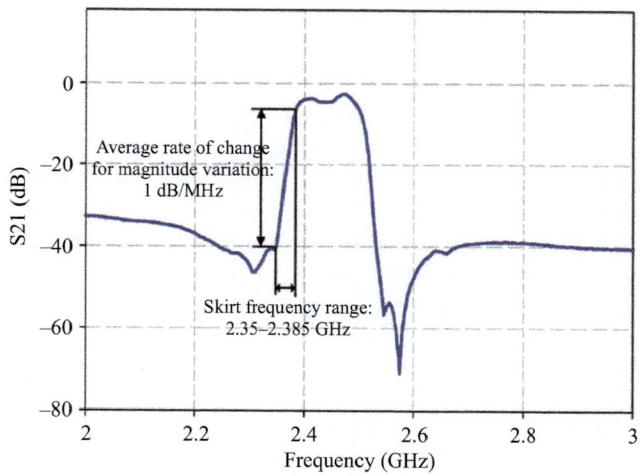

Figure 2.6 The performance of the SAW filter. © 2012 IEEE, Reprinted with permission from [19]

physiological signal is in skirt range of SAW filter. The average rate of change for magnitude variation of 0.6 dB/mm is obtained by the SAW filter within 20 mm distance. The output signal from the SAW filter can be converted to a DC voltage variation through an envelope detector.

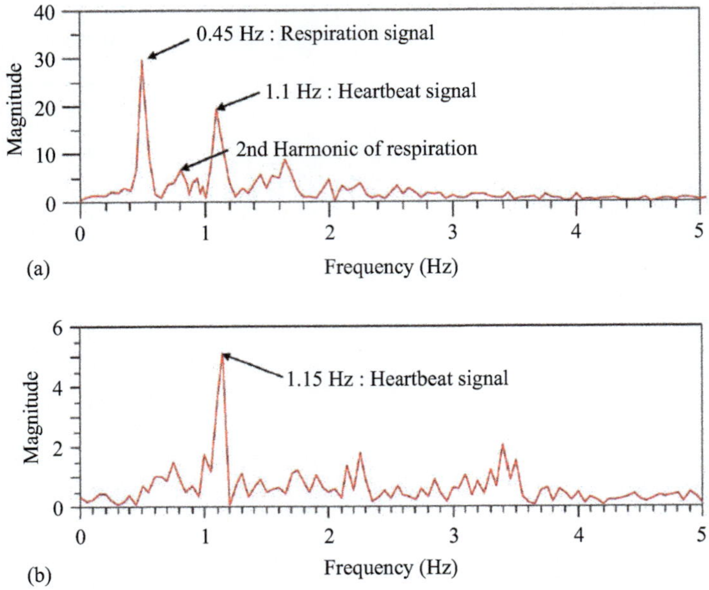

Figure 2.7 Measurement results. (a) Vital sign sensor with SAW filter. (b) Reference finger sensor. © 2012 IEEE, Reprinted with permission from [19]

By analyzing the signal passed through the operational amplifier, the heartbeat signal can be obtained. To validate the performance of the vital sign sensor using the SAW filter, the finger piezoelectric sensor is used. Figure 2.7(a) and (b) shows the measurement results of the vital sign sensor with SAW filter and reference finger sensor. In the result from the vital sign sensor using SAW filter, although the respiration signal has a larger magnitude than heartbeat signal, the heartbeat signal is clearly detected and it revealed excellent agreement with the result of the reference sensor.

2.2.2 PLL system

Although the vital sensor system using a SAW filter is simple and shows good performance, it has a few shortcomings. Since the frequency skirt range of the SAW filter is narrow, it is vulnerable to the frequency drift of the oscillator due to factors such as the aging of components, temperature variation, and so on. For the development of the stable system, the proximity vital sensor is coupled with a phase-locked loop (PLL) synthesizer. Figure 2.8 shows a block diagram of the sensing system based on a PLL synthesizer. The sensor consists of a voltage-controlled oscillator (VCO) with a planar-type circular resonator, a PLL, an operational amplifier, and a DSP stage. The interaction between a planar-type circular resonator and human body movement at the near-field range can be modeled

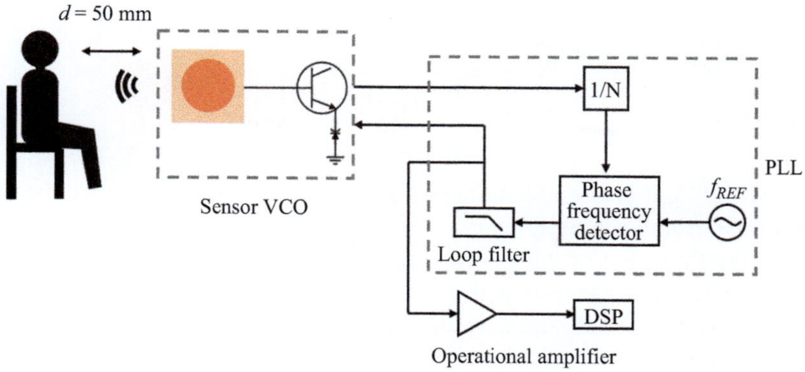

Figure 2.8 The architecture of the vital sign sensor using PLL

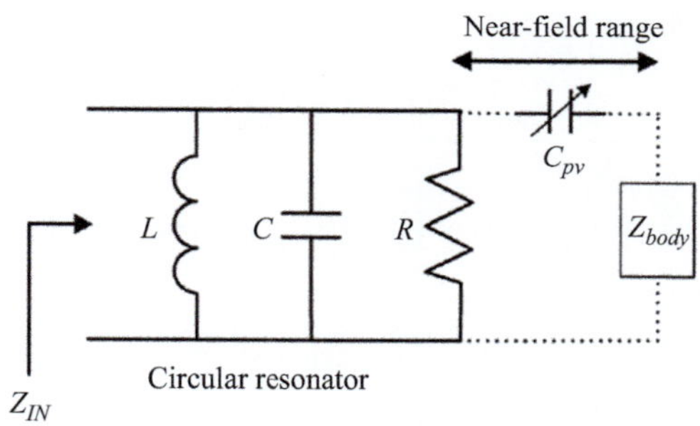

Figure 2.9 Circuit model of planar-type circular resonator associated with body proximity effect. © 2014 IEEE, Reprinted with permission from [21]

as a simplified circuit as illustrated in Figure 2.9. The input impedance of the resonator with the human body can be written as [22]:

$$|Z_{IN}(j\omega)| = \frac{R\|Z_b'}{\sqrt{1+\left[\omega_0\left(R\|Z_b'\right)\left(C\|C_{pv}'\left(\frac{\omega}{\omega_0}-\frac{\omega_0}{\omega}\right)\right)\right]}} \tag{2.3}$$

$$\phi(j\omega) = -\tan\left[\omega_0\left(R\|Z_b'\right)\left(C\|C_{pv}'\left(\frac{\omega}{\omega_0}-\frac{\omega_0}{\omega}\right)\right)\right] \tag{2.4}$$

$$Z_b' = \frac{1+\left(\omega C_{pv}Z_{\text{body}}\right)^2}{\left(\omega C_{pv}\right)^2 Z_{\text{body}}},\ C_{pv}' = \frac{C_{pv}}{1+\left(\omega C_{pv}Z_{\text{body}}\right)^2} \tag{2.5}$$

The change of the oscillation frequency is a significant factor in the sensor system. This can be represented by the stability factor, defined as [23]

$$S_F = \frac{\Delta\phi}{(\Delta\omega/\omega_0)} = \omega_0 \frac{d\phi}{d\omega}\bigg|_{\omega=\omega_0} = -2Q_L \tag{2.6}$$

$$\Delta\omega = \frac{\omega_0 \cdot \Delta\phi}{S_F} = \frac{-\Delta\phi \cdot \omega_0}{2Q_L} \tag{2.7}$$

In view of (2.7), it is desirable to design a low-Q_L resonator to improve the sensitivity of the overall sensor system.

The configuration of the designed resonator is presented in Figure 2.10. The planar-type circular resonator with a radius of 22.5 mm is printed on a substrate with a thickness of 0.787 mm, relative permittivity of 2.2, and loss tangent of 0.0009. The overall size of the resonator is 55 mm × 55 mm, and vertical interconnection via connecting to the active circuit is placed 12.5 mm away from the center of the resonator to obtain an appropriate Q_L value. The simulated and measured results for the characteristics of the designed resonator is shown in Figure 2.11. The results are well matched, and the measured Q_L is 25 with a 3 dB bandwidth of 102 MHz.

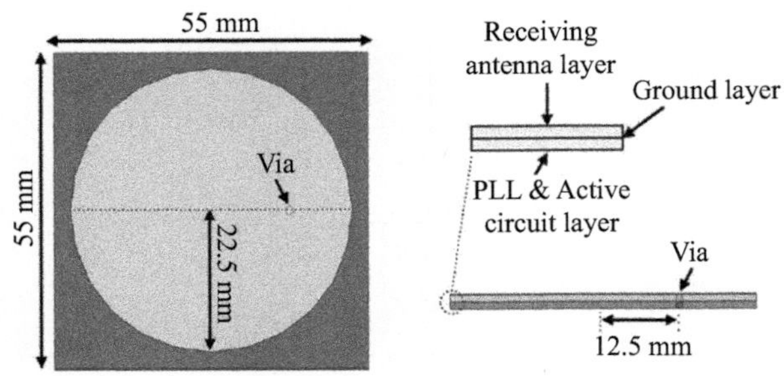

Figure 2.10 The layout of the planar-type circular resonator. © 2014 IEEE, Reprinted with permission from [21]

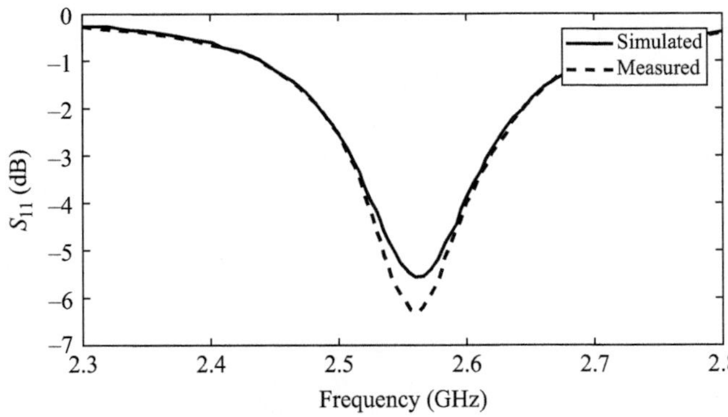

Figure 2.11 Simulated and measured S-parameter of the planar-type circular resonator

The planar-type circular resonator combined with a tunable negative resistance VCO. The VCO adopts series feedback topology, and the resonator is coupled to the base terminal of the BJT. Figures 2.12 and 2.13 represent the schematic of the VCO and the simulated and measured characteristics of the VCO circuit. The measured average gain of the VCO is 8.8 MHz/V (6–15 MHz/V) throughout the entire oscillation frequency range of 80 MHz.

When the human body approaches the planar-type circular resonator with VCO, the input impedance of the resonator is affected. For the analysis of the effect of the human body, simplified human body model is used. It contains fabric, skin, fat, and muscle layers and is located in front of the resonator. The human body model is simplified by a cuboid, with size of 300 mm × 300 mm × 35 mm. The detailed parameters of the human body model are listed in Table 2.1. The distance between resonator and human body model changes from 2 to 52 mm. Figure 2.14

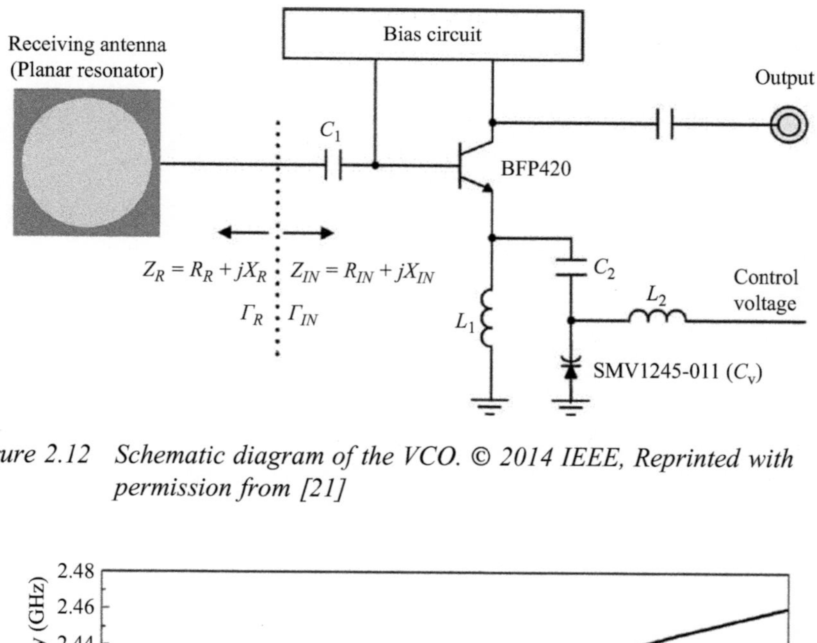

Figure 2.12 Schematic diagram of the VCO. © 2014 IEEE, Reprinted with permission from [21]

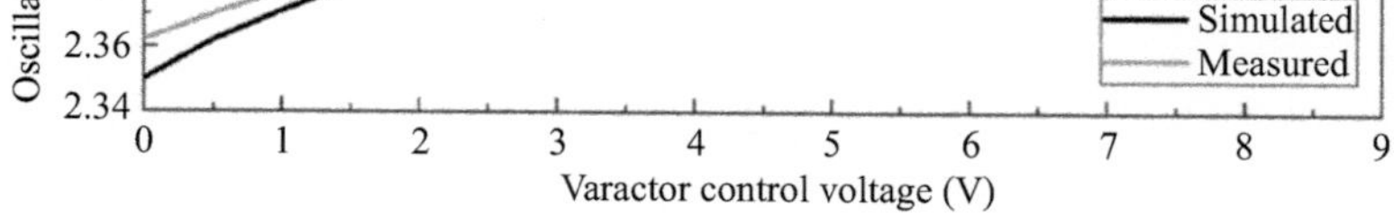

Figure 2.13 Simulated and measured results of the characteristics of VCO. © 2014 IEEE, Reprinted with permission from [21]

Table 2.1 Parameters of the simplified body model. © 2014 IEEE, Reprinted with permission from [21]

Tissues	Relative permittivity	Conductivity (S/m)	Thickness (mm)	Loss tangent
Fabric	1.52	—	3.20	0.012
Skin	38.0	1.46	1.30	—
Fat	10.8	0.27	10.5	—
Muscle	52.7	1.74	20.0	—

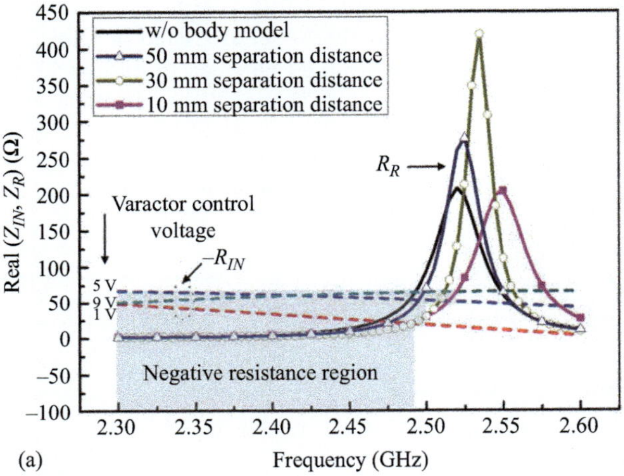

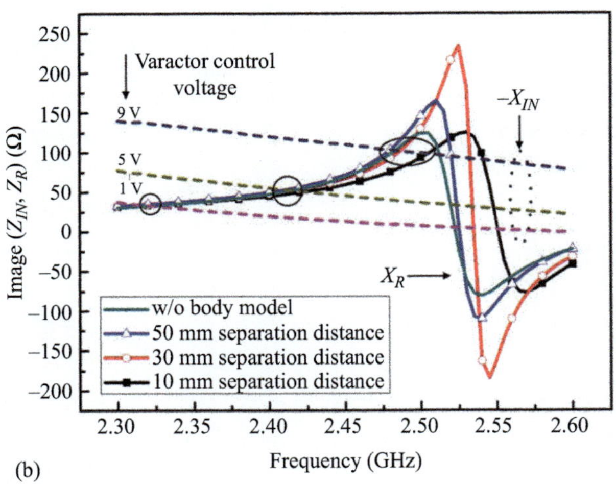

Figure 2.14 Simulated and measured results of the characteristics of VCO. © 2014 IEEE, Reprinted with permission from [21]

shows the simulation results. For the normal operation, the oscillation conditions must be satisfied for start-up [24]:

$$R_R + R_{IN} < 0 \tag{2.8}$$

$$X_R - X_{IN} = 0 \tag{2.9}$$

where R_{IN} and X_{IN} are, respectively, real and imaginary part of input impedance defined when we look into the base terminal of the BJT, while R_R and X_R are, respectively, the real and imaginary part of the input impedance of the resonator. The region of the negative resistance ($R_R + R_{IN} < 0$) is retained up to 2.52 GHz. Taking into account the given average gain of VCO, the amount of frequency deviation can be transformed into a voltage signal as a minimum value of 6.8 mV/mm and a maximum of 205 mV/mm. Now, the PLL is incorporated with the resonator and VCO for the stable sensor system. The PLL synthesizer operates as a frequency to time-domain converter, i.e. a frequency deviation detector.

For the validation of the sensor system with the PLL synthesizer, the piezoelectric finger sensor is used as a reference. A subject is seated at a distance away from the sensor fixed on a tripod. Figure 2.15 shows the experimental results for various distance and measurement position. The sensor system with PLL synthesizer shows excellent agreement with the reference sensor for all cases, front and back, up to 50 mm away from the subject.

2.2.3 Reflectometry system

This section introduces a vital sign sensor based on the reflection coefficient of an antenna. When the human body is placed in the near-field region of an antenna, the magnitude of the reflection coefficient of the antenna changes. Since vital sign can be detected by the changes in the magnitude of the reflection coefficient from an antenna, the sensor system can be designed as a parasite component in a conventional transceiver system. In other words, this system does not require additional signal source and antenna, so that it can be easily realized in RF Complementary metal–oxide–semiconductor (CMOS) Integrated Circuit (IC). It can be integrated into handheld communication systems such as smartphones, tablet PCs, etc.

Figure 2.16 shows a simplified configuration of a conventional wireless communication RF front end system. It consists of a power amplifier, a directional coupler, duplexer, and an antenna. The directional coupler is utilized to detect the transmitted power. The detector is located at the isolation port of the directional coupler to detect the reflection signal from the antenna. By comparing the reflected power and the transmitted power, it is possible to analyze the information contained in the reflected signal from the antenna and can be used to optimize the antenna performance. For this purpose, the directional coupler should have high directivity characteristics. Likewise, vital signs can be directly detected at the isolation port of the coupler, as shown in Figure 2.17, by detecting the magnitude of the reflected signal.

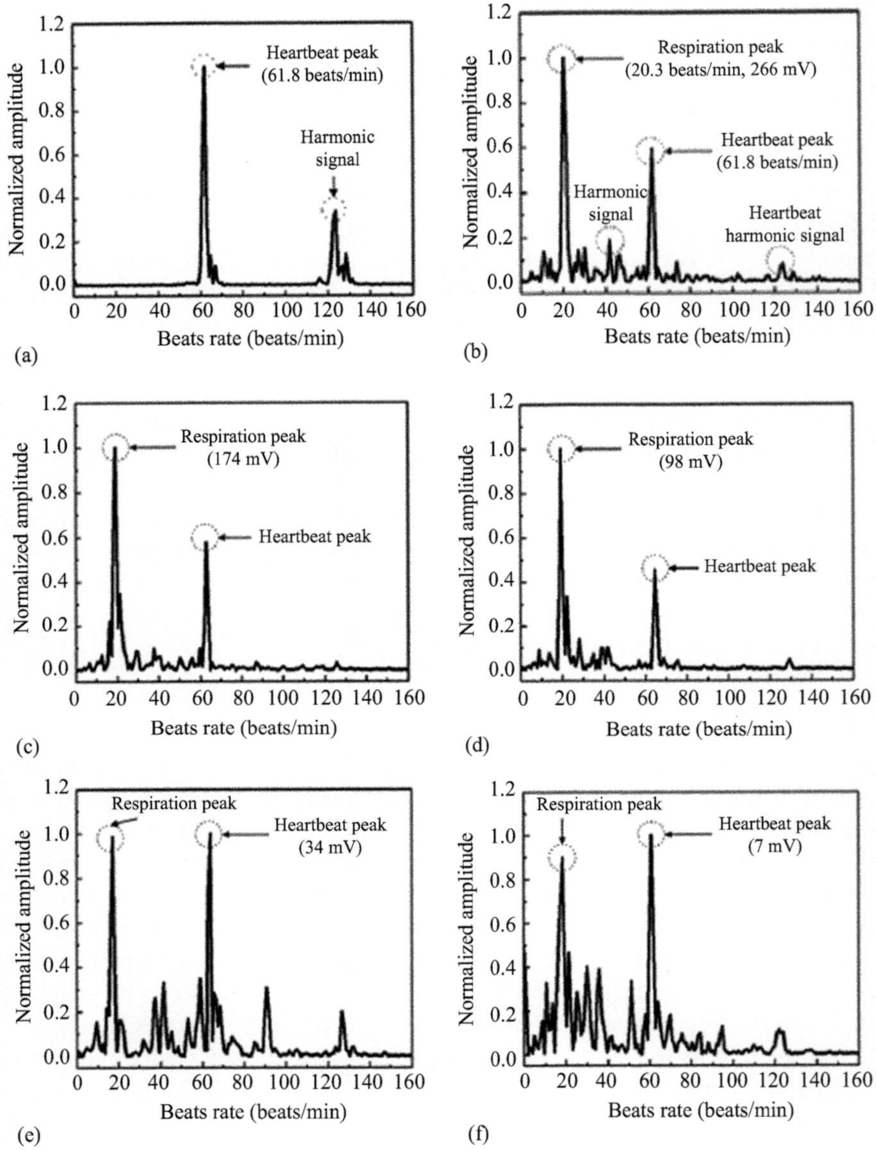

Figure 2.15 *Respiration and heart rates at the front and dorsal sides of a 30-year-old subject. (a) Reference signal. (b) At 10 mm from the front side. (c) At 20 mm from the front side. (d) At 50 mm from the front side. (e) At 30 mm from the dorsal side. (f) At 50 mm from the dorsal side. © 2014 IEEE, Reprinted with permission from [21]*

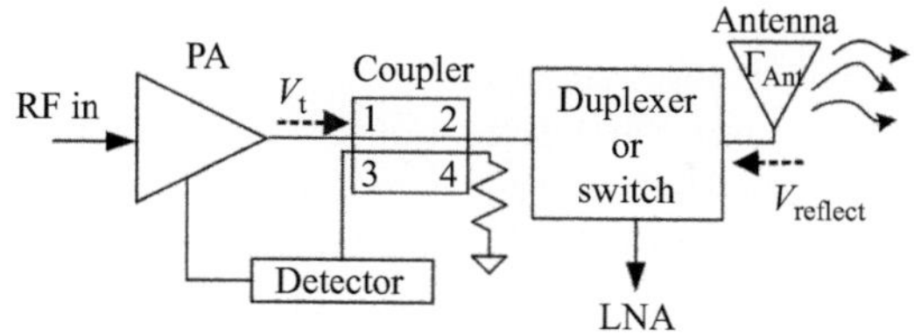

Figure 2.16 Conventional transmitter block diagram of the wireless communication system. © 2016 IEEE, Reprinted with permission from [25]

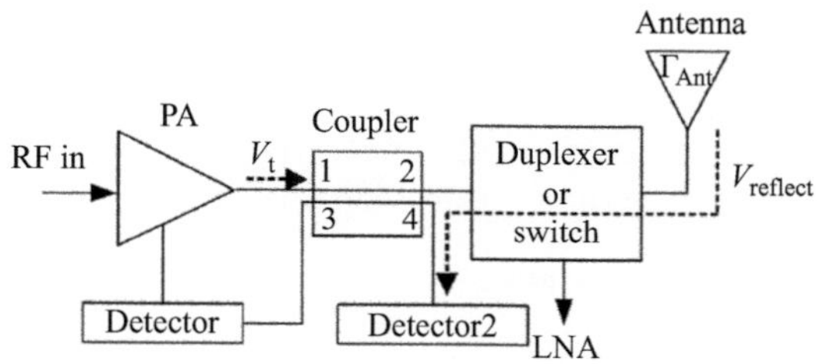

Figure 2.17 Wireless transmitter system having an additional RF power detector at port 4 for direct detection. © 2016 IEEE, Reprinted with permission from [25]

To model the effect of the reflection coefficient, the operating principle is summarized. When an antenna is located near the chest wall, the reflected signal from the antenna can be expressed as [26]:

$$V_{\text{reflect}} = \Gamma_{\text{ant}} \cdot V_t = |\Gamma_{\text{ant}}| \, e^{j(\angle\Gamma_{\text{ant}})} \cdot V_t = |\Gamma_{\text{a}}| \, |1+\gamma(t)| \, e^{j(\angle\Gamma_{\text{a}}+\angle\gamma(t))} \cdot V_t\sqrt{b^2-4ac} \tag{2.10}$$

where V_t is the transmitted signal, Γ_{ant} is the reflection coefficient of an antenna, Γ_{a} is a constant amplitude of the reflection coefficient for the fixed distance between the body and the antenna, and $\gamma(t)$ is a time-varying component of the reflection coefficient. Note that the magnitude and phase of the reflected signal depend on $\gamma(t)$. Since the magnitude and phase variations are very small and extremely low-frequency activities, they can be regarded as narrowband modulation. In a narrowband modulation signal, $\angle\gamma(t)$ is much smaller than unity. Thus, (2.10) can be simplified as

$$V_{\text{reflect}} = |\Gamma_{\text{a}}||1+\gamma(t)|[1+j(\angle\gamma(t))]e^{j(\angle\Gamma_{\text{a}})} \cdot V_t \tag{2.11}$$

In (2.11), $\angle\gamma(t)$ can be considered as a vital signal. That is, V_{reflect} is clearly amplitude-modulated by the magnitude and phase terms of the reflection coefficient. To concentrate on the effect of the magnitude variation of $\gamma(t)$, $\angle\gamma(t)$ is considered as a constant. Consequently, the reflected signal, which contains vital sign signal, can be expressed as

$$V_{\text{reflect}} = |\Gamma_{\text{a}}||1+\gamma(t)|e^{j(\angle\Gamma_{\text{a}})} \cdot V_t \tag{2.12}$$

Since the reflected signal is much smaller than that of the transmitted signal, the directivity of the coupler plays significant rule. However, since the conventional coupler has finite directivity, the detection of the reflection coefficient is restricted by its finite directivity. Assuming that the magnitude of the reflection coefficient from the antenna is −10 dB with ±0.05 dB variation due to heartbeat, the vital sign signal level has to be larger than the leakage signal from the power amplifier. The isolation of the coupler should be larger than the coupling level. Thus, 20 dB microstrip coupler having about 50 dB directivity at 2.4 GHz is utilized for reliable detection [27].

The reflected signal from the antenna, which is coupled to port 4 of the coupler as shown in Figure 2.17, can be expressed as

$$V_{\text{isolation}} = \sqrt{C}|\Gamma_{\text{a}}||1+\gamma(t)|e^{j(\angle\Gamma_{\text{a}})} \cdot V_t \tag{2.13}$$

where C is the coupling coefficient. The output from the RF detector can be modeled when it is operated in a linear range:

$$V_{\text{out}} = k \cdot 20 \cdot \log_{10}(u) + V_{\text{initial}} \tag{2.14}$$

where k is the slope of the detector output, V_{initial} is a constant value when the input is at the minimum detectable power signal, and u is the input parameter of the power detector. Due to the variation of the reflection level from the antenna, the output of the detector can be expressed:

$$V_{\text{out}} = k \cdot 20 \cdot \log_{10}\left(\sqrt{C}|\Gamma_{\text{a}}||1+\gamma(t)|e^{j(\angle\Gamma_{\text{a}})} \cdot V_t\right) + V_{\text{initial}} \tag{2.15}$$

Therefore, the sensitivity of the direct detection can be defined as the slope of the detector output k.

When the direct detection scheme is used, the vital sign sensor has very low sensitivity. For reliable vital sign detection, it is important to enhance the sensitivity of the sensing method. It is important to identify the major factor influencing the sensitivity of the system. As mentioned in the preceding formula, the reflection coefficient of an antenna is separated into two parts: Γ_{a}, which is a large value and $\gamma(t)$, which is a relatively small time-variant value. The output signal of a detector can be formulated with small signal analysis as

$$V_{\text{out}}(u) = V_{\text{OUT}}(u_0) + v_{\text{out}} = k \cdot 20 \cdot \log_{10}(u_0) + V_{\text{initial}} + \left.\frac{\partial V_{\text{out}}}{\partial u}\right|_{u=u_0} \Delta u \tag{2.16}$$

where u is an arbitrary variable, u_0 can be considered as a large signal, and Δu is designated as a small signal. Similarly, V_{OUT} is the output voltage determined by a large signal, while v_{out} is the output voltage of a small signal. Now, the small signal of the detector output v_{out} can be calculated as

$$v_{\text{out}}(u) = \frac{\partial V_{\text{out}}}{\partial u}\Delta u = \frac{k \cdot 20 \cdot u'}{\ln 10 \cdot u}\Delta u \tag{2.17}$$

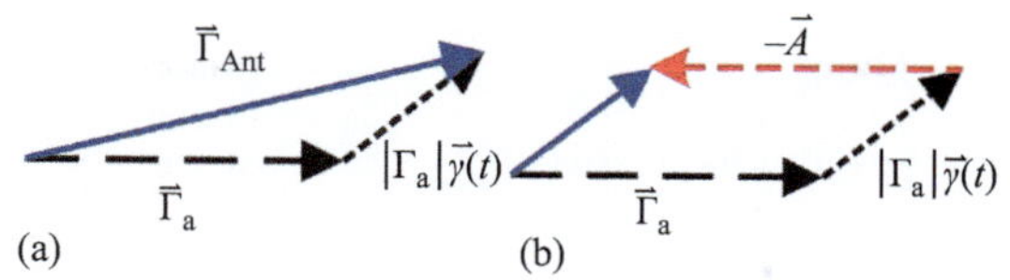

Figure 2.18 Signal vector diagram. (a) Narrowband modulated signal. (b) Demodulation concept of the subtract method. © 2016 IEEE, Reprinted with permission from [25]

If u is a first-order polynomial, v_{out} becomes larger as u become smaller, meaning that a small input signal becomes a large output signal. This implies that the magnitude of the reflected signal from an antenna should be reduced to improve the sensitivity of the sensor system. In other words, it can be considered to use an attenuator to reduce the signal level. However, the attenuator reduces not only the signal level, but also the information contained in a small signal, which is a vital sign. Thus, it is necessary to reduce only the large carrier signal component in the reflection signal for the enhancement of the system sensitivity. In the vector diagram as shown in Figure 2.18, the complex signal of reflected signal can be demodulated into the large unwanted vector Γ_a, which is a carrier signal, and the small vector $\Gamma_a \cdot \gamma(t)$, which is the information-bearing part. When the magnitude of Γ_a is subtracted from the overall reflection Γ_{ant}, the rest of the signal can be only the information bearing signal with a significantly reduced power level. The carrier subtraction method is applied to (2.15) to acquire:

$$V_{out} = k \times 20 \times \log_{10}\left(A + |\Gamma_a||1+\gamma(t)|e^{j(\angle\Gamma_a)}\right) + k \times 20 \times \log_{10}\left(\sqrt{C}\right) + V_{\text{initial}} \tag{2.18}$$

where A is an adjustable constant to eliminate the unwanted carrier signal. When A is nearly matched to $-|\Gamma_a|$ with $\angle\Gamma_a$ close to π, the first term of (2.18) will have only the wanted signal $\gamma(t)$. In addition variable u and Δu are substituted to $(A - |\Gamma_a| \cdot |1+\gamma(t)|)$ and $\gamma(t)$, respectively, in the small signal output expression to give:

$$v_{\text{out}} = \frac{\partial V_{\text{out}}}{\partial \Gamma_a}\gamma(t) = \frac{k \cdot 20 \cdot (1+\gamma(t))}{\ln 10(|\Gamma_a| \cdot |1+\gamma(t)| - A)}\gamma(t) \tag{2.19}$$

In (2.19), the variation in the small signal component can be analyzed by a differential function of the detector output. The validity of the above method is ensured by calculation of the detector output power as a function of $\sqrt{C} \cdot \Gamma_a$ in the range of -31 to -29 dB as given in Figure 2.19. The dotted line in Figure 2.19 shows the output power of the direct detection method. Compared to the direct detection method, the sensitivity enhancement method reveals drastically improved results as Γ_a approaches to A. In other words, the slope of the solid line becomes much steeper, and it even becomes infinite theoretically when A approaches to Γ_a.

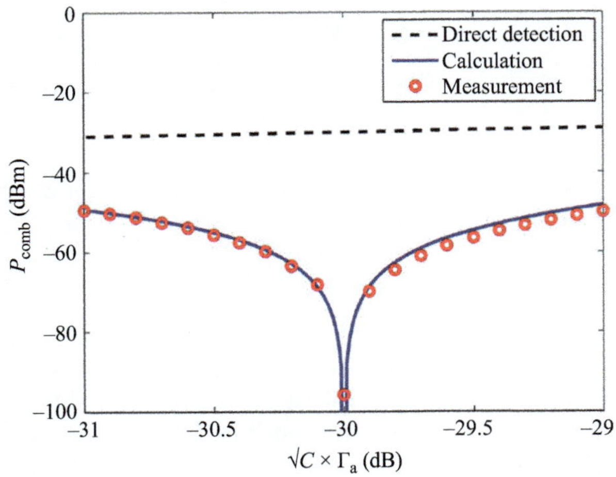

Figure 2.19 Comparing output power between detector only as direct detection from coupler and level of the power combiner output with sensitivity enhancement calculation and measurement results. © 2016 IEEE, Reprinted with permission from [25]

In actual circuit implementation, the errors of both amplitude and phase differences can occur. Also, the amplitude difference from the phase shifter and the phase difference from the attenuator may not be perfectly adjusted. Furthermore, the reflection coefficient of an antenna can be influenced by a position as well as skin conditions. Thus, the effects of the amplitude and phase mismatch need to be analyzed. The real part of the output signal from the proposed system can be written as

$$V_{\text{out}} = k \cdot 20 \cdot \log_{10}\big(A + |\Gamma_{\text{a}}| \cos\big(\angle\Gamma_{\text{a}} + \phi_{ps}\big)\big) \tag{2.20}$$

where ϕ_{ps} is the phase variation from the phase shifter. If ϕ_{ps} satisfies $\phi_{ps} = \pi - \angle\Gamma_{\text{a}}$, then condition (2.20) becomes an ideal detection condition. Let us define the phase difference θ_d as $\theta_d = \angle\Gamma_{\text{a}} + \phi_{ps}$. Then the slope of the detector output is determined as

$$\frac{\partial V_{\text{out}}}{\partial \Gamma_{\text{a}}} = \frac{k \cdot 20}{\ln 10} \frac{\cos(\theta_d)}{A + |\Gamma_{\text{a}} \cos(\theta_d)|} \tag{2.21}$$

The system sensitivity is a function of the amplitude (A and $|\Gamma_{\text{a}}|$) and the phase θ_d. Figure 2.20 shows that the sensitivity enhancement depends not only on the magnitude matching performance ($|\Gamma_{\text{a}}|/A$) but also on the phase variation (θ_d). It is clear that the maximum sensitivity enhancement can be obtained at each corresponding amplitude ratio $|\Gamma_{\text{a}}|/A$ when the phase changes. This analysis shows that the amplitude ratio is a dominant factor influencing the sensitivity rather than the phase difference θ_d. In view of this observation, phase variation θ_d of the system is set to near 180 degrees.

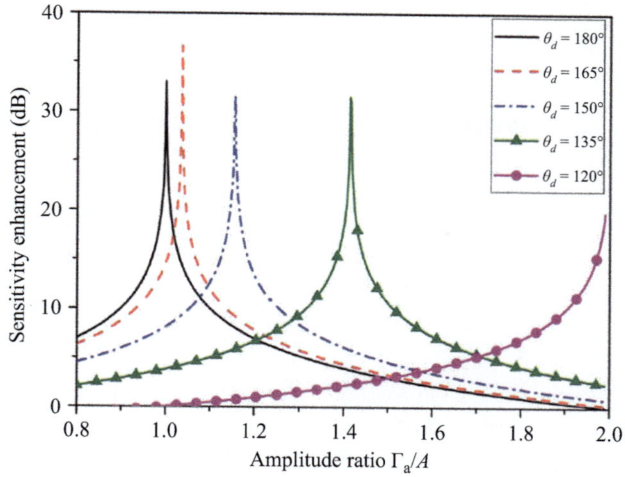

Figure 2.20 Sensitivity enhancement calculated by amplitude ratio variation, and the phase difference changed between the attenuated signal and the reflected signal. © 2016 IEEE, Reprinted with permission from [25]

Figure 2.21(a) shows a block diagram of the sensitivity-enhanced vital signal sensor system. It consists of a power amplifier, antenna, 20 dB directional coupler, phase shifter, and the attenuator. The variable attenuator is connected to port 3, and the phase shifter is connected to port 4 to control the delay of the reflected signal. The combined signals from the variable attenuator and the phase shifter are amplified to increase the signal level above the minimum detectable level. Then, the vital sign signals are detected by a power detector placed at the output of the power combiner. The system is suitable for energy-efficient, lower-cost, and lower-noise-level system. In Figure 2.21(b), slightly different circuit topology is illustrated. This system can significantly reduce the noise level. In conventional communication system, low-noise amplifier is generally placed in front of the receiver to reduce noise factor. In this system, the variable amplifier composed of low-noise amplifier and variable attenuator. In addition, the phase shifter is located at port 3 of the coupler and the RF amplifier located after the power combiner as shown in Figure 2.21(a) is removed for the same power consumption.

The sensitivity-enhanced systems are implemented with off-the-shelf components, excluding the power amplifier, which is replaced with a signal generator for consistent measurement. The coupler and a 2.4 GHz dipole antenna are implemented on a printed circuit board. An analog voltage-controlled variable attenuator is utilized to control variable A, and a voltage-controlled phase shifter consisting of varactor diodes is utilized to control variable ϕ_{ps} [28]. A Wilkinson power combiner is used to add signals, and an RF amplifier is connected to the output of the power combiner for amplification as shown in Figure 2.21(a) and connected to port 4 as shown in Figure 2.21(b). An rms power detector is utilized to detect power containing vital sign information.

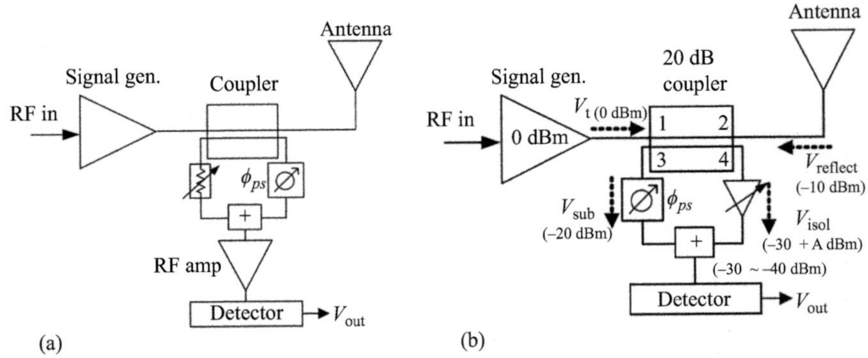

Figure 2.21 Proposed system topology. (a) Utilizing variable attenuator and phase shifter. (b) Utilizing variable amplifier and phase shifter for comparing sensitivity. © 2016 IEEE, Reprinted with permission from [25]

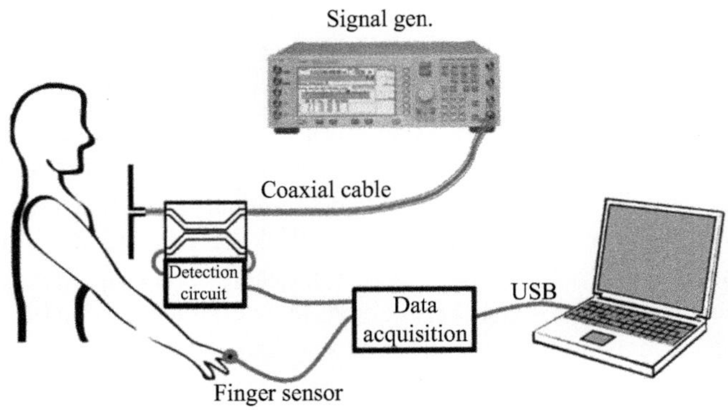

Figure 2.22 Measurement setup of the vital sign detection system. © 2016 IEEE, Reprinted with permission from [25]

Figure 2.22 shows the configuration of the experimental setup for detection of the vital signs, and Figure 2.23 presents the results of the abovementioned methods. Figure 2.23(a) shows the result of the direct detection method, and the peak value of the heartbeat frequency is about 0.4 mV. Because the sensitivity of the detector is 36 mV/dB, the magnitude variation of the reflection coefficient is in the range of 0.1 dB due to heartbeat activity. On the other hand, the peak value of the heartbeat frequency is about 12 mV for the sensitivity-enhanced method using variable attenuator and phase shifter, as shown in Figure 2.23(b). It shows the output voltage of the system is 10 to 30 times larger than the direct detection method, revealing the superior performance of the improved system. Several peaks in other frequencies are induced by the arbitrary motion of the human subject. Figure 2.23(c) shows the result of the sensitivity-enhanced system using an LNA operating as the variable amplifier, revealing the superior performance.

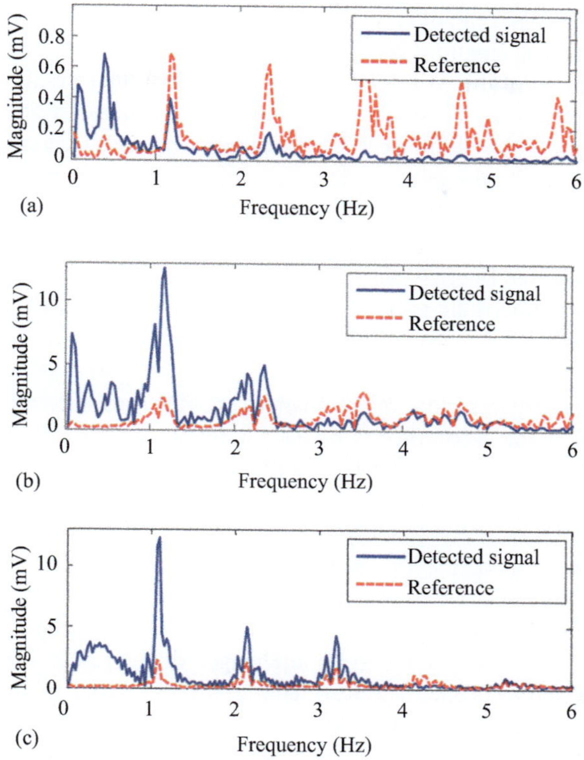

Figure 2.23 Measured results for the vital signs. (a) Direct detection method. (b) Sensitivity-enhanced method using a variable attenuator and phase shifter. (c) Sensitivity-enhanced method using a variable amplifier and phase shifter. © 2016 IEEE, Reprinted with permission from [25]

2.3 Wrist pulse

Unlike the measurement in the vicinity of a chest described in the previous section, this chapter deals with the sensor systems that measure the pulse from the wrist. It detects the change of the input impedance of an antenna or resonator due to the changes in the diameter of the blood vessel caused by the pulse similarly to the operating principle near a chest. If reliable wrist pulse measurement is possible, it can be applicable to wearable devices. Three different methods will be detailed as follows.

2.3.1 Injection-locked PLL

The system introduced in this section is a system that detects vital signs via changes in the diameter of arterial blood vessels in the wrist due to heart pulses. Unlike the

PLL method described previously, the sensor oscillator is separated from the VCO and the PLL circuit. Figure 2.24 shows a configuration of the wrist pulse sensor. For the application of the wearable device on the wrist, an interdigital electrode is employed as an RF planar resonator that acts as a transducer, transforming changes in blood volume in the radial artery to frequency and phase variations. The interdigital electrode has a finger-like, periodic, and parallel pattern.

When the interdigital electrode is placed near the wrist, the fringing fields of the interdigital electrode are affected by the wrist pulse of the radial artery. This near-field proximity effect changes the characteristics of the interdigital electrode. In other words, the periodic pulse of the radial artery near the wrist causes periodic variations in the *S*-parameters of the interdigital electrode, and a free-running oscillator is constructed with the interdigital electrodes as a band-pass filter. In other words, the *S*-parameters of the interdigital electrode determines the oscillation condition. If the *S*-parameters of the interdigital electrode varies due to the wrist pulse, the oscillation frequency will change periodically. Throughout the remainder of this section, this free-running oscillator is termed as the sensor oscillator.

The operating principle of the interdigital electrode as a transducer and its equivalent circuit model are well known. This sensor oscillator is placed near the wrist, where the radial artery passes, so that the fringing fields of the interdigital electrodes are influenced by the volume changes of the radial artery. The output signal from the sensor oscillator is injected into the VCO via a micro-coaxial cable, and the frequency of the VCO is synchronized with the frequency of the sensor oscillator by the injection-lock phenomenon, under the condition that the frequency difference between the two oscillators is within the injection-lock range. The frequencies of both oscillators change with the periodic volume changes of the radial artery. Finally, the PLL synthesizer stabilizes the VCO frequency by applying a

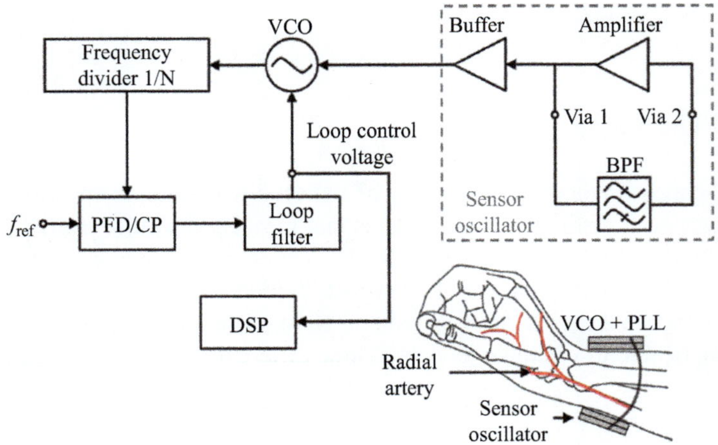

Figure 2.24 Block diagram of the wrist pulse sensor. © 2016 IEEE, Reprinted with permission from [29]

loop-control voltage, and the physiological information can be detected from voltage deviation from the loop filter.

When a human body is located within the near-field range of the interdigital electrodes, the frequency response of the interdigital electrode generally shifts to a lower-frequency region. To confirm the proximity effect of a human body, a wrist model is designed for full-wave electromagnetic simulation, as illustrated in Figure 2.25. The detailed electrical characteristics of the wrist model are provided in Table 2.2. Compared with the free-space condition, the resonance frequency of the interdigital electrode shifts to a lower-frequency region about 360 MHz, as illustrated in Figure 2.26. In the design of the sensor oscillator, this proximity effect has to be carefully considered and the oscillation conditions have to be satisfied in the proximity condition rather than in the free-space condition.

Then, an amplifier using a BJT is combined with the interdigital electrodes to form a loop. The open-loop characteristics of the sensor oscillator are examined by hybrid simulation (full-wave electromagnetic simulation), as shown in Figure 2.27; the diameter of the blood vessel in the simplified wrist model is varied. The volume changes of the radial artery clearly affect the open-loop characteristics, and subsequently results in a frequency deviation satisfying the oscillation condition (for phase) about 0.7 MHz. Figure 2.28 illustrates the measured output spectrum of the sensor oscillator with the interdigital electrode attached to the wrist above the radial artery. The output power of the sensor oscillator is about 7 dBm, and the oscillation

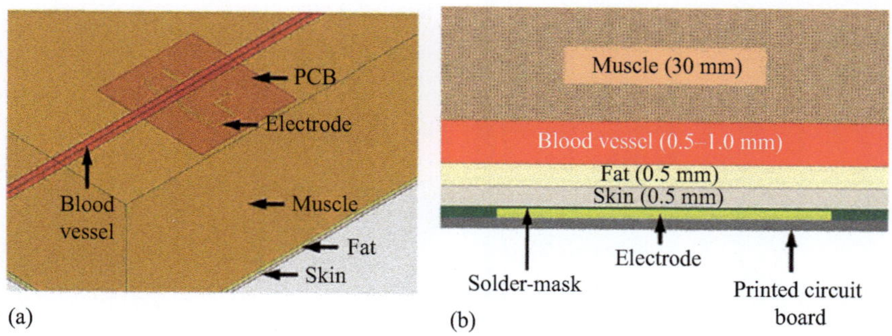

Figure 2.25 (a) Configuration for full-wave electromagnetics (EM) simulation. (b) Side view of the configuration. © 2016 IEEE, Reprinted with permission from [29]

Table 2.2 Electrical characteristics of human tissue. © 2016 IEEE, Reprinted with permission from [29]

Tissues	**Relative permittivity**	**Conductivity (S/m)**
Skin	44.6	1.70
Fat	4.40	0.08
Muscle	54.0	2.20
Blood	53.0	2.70

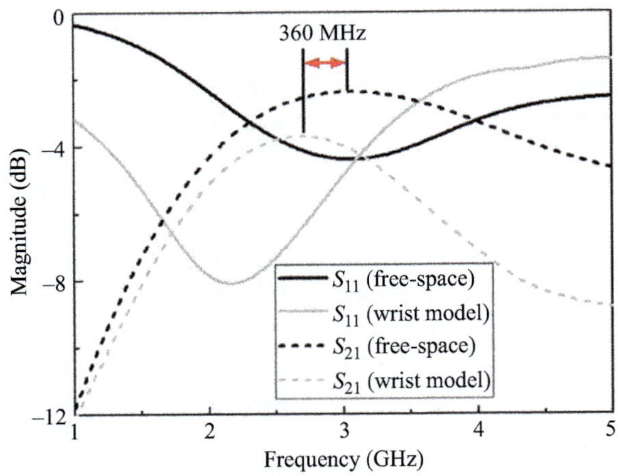

Figure 2.26 S-parameters comparison between the interdigital electrodes in free-space condition and wrist model condition. © 2016 IEEE, Reprinted with permission from [29]

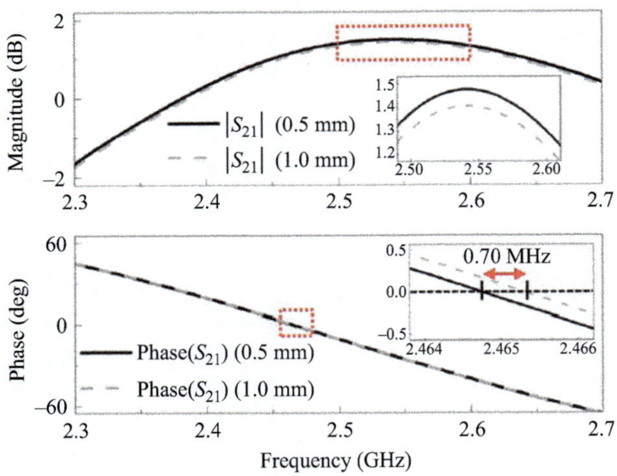

Figure 2.27 Open-loop characteristic with varying diameter of the radial artery. © 2016 IEEE, Reprinted with permission from [29]

frequency varies (low to high) about 0.74 MHz with the period of the wrist pulse. This result shows excellent agreement with the results of the open-loop simulation shown in Figure 2.27. Note that the sensor oscillator is followed by a buffer amplifier to minimize the load-pull effect due to the cascade components, micro-coaxial cable, and PLL synthesizer.

The injection-locking and injection-pulling phenomena of this oscillatory system have been well documented. The usual oscillators in RF systems have been phase-locked by the PLL synthesizer to a desired frequency. The system architecture can be

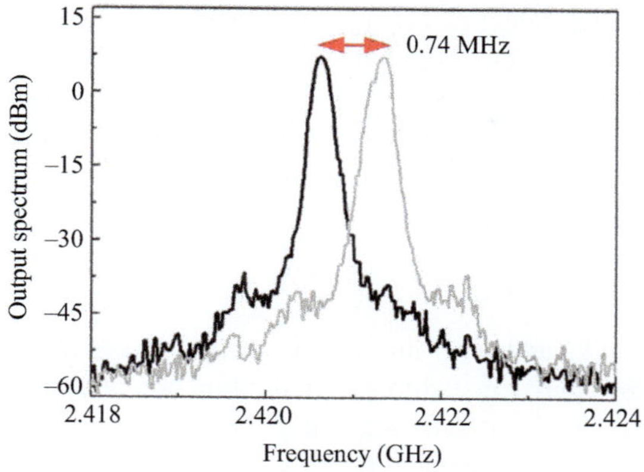

Figure 2.28 Measured output spectrum of the sensor oscillator. © 2016 IEEE, Reprinted with permission from [29]

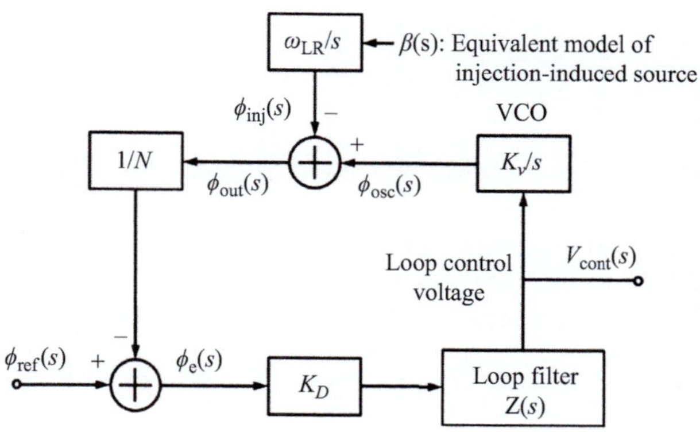

Figure 2.29 Linear frequency-domain model of the proposed sensor. © 2016 IEEE, Reprinted with permission from [29]

transformed into a linear frequency-domain model as presented in Figure 2.29. The variables $V_{cont}(s)$ and $\beta(s)$ denote the loop-control voltage and the injection-induced source, respectively, and the injection signal from the sensor oscillator is phase-modulated by the periodic change of the blood volume at the wrist.

The relation between $V_{cont}(s)$ and $\beta(s)$ can then be derived as

$$V_{cont}(s) = \frac{K_D \omega_{LR} Z(s)/N}{s + (K_D K_V/N)Z(s)} \beta(s) \tag{2.22}$$

where

$$\omega_{LR} \approx \frac{\omega_{osc}}{2Q} \frac{E_{inj}}{E_{osc}} = \frac{\pi f_{syn}}{Q} \sqrt{\frac{P_{inj}}{P_{osc}}} \tag{2.23}$$

is the single side-lock range of the injection-locked oscillator under weak injection. Here, P_{inj} is the power of the injection signal, i.e. the output signal from the sensor oscillator, and P_{osc} is the power of the VCO, while Q is the quality factor. Substituting (2.23) into (2.22) and assuming s is small, (2.22) can be approximated as

$$V_{\text{cont}}(s) \approx \frac{\pi f_{syn}}{Q \cdot K_V} \sqrt{\frac{P_{\text{inj}}}{P_{\text{osc}}}} \beta(s) \tag{2.24}$$

$\beta(t)$ is a phase-modulated term due to the change in the diameter of the blood vessel, so it can be approximated to a periodic function with the periodicity of the wrist pulse. If we simply assume that the $\beta(t)$ is a sinusoidal function having a frequency f, which is the frequency of the wrist pulse:

$$\beta(t) \approx A \sin(2\pi f t) \tag{2.25}$$

then $V_{\text{cont}}(t)$ also becomes a periodic function having the frequency of the wrist pulse:

$$V_{\text{cont}}(t) \approx \frac{A \cdot \pi f_{syn}}{Q \cdot K_V} \sqrt{\frac{P_{\text{inj}}}{P_{\text{osc}}}} \sin(2\pi f t) \tag{2.26}$$

Thus, the loop-control voltage can be viewed as directly correlated with the wrist pulse of a human subject.

Figure 2.30 shows the experimental setup for detection of vital sign signals, and Figure 2.31 shows the experimental results. It reveals that the result of the wrist pulse sensor based on injection-locked PLL system shows excellent agreement with the reference sensor.

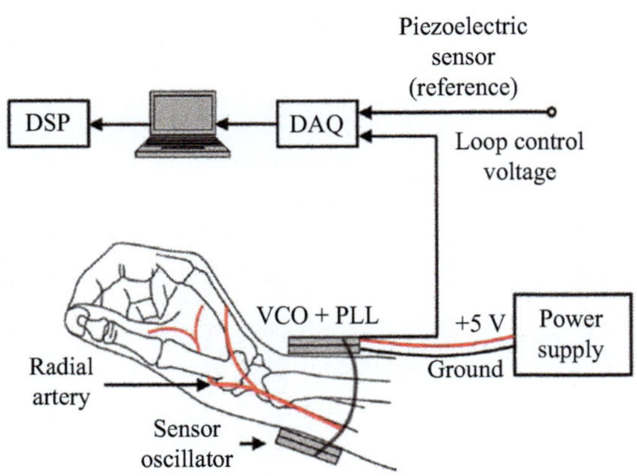

Figure 2.30 The configuration of the measurement setup. © 2016 IEEE, Reprinted with permission from [29]

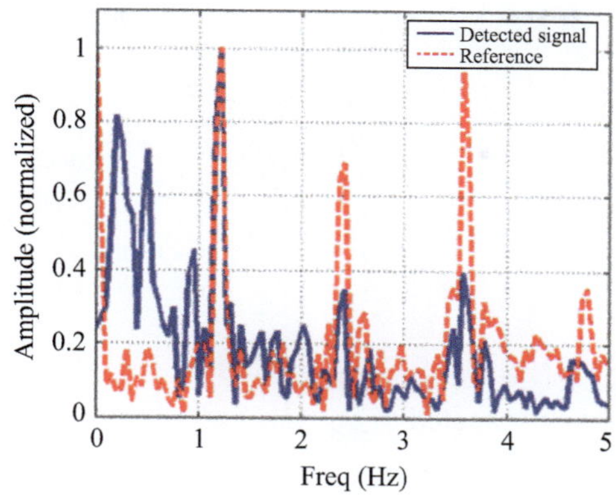

Figure 2.31 Measured results from the sensor using injection-locked PLL and piezoelectric finger sensor. © 2016 IEEE, Reprinted with permission from [29]

2.3.2 Reflectometry system with array resonator

For wearable applications, it is assumed that a resonator can be embedded under a wrist strap of a smartwatch and designed with a few considerations. First, the size of the resonator should be reasonable. The maximum dimension is restricted to the width of a watch strap. Meanwhile, the strap should be flexible to be worn on the wrist. A resonant-type antenna facing inside of the strap is used to induce large frequency deviation owing to the volume changes of the radial artery. So that it is only affected by the motion of the radial artery, not by outside motion. In this section, 0.25 mm thick Teflon is used as a substrate and a patch-type planar inverted-F antenna structure is employed to minimize the size of the resonator. The resonant frequency is in general shifted to lower-frequency region when it is placed near a human body. To verify the effect of a human body with the resonator, 3D full-wave electromagnetic simulation has been performed. For simulation of human skin, a box with tissue property is located in front of the resonator with distance ranging from 0.1 to 0.6 mm. As the resonator is placed closer to the skin having much higher dielectric constant than Teflon, the effective dielectric constant seen from the resonator becomes much larger, so that the size of the resonator has to be adjusted to resonate at 2.4 GHz, becoming about 19.8 mm. The designed resonator is presented in Figure 2.32, and the simulation results are illustrated in Figure 2.33(a). The L_s is one of the control variables of the resonator which is related to matching frequency and quality factor, and the L_s is selected to be 3.8 mm. The simulation results show that the resonator is well matched with distance ranging from 9.2 to 9.3 mm. Figure 2.33(b) also shows the measured S_{11} of the flexible resonator. To change the distance between the resonator and skin, the resonator is placed on the wrist and pressure is applied from outside. The tendency of the resonance frequency variation with different pressure shows very similar behavior with the simulation result as shown in Figure 2.33(a).

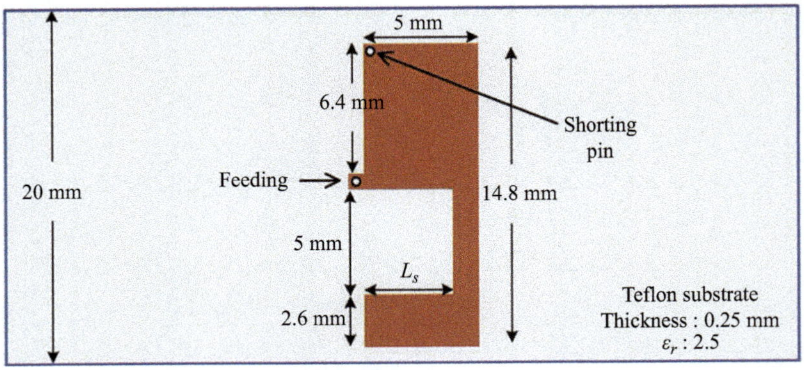

Figure 2.32 Designed resonator structure and dimensions. © 2016 IEEE, Reprinted with permission from [30]

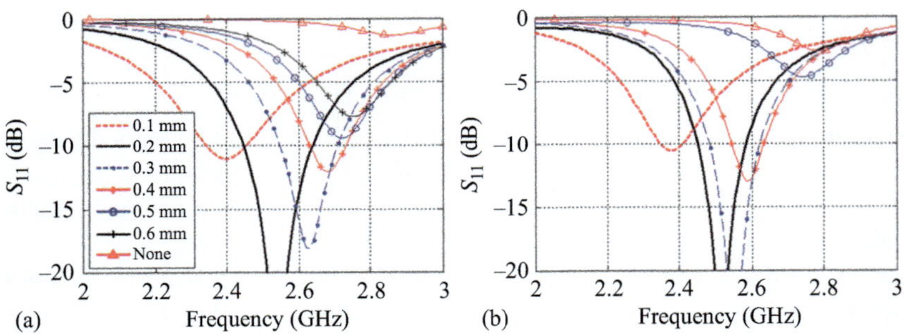

Figure 2.33 Simulated and measured S-parameters. (a) 3D EM simulation using a skin model with varying of the distance. (b) Resonator on the wrist with varying tightness. © 2016 IEEE, Reprinted with permission from [30]

On the wrist, pulse can be easily detected on several positions where the arteries are located. However, since the shape and position of a wrist strap could be slightly different for different types of equipment and since a strap can rotate on the wrist during movement, the position of the resonator needs to be flexible. For a minimized form factor, the size of the proposed resonator is selected to be less than 20 mm, and the detection areas are relatively small with a single resonator. To overcome this issue, array resonator geometry is proposed to extend the detection area to as large as possible. In addition, a multiple detection scheme utilizing the array resonator is also proposed. As shown in Figure 2.34, the proposed array resonators are placed 5 mm apart from one another, extending the detection coverage area to about 35 mm, which is sufficient to sense at least one artery of the wrist.

The array resonator is connected to the detection system, with an absorptive single-pole four-throw (SP4T) RF switch, and the switch is controlled by a 2-bit binary signal. The electrical spacing between the resonators (edge-to-edge) is about (5 mm) to ensure reasonable isolation between the resonators. Note that while such

Figure 2.34 Photograph of the flexible array resonator with 5 mm spacing. © 2016 IEEE, Reprinted with permission from [31]

isolation is generally an important factor in array antenna design, it might be relatively negligible factor in this system. It should be emphasized that the proposed detection system can be operated with one resonator at a time and that any leakage to the other resonators is absorbed by the absorptive off-state port of the switch. This absorptive characteristic provides high isolation level between output ports of the switch and does not disturb impedance matching performance.

To enhance the sensitivity of the sensor, an active isolation technique is utilized. This method allows detection of the reflected signal from the resonator with much higher sensitivity compared to using a system with a single detector.

The reflected signal from the resonator can be expressed as

$$V_{\text{reflect}} = \Gamma_{\text{res}} \cdot V_t = |\Gamma_L||1 + \gamma(t)|e^{j(\angle\Gamma_L + \angle\gamma(t))} \cdot V_t \tag{2.27}$$

where Γ_{res} is the reflection coefficient of the resonator, V_t is the transmitted signal, Γ_L is the constant component as a large signal of the reflection coefficient, and $\gamma(t)$ is the time-varying pulse information as the small signal component. A coupler is utilized to detect the reflected signal. Thus, the reflected signal (V_{reflect}) is coupled to $\sqrt{C} \cdot V_{\text{reflect}}$, where C is the coupling coefficient of the coupler. If RF detector is used, the output can be expressed as

$$V_{\text{out}} = k \cdot 20 \log_{10}\left(\sqrt{C} \cdot V_{\text{reflect}}\right) + V_{\text{initial}} \tag{2.28}$$

where k is the detector coefficient, and V_{initial} is the initial output voltage when the detector input is zero. To obtain the conversion gain slope of the detector, the output voltage in (2.28) is partially differentiated with Γ_L to have:

$$\frac{\partial V_{\text{out}}}{\partial \Gamma_L} = \frac{k \cdot 20}{\ln 10 \cdot V_{\text{reflect}}} \frac{\partial V_{\text{reflect}}}{\partial \Gamma_L} = \frac{k \cdot 20}{\ln 10 \cdot |\Gamma_L|} \tag{2.29}$$

The above (2.29) shows the sensitivity of the proposed system. To increase the sensitivity, Γ_L should be minimized; thereby, the active isolator concept can be best suited to increase sensitivity by reducing the constant component of V_{reflect}.

A block diagram of the proposed system is shown in Figure 2.35. V_t is transmitted into port 2 of the coupler and most of the signal reaches the RF switch which is connected to port 1 of the coupler. Then the RF signal is relayed to each resonator, depending on the RF switch status. As mentioned in Section 2.2, the resonator

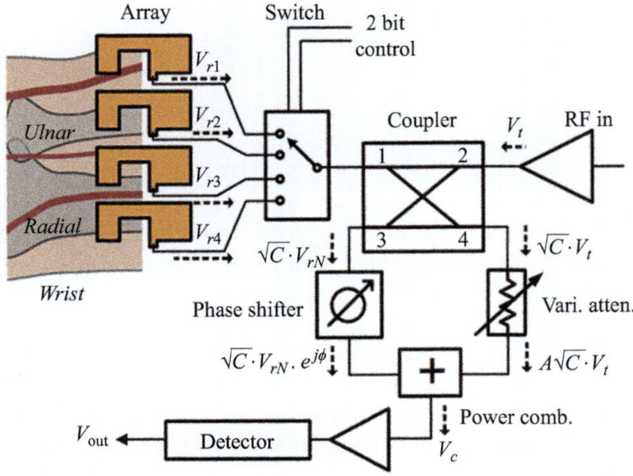

Figure 2.35 Block diagram of the proposed system. © 2016 IEEE, Reprinted with permission from [31]

elements of the array have a reflection coefficient dependent on the near-field variation due to physiological movement. The reflected signal (V_{rn}) can be expressed as

$$V_{rn}(t) = \begin{cases} V_{r1} = \Gamma_1(t) \cdot V_t(t) & \text{if } Sn|_{n=1} = \text{“00”} \\ V_{r2} = \Gamma_2(t) \cdot V_t(t) & \text{if } Sn|_{n=2} = \text{“01”} \\ V_{r3} = \Gamma_3(t) \cdot V_t(t) & \text{if } Sn|_{n=3} = \text{“10”} \\ V_{r4} = \Gamma_4(t) \cdot V_t(t) & \text{if } Sn|_{n=4} = \text{“11”} \end{cases} \tag{2.30}$$

where Γ_n is the reflection coefficient of the resonator elements, and Sn is the 2-bit binary signal to control the RF switch. Now, the reflected signal from the resonator is coupled to port 3 of the coupler and becomes $\sqrt{C} \cdot V_{rn}$. In the same way, the reflected signal is coupled to port 4 and can be expressed. For phase synchronization, a phase shifter is used to the reflected signal, and the signal becomes $\sqrt{C} \cdot V_{rn} \cdot e^{j\phi}$, where $e^{j\phi}$ is the shifted phase. In addition, a variable attenuator is utilized to adjust the magnitude of the transmitted signal at port 4, and the adjusted signal becomes $A \cdot \sqrt{C} \cdot V_t$. These two signals are combined at the power combiner, and the combined signal V_c can be expressed as follows:

$$V_c(t) = \sqrt{C} \cdot V_t(t)\left(\Gamma_n \cdot e^{j\phi} + A\right) \tag{2.31}$$

The output of the detector is now differentiated with respect to Γ_{nL}. Thus, the sensitivity of the system becomes:

$$\frac{\partial V_{\text{out}}}{\partial \Gamma_{nL}} = \frac{k \cdot 20}{\ln 10 \cdot V_c} \frac{\partial V_c}{\partial \Gamma_{nL}} = \frac{k \cdot 20 \cdot |1 + \gamma_n(t)| e^{j(\angle\Gamma_{nL} + \phi)}}{\ln 10 \cdot (A + |\Gamma_{nL}| \cdot |1 + \gamma_n(t)| e^{j(\angle\Gamma_{nL} + \phi)})} \tag{2.32}$$

To increase the slope, i.e. the sensitivity of the combined signal, the denominator in (2.32) should be as small as possible. For sensitivity enhancement, the phase shifter and the variable attenuator should be properly set as

$$\angle\Gamma_{nL} + \phi = \pi, |\Gamma_{nL}| = A \tag{2.33}$$

It is necessary to note that the reflection coefficient of each resonator is slightly different. As a result, these phase and magnitude differences could degrade the detection sensitivity. Figure 2.36 depicts the calculated results based on (2.32). The figure shows that the sensitivity enhancement curves are dependent on the magnitude and phase differences and reveals the maximum sensitivity when the ideal conditions are satisfied as in (2.33). It is clear that the sensitivity can be increased by about 40 dB, as shown in Figure 2.36. Note that the amount the sensitivity enhancement varies with respect to both the magnitude ratio ($|\Gamma_{nL}|/A$) of the reflected and the attenuated signal and the phase difference ($\angle\Gamma_{nL} + \phi$).

As shown in Figure 2.36, if the phase difference is fixed to 180°, the sensitivity enhancement is symmetrically decreased, while $|\Gamma_{nL}|/A$ is shifted away from the highest sensitivity enhancement value at the center point $|\Gamma_{nL}|/A = 1$. However, when the phase difference is shifted from 180 degrees, the center point of the symmetry is shifted to $|\Gamma_{nL}|/A > 1$. The center point is determined by using $|\Gamma_{nL}|\cos(\angle\Gamma_{nL} + \phi) = -A$, and consequentially the sensitivity enhancement region is largely spread on the $|\Gamma_{nL}|/A > 1$ condition. In a case where $|\Gamma_{nL}| > A$, the sensitivity also increases to near $|\Gamma_{nL}|\cos(\angle\Gamma_{nL} + \phi) = -A$ points. Therefore, when using a smaller magnitude of A than $|\Gamma_{nL}|$, the sensitivity enhancement range is extended on the phase difference domain in Figure 2.36. This means that if the condition of the array system is set to $|\Gamma_{nL}| > A$, some of the phase mismatches can be accepted for sensitivity enhancement. According to Figure 2.34, at least 10 dB the sensitivity can be increased at least 10 dB, where $|\Gamma_{nL}|/A = 1.1$ with measured

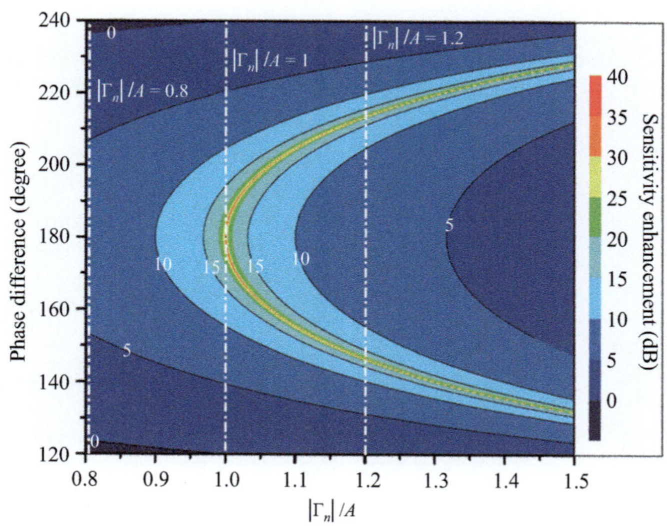

Figure 2.36 Calculated result of the sensitivity enhancement depending on the magnitude and phase difference. © 2016 IEEE, Reprinted with permission from [31]

phase differences. Therefore, the array detection system can provide high sensitivity without controlling the phase shifter and variable attenuator for each resonator.

Figure 2.37 shows the experimental setup for detecting vital sign signals, and Figure 2.38 shows the experimental results. The switch control binary bits are changed every 25 ms so that each resonator is selected 10 times/s. To reduce the complexity of the signal processing, an identical 10 sample/s sampling rate is used. It is clear that all four resonators can detect a heartbeat pulse with different signal levels, while S1 and S4 reveal excellent pulse waveforms. To demonstrate this in detail, all the sampled signals are interpolated with a sinc function. In accordance with the Nyquist theorem, signals detected via the 10 samples/s per channel can be

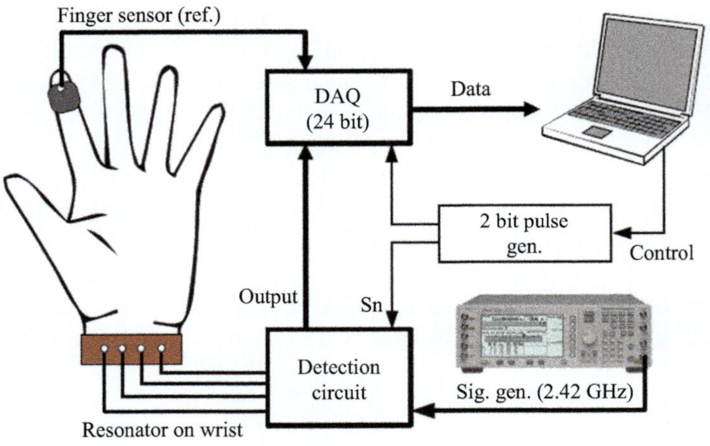

Figure 2.37 Measurement setup of the proposed wrist pulse detection system. © 2016 IEEE, Reprinted with permission from [31]

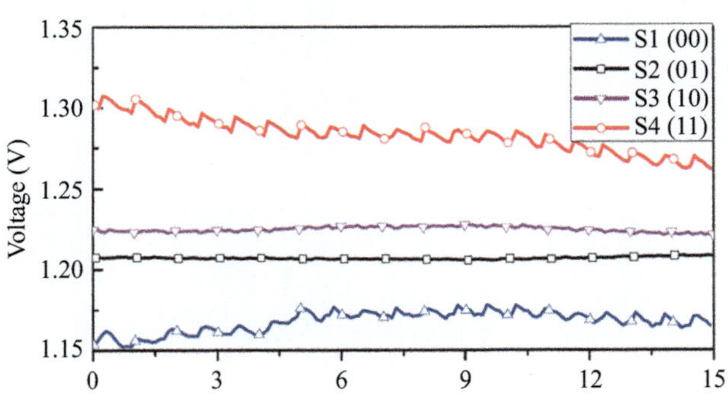

Figure 2.38 Measurement result with simultaneous switching at 10 samples/s. © 2016 IEEE, Reprinted with permission from [31]

recovered up to 5 Hz, and this renders to resolve the second peak in the pulse from the sampled result.

2.3.3 Interferometry system

Figure 2.39 shows the block diagram of the RF interferometer system, utilizing variations in transmission coefficient [32]. To minimize the unwanted power loss, the system employs the Wilkinson power divider instead of a directional coupler. The divided signal experiences two different paths. One path is composed of a two-port resonator, and the other path is composed of a variable attenuator and a phase shifter. The combined signal is amplified and fed to the power detector. The operating circuit topology is similar to the method, except that the input signal is transmitted signal, not reflected one.

The concept of the wrist pulse detection method is same as previous wrist pulse sensors. The resonator elements are positioned near the radial artery and facing the radial artery. Large invariant signal is always transmitted independent of the movement of the radial artery while the small movement of the radial artery causes variations in the transmission coefficient between the resonators. To validate the performance of the sensor system, a 3D electromagnetic simulation is conducted to optimize the transmission characteristics of the resonators with human skin model, which has a relative permittivity of 37 and conductivity of 1.7 S/m. The space between the resonator and the skin is filled with a preventive cover layer, which has a relative permittivity of 3.2, and thickness of 0.25–1.0 mm to prevent direct contact between the resonator, conductor and the skin. The simulation and measurement results are illustrated in Figure 2.40. The maximum magnitude of S21 is about −10 dB, and the peak frequency shifts to a higher-frequency region with smaller spacing due to the increased effective dielectric constant.

Figure 2.41 shows the measured results with and without an interferometer system. From both experiments, the wrist pulse can be clearly detected on the basis of the transmitted signal between the two resonators. Note that the magnitude of the wrist pulse signal at the fundamental frequency from the conventional circuit topology is about 0.34 mV, whereas the magnitude of the wrist pulse signal at the fundamental

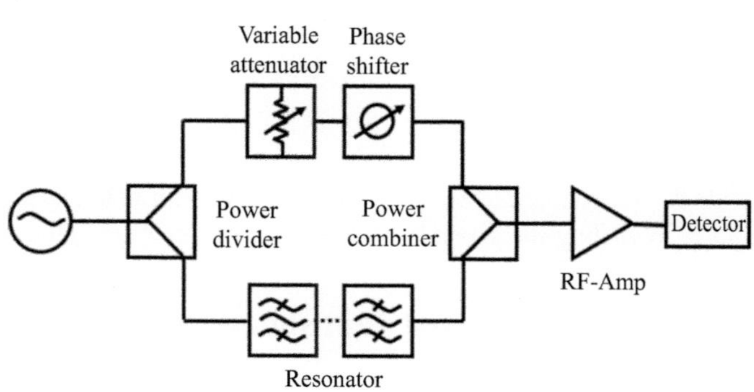

Figure 2.39 Block diagram of the RF interferometer system

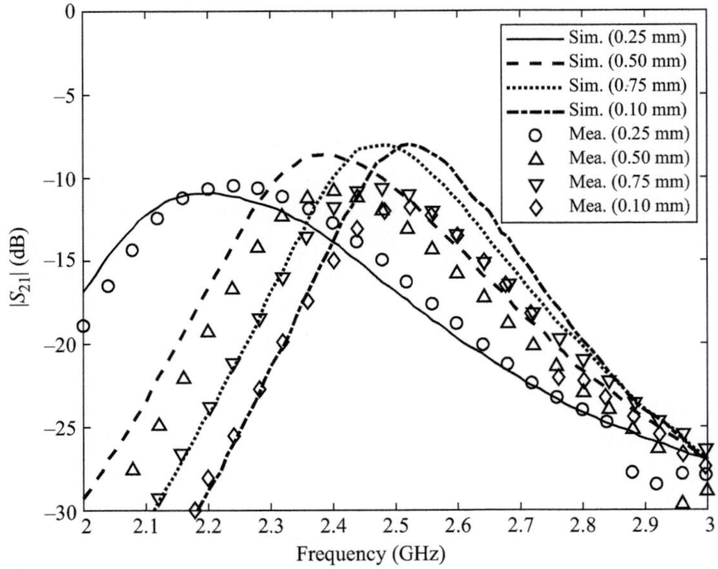

Figure 2.40 Simulated and measured results for the transmission coefficient of the two-port resonator with various distances

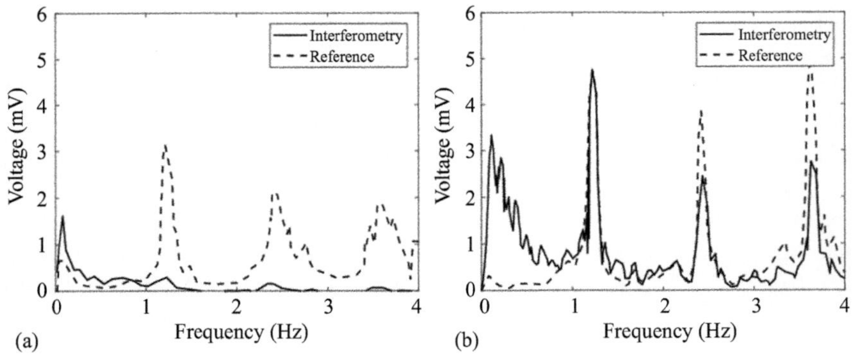

Figure 2.41 Measurement result. (a) Direct detection system. (b) Interferometer system

frequency with the proposed interferometry circuit reveals much larger signal level with lower noise, revealing the superior performance of the improved system.

2.4 AR method for improved vital sign estimation

Nonparametric techniques, such as Fourier-based methods, provide spectral estimation where no prior model is assumed, in the sense that no assumptions are made concerning the physical process that generated a given data. They are also known as the classical methods of spectral estimation. Although this approach of signal parameter estimation is computationally efficient, it has limited frequency resolution. These methods also suffer from spectral leakage effects that often mask weak signals. Prominent conclusions from these nonparametric techniques are that there is always a compromise in the bias-variance trade-off because both of these errors cannot be minimized simultaneously. The parametric or model-based methods of spectral estimation assume that the signal satisfies a generating model with known functional form, and then proceed by estimating the parameters in the assumed model. The signal's spectral characteristics of interest are then derived from the estimated model. In those cases where the assumed model is a close approximation to the reality, it is no wonder that the parametric methods provide more accurate spectral estimation than the nonparametric techniques. The nonparametric approach to power spectral density estimation remains useful, though, in applications where there is little or no information about the signal in question [33,34].

The AR method is the simplest of the parametric-based technique. For the rational transfer function, the system model can be written in the time domain as

$$y(t) = (1/A(z))e(t) \tag{2.34}$$

where $y(t)$ is the filter output, $1/A(z)$ is the transfer function, and $e(t)$ is white noise of variance equal to σ^2. It can be written as where a_i is the AR coefficient. After multiplying (2.35) by $y^*(t-k)$, the equation can be solved by the Yule–Walker method in matrix form. The Yule–Walker equations can be written as

$$y(t) + \sum_{i=1}^{n} a_i y(t-i) = e(t) \tag{2.35}$$

$$\begin{bmatrix} r(0) & r(-1) & \cdots & r(-n) \\ r(1) & r(0) & \cdots & r(-n+1) \\ \vdots & \vdots & \ddots & \vdots \\ r(n) & r(n-1) & \cdots & r(0) \end{bmatrix} \begin{bmatrix} 1 \\ a_1 \\ \vdots \\ a_n \end{bmatrix} = \begin{bmatrix} \sigma^2 \\ 0 \\ \vdots \\ 0 \end{bmatrix} \tag{2.36}$$

Excluding the first row from (2.36), the AR coefficient can be derived by

$$\mathbf{\Phi} = -\mathbf{R}_n^{-1}\mathbf{r}_n \tag{2.37}$$

where

$$\begin{aligned} \mathbf{\Phi} &= [a_1 \quad a_2 \quad \cdots \quad a_n], \\ \mathbf{r}_n &= [r(1) \quad r(2) \quad \cdots \quad r(n)], \\ \mathbf{R}_n &= \begin{bmatrix} r(0) & r(-1) & \cdots & r(-n) \\ r(1) & r(0) & \cdots & r(-n+1) \\ \vdots & \vdots & \ddots & \vdots \\ r(n) & r(n-1) & \cdots & r(0) \end{bmatrix} \end{aligned} \tag{2.38}$$

To validate the performance of the AR method in vital sign detection, it is applied to the incoming time-domain signal of the proximity sensor. The received signal by the SAW filter method is shown in Figure 2.42(a) [35]. The received signal is

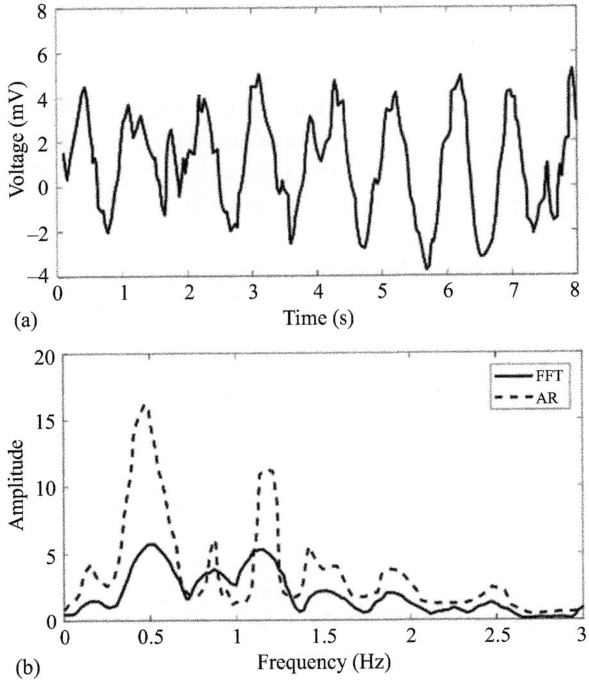

Figure 2.42 (a) Measured heartbeat signal from the SAW sensor. (b) Comparison between the results of the FFT and AR methods

converted to frequency-domain signal by two methods, i.e. fast Fourier transform (FFT) and AR methods, as presented in Figure 2.42(b). Under practical application, a proper DSP for a short data sequence should be applied to analyze and eliminate the frequency component from the artificial noise caused by the random body movement. The FFT method shows low resolution in the frequency spectrum for short data sequences, and it makes the vital sign detection difficult. On the other hand, the AR method provides a higher frequency spectrum resolution than the FFT method. Thus, it is clear that AR method is one of the appropriate signal processing methods for reliable vital sign detection from the artificial noise such as motion artifacts. Of course, we can apply various spectral estimation methods to improve time-domain vital signal processing for more reliable and accurate vital sign estimation.

2.5 Conclusion

In this chapter, we have reviewed various vital sensor based on the proximity effect of the antenna (or resonator). The basic operational principle is to measure the impedance change of the antenna. The measurement system includes a combination of an oscillation circuit and an antenna, and a reflectometry or interferometry system. Unlike conventional PPG or ECG sensors, measurement is possible without direct contact with the skin (even on your clothes). Although the measurement distance is shorter than that of the Doppler radar method, there is no need to consider intersensor interference with the methods introduced here. It is worth to mention that the method of detecting movement of the chest wall can be applied to smart bed or smart chair, while the method of detecting a wrist pulse can be applied to a smartwatch application. For commercialization, further development of countermeasures against motion artifact and system integration is necessary.

References

[1] Choi H, Naylon J, Luzio S, *et al.* Design and in vitro interference test of microwave noninvasive blood glucose monitoring sensor. IEEE Transactions on Microwave Theory and Techniques. 2015;63(10):3016–3025.
[2] Buxi D, Redouté JM, and Yuce MR. Blood pressure estimation using pulse transit time from bioimpedance and continuous wave radar. IEEE Transactions on Biomedical Engineering. 2017;64(4):917–927.
[3] Yilmaz T, Foster R, and Hao Y. Detecting vital signs with wearable wireless sensors. Sensors. 2010;10(12):10837–10862.
[4] Fukushima H, Kawanaka H, Bhuiyan MS, *et al.* Estimating heart rate using wrist-type photoplethysmography and acceleration sensor while running. In: Engineering in Medicine and Biology Society (EMBC), 2012 Annual International Conference of the IEEE. IEEE; 2012. p. 2901–2904.
[5] Gil E, Orini M, Bailón R, *et al.* Photoplethysmography pulse rate variability as a surrogate measurement of heart rate variability during non-stationary conditions. Physiological Measurement. 2010;31(9):1271.
[6] Zhang Z. Photoplethysmography-based heart rate monitoring in physical activities via joint sparse spectrum reconstruction. IEEE Transactions on Biomedical Engineering. 2015;62(8):1902–1910.
[7] Lu G, Yang F, Taylor J, *et al.* A comparison of photoplethysmography and ECG recording to analyse heart rate variability in healthy subjects. Journal of Medical Engineering & Technology. 2009;33(8):634–641.
[8] Allen J. Photoplethysmography and its application in clinical physiological measurement. Physiological Measurement. 2007;28(3):R1.
[9] Zijlmans M, Flanagan D, and Gotman J. Heart rate changes and ECG abnormalities during epileptic seizures: prevalence and definition of an objective clinical sign. Epilepsia. 2002;43(8):847–854.
[10] Jung SJ, Shin HS, and Chung WY. Driver fatigue and drowsiness monitoring system with embedded electrocardiogram sensor on steering wheel. IET Intelligent Transport Systems. 2014;8(1):43–50.
[11] Park BK, Boric-Lubecke O, and Lubecke VM. Arctangent demodulation with DC offset compensation in quadrature Doppler radar receiver systems. IEEE Transactions on Microwave Theory and Techniques. 2007;55(5):1073–1079.
[12] Lohman B, Boric-Lubecke O, Lubecke V, *et al.* A digital signal processor for Doppler radar sensing of vital signs. IEEE Engineering in Medicine and Biology Magazine. 2002;21(5):161–164.

[13] Massagram W, Lubecke VM, HØst-Madsen A, *et al.* Assessment of heart rate variability and respiratory sinus arrhythmia via Doppler radar. IEEE Transactions on Microwave Theory and Techniques. 2009;57(10):2542–2549.
[14] Xiao Y, Li C, and Lin J. A portable noncontact heartbeat and respiration monitoring system using 5-GHz radar. IEEE Sensors Journal. 2007;7(7): 1042–1043.
[15] Li C, Xiao Y, and Lin J. Experiment and spectral analysis of a low-power Ka-band heartbeat detector measuring from four sides of a human body. IEEE Transactions on Microwave Theory and Techniques. 2006;54(12):4464–4471.
[16] Chen KM, Misra D, Wang H, *et al.* An X-band microwave life-detection system. IEEE Transactions on Biomedical Engineering. 1986;(7):697–701.
[17] Bakhtiari S, Elmer TW, Cox NM, *et al.* Compact millimeter-wave sensor for remote monitoring of vital signs. IEEE Transactions on Instrumentation and Measurement. 2012;61(3):830–841.
[18] Stoica P, and Moses RL. Spectral analysis of signals. Upper Saddle River, NJ: Prentice Hall, 2005.
[19] Kim SG, Yun GH, and Yook JG. Compact vital signal sensor using oscillation frequency deviation. IEEE Transactions on Microwave Theory and Techniques. 2012;60(2):393–400.
[20] Gabriel C, Gabriel S, and Corthout E. The dielectric properties of biological tissues: I. Literature survey. Physics in Medicine & Biology. 1996;41(11):2231.
[21] Hong Y, Kim SG, Kim BH, *et al.* Noncontact proximity vital sign sensor based on PLL for sensitivity enhancement. IEEE Transactions on Biomedical Circuits and Systems. 2014;8(4):584–593.
[22] Hong Y, Kim SG, Kim BH, *et al.* Advanced non-contact near-field proximity vital sign sensor using phase locked loop. In: Microwave Conference (EuMC), 2013 European. IEEE; 2013. p. 605–608.
[23] Gonzalez G. Foundations of Oscillator Circuit Design. Boston: Artech house; 2007.
[24] Gonzalez G, and Sosa OJ. On the design of a series-feedback network in a transistor negative-resistance oscillator. IEEE transactions on microwave theory and techniques. 1999;47(1):42–47.
[25] An YJ, Yun GH, and Yook JG. Sensitivity enhanced vital sign detection based on antenna reflection coefficient variation. IEEE Trans Biomed Circuit sand Systems. 2016;10(2):319–327.
[26] Proakis JG, Salehi M, Zhou N, *et al.* Communication Systems Engineering. Vol. 2. Prentice Hall, New Jersey; 1994.
[27] Lee S, and Lee Y. A design method for microstrip directional couplers loaded with shunt inductors for directivity enhancement. IEEE Transactions on Microwave Theory and Techniques. 2010;58(4):994–1002.
[28] Kim H, Kozyrev AB, Karbassi A, *et al.* Linear tunable phase shifter using a left-handed transmission line. IEEE Microwave and Wireless Components Letters. 2005;15(5):366–368.
[29] Kim BH, Hong Y, An YJ, *et al.* A proximity coupling RF sensor for wrist pulse detection based on injection-locked PLL. IEEE Transactions on Microwave Theory and Techniques. 2016;64(5):1667–1676.

[30] An YJ, Yun GH, Kim SW, *et al.* Wrist pulse detection system based on changes in the near-field reflection coefficient of a resonator. IEEE Microwave and Wireless Components Letters. 2014;24(10):719–721.

[31] An YJ, Kim BH, Yun GH, *et al.* Flexible non-constrained RF wrist pulse detection sensor based on array resonators. IEEE Transactions on Biomedical Circuits and Systems. 2016;10(2):300–308.

[32] Kim BH, An YJ, Yun GH, *et al.* RF interferometric wearable wrist pulse detection system. IET Microwaves, Antennas & Propagation. 2017;12(2): 167–172.

[33] Tarassenko L, Villarroel M, Guazzi A, *et al.* Non-contact video-based vital sign monitoring using ambient light and auto-regressive models. Physiological Measurement. 2014;35(5):807.

[34] Nagae D, and Mase A. Measurement of heart rate variability and stress evaluation by using microwave reflectometric vital signal sensing. Review of Scientific Instruments. 2010;81(9):094301.

[35] Kim SG, Yun GH, and Yook JG. Wireless RF vital signal sensor using autoregressive spectral estimation method. IEEE Antennas and Wireless Propagation Letters. 2012;11:535–538.

Chapter 3

Wi-Fi-based sensing for gesture control applications

Mu-Cyan Tang[1], Yi-Chen Lai[1], Min-Hui Lin[1], Chien-Lun Chen[1], Chuan-Chi Chou[1], Fu-Kang Wang[1], Chia-Hung Yeh[1,2], Chien-Jung Li[3], and Tzyy-Sheng Jason Horng[1]

This chapter introduces a Wi-Fi-based sensing technology that is able to detect and recognize hand gestures with Wi-Fi signals. It covers the underlying theory and state-of-the-art sensor systems. It also demonstrates several gesture control applications with the additional use of machine learning and sensor fusion techniques.

3.1 Introduction

Touchless control using gestures for mobile devices, game stations, smart vehicles, and homes is gaining popularity and has attracted much attention from the global information and communications technology industry. Most gesture sensors are of the contact type; they include inertia sensors in the Wii system [1], bend sensors in the CyberGlove system [2], and electromyographic sensors in the Myo system [3]. However, noncontact gesture sensors are favored because they allow more freedom in the making of the gestures that they can sense. Well-known examples include Microsoft Kinect, Intel RealSense, and Leap Motion, which track a user's gestures using three-dimensional (3D) cameras [4]. However, a touchless control device with a 3D camera consumes considerable power and is affected by ambient light. Such a device cannot detect through obstacles. Other infrared (IR)-based [5] and electric-near-field-based [6] gesture sensors may solve the problems of power consumption and background light, but only at the expense of a limited field-of-view (FOV) or reduced sensing range, and they also cannot detect through obstacles.

Ultrasonic [7–9] and microwave [10–13] Doppler radar sensors consume much less power and respond much faster to gestures than do the aforementioned

[1]Department of Electrical Engineering, National Sun Yat-Sen University, Taiwan
[2]Department of Electrical Engineering, National Taiwan Normal University, Taiwan
[3]Sivann Inc., Taiwan

non-Doppler-based sensors. Moreover, these radars can sense gestures when they are concealed by clothes, bags, or even walls. Nonetheless, interference between such radars, which all use similar frequencies, is problematic as it degrades the captured Doppler signal. Ambient noise and interferers also degrade Doppler signals. This problem is particularly evident for microwave Doppler radar systems that operate in Industrial Scientific Medical (ISM) bands because these radars may be overwhelmed by a large number of wireless network devices that are nearby, such as Wi-Fi and Bluetooth. Unfortunately, microwave Doppler radar systems have not been regulated to resolve radio interference with wireless network devices.

Passive radar is a receive-only radar that exploits radio wave from the environment as its transmission source. Over the last decade or more, many communication signals have been utilized as "illuminators of opportunity" for this radar, including television, direct-broadcast satellite, FM radio, global system for mobile communication (GSM), worldwide interoperability for microwave access (WiMAX), digital audio/video broadcast (DAB/DVB), and Wi-Fi signals [14–21]. A conventional passive radar comprises two receivers with their own antennas to receive the direct-path reference signal from the opportunistic transmitter and the echo signal that is reflected from the target. Then, the Doppler information can be processed by correlating/mixing these two received signals. However, such a signal-reception mechanism is difficult to implement indoors because of multipath reflections.

Recently, based on the concept of passive radar, gesture sensors that use wireless communication signals have been extensively reported upon. Pu *et al.* presented a WiSee system to sense and identify human gestures and motions by leveraging Wi-Fi signals in an indoor environment [22]. Kellogg *et al.* proposed an Allsee system that extracts Doppler information from the amplitude of wireless communication signals, such as TV and radio frequency identification (RFID) transmissions, for use in sensors with low complexity and power consumption, to enable always-on gesture recognition by smartphones and tablets [23]. Wang *et al.* demonstrated a Doppler radar sensor that is based on an injection-locked quadrature receiver (ILQR) for sensing gestures with ambient wireless communication signals [24]. The radar by Wang *et al.* [24] uses a concept that is similar to that of the passive radar, which detects moving objects by processing echoes from opportunistic transmitters. However, it differs from the passive radar by the addition of a retransmitting and coupling mechanism to separate the received echo signal from the received opportunity signal. Based on prior research [24–26], this chapter introduces a Wi-Fi-based sensing technology for detecting and identifying hand gestures using Wi-Fi signals from an access point (AP) or an ambient environment. The chapter is organized as follows.

Section 3.2 introduces the underlying injection-locking theory and uses it to model the local oscillator (LO) of the ILQR. It derives the generalized locking equation for an oscillator with an injected modulated signal (Section 3.2.1) and describes the crucial characteristics of injection-locking and injection-pulling phenomena, such as locking range and lock-in time (Section 3.2.2), frequency pulling (Section 3.2.3), and synchronization (Section 3.2.4). This section also provides a discrete-time method for solving the derived differential equation that is applicable to injection-locking analysis (Section 3.2.5). Section 3.3 presents a passive radar that uses Wi-Fi/long term evolution (LTE) signals for sensing gestures, including its system architecture (Section 3.3.1), principles of sensing (Section 3.3.2), system performance simulation and verification (Section 3.3.3), and experimental results (Section 3.3.4). Section 3.4 presents applications of a Wi-Fi-based passive radar to gesture control in, for example, swiping pages (Section 3.4.1), and playing games (Section 3.4.2) using hand gestures on a tablet. This section also addresses the use of machine learning to recognize hand gestures (Section 3.4.3) and describes a sensor fusion technique that combines the Wi-Fi-based passive radar with a camera to track a 3D hand trajectory (Section 3.4.4).

3.2 Injection locking with a modulated signal

This section begins by deriving the generalized locking equation for an oscillator with a modulated injection signal. Background information on injection locking and pulling is provided. A discrete-time solution to the derived locking equation is given to analyze the behavior of an oscillator under the injection of a modulated signal with a time-varying envelope.

3.2.1 Generalized locking equation

Adler [27] and many others [28–34] have studied the synchronization of an oscillator with the injection of a sinusoidal signal. Adler used a vector diagram to describe the relation between an oscillator output signal and an injection signal; he derived a phase differential equation that accounts for many observed phenomena of synchronization due to the injection of a weak signal into a free-running oscillator. In 1947, Huntoon and Weiss [28] developed a synchronization equation that was similar to Adler's equation by considering a fictitious change of load impedance subject to injections. In 1965, Paciorek [30] extended Adler's work to the application of stronger injection signals. In the 2000s, Razavi [33] and Mirzaei [34] derived a locking equation that is identical to that of Adler and Paciorek using a different approach. Although a great deal of attention has been paid to sinusoidal signal injection, a derivation of a locking equation for modulated injection signals is lacking. Based on Adler's approach, this subsection develops a detailed model of an oscillator with modulation signal injections [35–37]. This model will be used to characterize the LO of the ILQR in the passive radar that is described in Section 3.3.

Figure 3.1 displays the block diagram of an oscillator, which consists of an amplifier-limiter, a tank circuit, and a feedback path. The amplifier-limiter is an

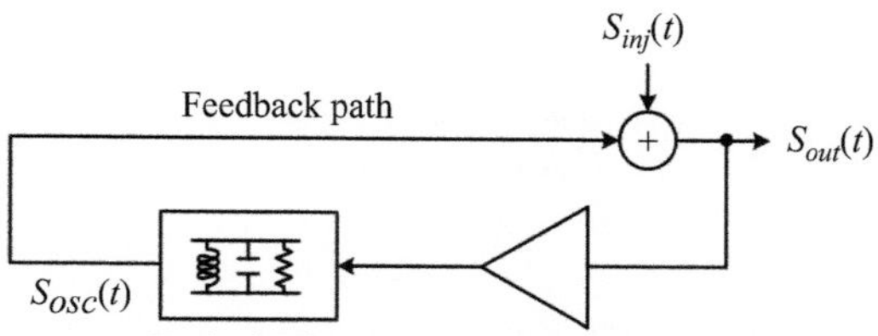

Figure 3.1 Block diagram of an oscillator under injection

amplifier with a nonlinear element to stabilize the amplitude of a self-excited oscillator. The tank circuit has a resonant frequency that determines the frequency of a free-running oscillator. The feedback path forms a loop to establish the oscillation conditions. Notably, the feedback path includes a summing point to allow the injection of signals. In Figure 3.1, $S_{osc}(t)$ represents the inherent oscillation signal, $S_{inj}(t)$ represents the injection signal, and $S_{out}(t)$ is the resultant oscillator output signal under injection.

First, consider no injection of an external signal into an oscillator that is free-running at a frequency that is determined by the tank circuit. Under this condition, $S_{out}(t)$ and feedback signal $S_{osc}(t)$ must coincide for the oscillator to maintain oscillation at the resonant frequency of the tank circuit. Next, consider the injection of a sinusoidal signal of a slightly different frequency into the oscillator. As shown in Figure 3.1, the feedback signal $S_{osc}(t)$ combines with the injection signal $S_{inj}(t)$ to output $S_{out}(t)$. The addition of $S_{inj}(t)$ induces a phase shift between $S_{osc}(t)$ and $S_{out}(t)$, causing the free-running oscillator to change to a new oscillation frequency ω_{inj} for introducing a necessary phase shift in the tank circuit to satisfy the new oscillation condition.

Figure 3.2 provides a vector representation of the signals in Figure 3.1. First, consider the modulated injection signal $S_{inj}(t)$ with a time-varying amplitude $E_{inj}(t)$ and an instantaneous frequency $\omega_{inj}(t)$, and $S_{inj}(t)$ is at rest at any instant with respect to the observer. The instantaneous frequency of the injection signal is given by

$$\omega_{inj}(t) = \omega_{inj} + \frac{d\theta_i(t)}{dt} \tag{3.1}$$

where ω_{inj} is the center frequency of $S_{inj}(t)$ and $d\theta_i(t)/dt$ denotes the instantaneous modulation frequency of $S_{inj}(t)$. The inherent oscillation signal $S_{osc}(t)$ has a constant amplitude E_{osc} without perturbation and the oscillator is free-running at frequency ω_{osc}. The resultant oscillator output signal $S_{out}(t)$ is also considered to have a constant amplitude E_{out} and an instantaneous output frequency $\omega_{out}(t)$. As depicted

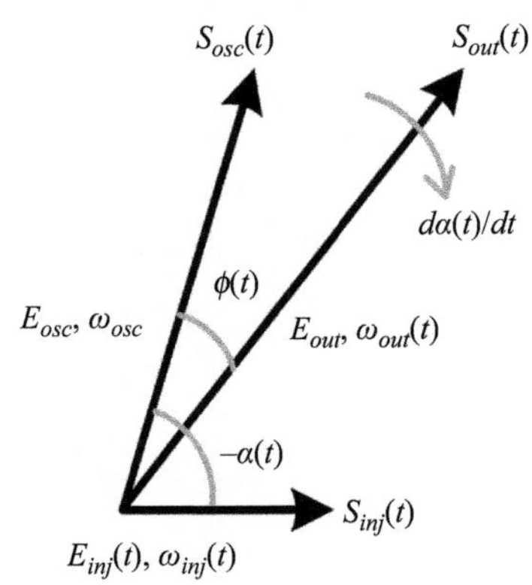

Figure 3.2 Vector diagram of the signals in an oscillator under injection

in Figure 3.2, $S_{out}(t)$ can also be regarded as a vector that is rotating clockwise with a beat frequency $d\alpha(t)/dt$ with respect to $S_{inj}(t)$, so that the instantaneous frequency $\omega_{out}(t)$ can be expressed as

$$\omega_{out}(t) = \omega_{inj}(t) + \frac{d\alpha(t)}{dt} \tag{3.2}$$

From the vector diagram in Figure 3.2, the following relations at any instant can be derived:

$$E_{out} \sin \phi(t) = E_{inj}(t)\sin(-\alpha(t)) \tag{3.3}$$

$$E_{out} \cos \phi(t) = E_{osc} + E_{inj}(t)\cos(-\alpha(t)) \tag{3.4}$$

where the phase difference $\phi(t)$ between $S_{osc}(t)$ and $S_{out}(t)$ arises from the addition of $S_{inj}(t)$. Dividing (3.3) by (3.4) yields:

$$\tan \phi(t) = -\frac{E_{inj}(t)\sin \alpha(t)}{E_{osc} + E_{inj}(t)\cos \alpha(t)} \tag{3.5}$$

Substituting (3.1) and (3.2) into the phase shift for a tank circuit [38] yields:

$$\tan \phi(t) = \frac{2Q}{\omega_{osc}}[\omega_{out}(t) - \omega_{osc}] = \frac{2Q}{\omega_{osc}}\left\{\frac{d\alpha(t)}{dt} - \left[\omega_{osc} - \omega_{inj} - \frac{d\theta_i(t)}{dt}\right]\right\} \tag{3.6}$$

where Q is the quality factor of the tank circuit. Setting (3.5) equal to (3.6) results in

$$\frac{d\alpha(t)}{dt} = -B(t)\sin \alpha(t) + \Delta\omega_{osc} - \frac{d\theta_i(t)}{dt} \tag{3.7}$$

where

$$B(t) = \frac{\omega_{osc}}{2Q}\frac{E_{inj}(t)}{E_{osc}}\frac{1}{1 + \frac{E_{inj}(t)}{E_{osc}}\cos \alpha(t)} \tag{3.8}$$

and

$$\Delta\omega_{osc} = \omega_{osc} - \omega_{inj} \tag{3.9}$$

is the undisturbed beat frequency that is the spontaneous frequency separation between the oscillator inherent frequency and the injection frequency. Unlike in other studies [27,28], the beat frequency $d\alpha(t)/dt$ in (3.7) evidently relates to the instantaneous modulation frequency $d\theta_i(t)/dt$ and the injection amplitude $E_{inj}(t)$. Equation (3.7) is the generalized locking equation that describes the phase dynamics of an oscillator under the influence of a modulated injection signal with time-varying envelope and phase. Substituting (3.7) into (3.2) yields the instantaneous frequency of the resultant oscillator output signal, given by

$$\omega_{out}(t) = \omega_{osc} - B(t)\sin \alpha(t) \tag{3.10}$$

Equation (3.10) indicates an amplitude-modulation to frequency-modulation (AM-FM) effect of a modulated injection signal with a time-varying envelope. This effect induces frequency distortion of an oscillator that is synchronized to a modulated injection signal.

Considering a sinusoidal injection signal with constant amplitude and frequency, $E_{inj}(t) = E_{inj}$ and $d\theta_i(t)/dt = 0$, (3.7) becomes:

$$\frac{d\alpha(t)}{dt} = -\frac{\omega_{osc}}{2Q}\frac{E_{inj}}{E_{osc}}\frac{\sin\,\alpha(t)}{1+\dfrac{E_{inj}}{E_{osc}}\cos\,\alpha(t)} + \Delta\omega_{osc} \tag{3.11}$$

which is identical to the result of Paciorek [30]. Substituting a current ratio I_{inj}/I_{osc} for the voltage ratio E_{inj}/E_{osc} in (3.11) yields the generalized Adler's equation that was derived by Mirzaei [34]. Consider a sinusoidal injection signal of low amplitude. Now, (3.11) can be simplified to the well-known Adler's equation [27]:

$$\frac{d\alpha(t)}{dt} = -\frac{\omega_{osc}}{2Q}\frac{E_{inj}}{E_{osc}}\sin\,\alpha(t) + \Delta\omega_{osc} \tag{3.12}$$

3.2.2 Locking range and lock-in time

The locking range ω_{LR} is defined as the maximum undisturbed beat frequency for an oscillator to be locked with a sinusoidal injection signal. Upon locking, the phase difference $\alpha(t)$ approaches a fixed value α_{ss} as time increases. This value α_{ss} represents the phase difference between $S_{osc}(t)$ and $S_{inj}(t)$ in the steady state; it is obtained by making (3.11) approach zero:

$$\Delta\omega_{osc} \approx \frac{\omega_{osc}}{2Q}\frac{E_{inj}}{E_{osc}}\frac{\sin\,\alpha_{ss}}{1+\dfrac{E_{inj}}{E_{osc}}\cos\,\alpha_{ss}} \tag{3.13}$$

Differentiating $\Delta\omega_{osc}$ with respect to α_{ss} reveals that the undisturbed beat frequency is maximal when $\cos\,\alpha_{ss} = -E_{inj}/E_{osc}$. Substituting this condition back into (3.13) leads to an equation for locking range that can be found elsewhere [30]:

$$\omega_{LR} = \max(\Delta\omega_{osc}(\alpha_{ss})) = \frac{\omega_{osc}}{2Q}\frac{E_{inj}}{E_{osc}}\frac{1}{\sqrt{1-\left(\dfrac{E_{inj}}{E_{osc}}\right)^2}} \tag{3.14}$$

When the injection level is low, (3.14) can then be simplified as [27]:

$$\omega_{LR} \approx \frac{\omega_{osc}}{2Q}\frac{E_{inj}}{E_{osc}} \tag{3.15}$$

Note that ω_{LR} is approximately proportional to the voltage amplitude ratio E_{inj}/E_{osc} or the square root of the power ratio $\sqrt{P_{inj}/P_{osc}}$, where P_{inj} and P_{osc} are the inherent oscillation output power and the injection signal power, respectively. When $\Delta\omega_{osc}$ is less than ω_{LR}, the pull-in process [27] pulls the oscillator output frequency $\omega_{out}(t)$ to lock on to the injection frequency ω_{inj}, so the oscillator is under the injection-locking condition.

In the steady state of an oscillator with weak injection, $\alpha(t)$ in (3.12) approaches a constant value α_{ss}, which can be obtained by setting (3.12) to zero:

$$\alpha_{ss} = \sin^{-1}\left(\frac{\Delta\omega_{osc}}{\omega_{LR}}\right) \tag{3.16}$$

A lock-in time τ_l is defined as the partial derivative of α_{ss} with respect to $\Delta\omega_{osc}$:

$$\tau_l \triangleq \frac{\partial(\alpha_{ss})}{\partial(\Delta\omega_{osc})} = \frac{1}{\sqrt{\omega_{LR}{}^2 - \Delta\omega_{osc}{}^2}} \tag{3.17}$$

3.2.3 Frequency pulling

When a signal outside the locking range is injected into the oscillator, $|\Delta\omega_{osc}| > \omega_{LR}$, pull-in may never be achieved and $\omega_{out}(t)$ will be modulated with $d\alpha(t)/dt$, which has an average beat frequency ω_b. In this situation, the oscillator undergoes injection pulling and the average beat frequency is given by [33]:

$$\omega_b = \sqrt{\Delta\omega_{osc}^2 - \omega_{LR}^2} \tag{3.18}$$

Figure 3.3 shows the injection-pulling behavior of an oscillator using a series of correlated graphs. Injection pulling can be divided into fast-beat and quasi-lock [33], as depicted in Figure 3.3(a) and (b), respectively. In each figure, the top graph plots

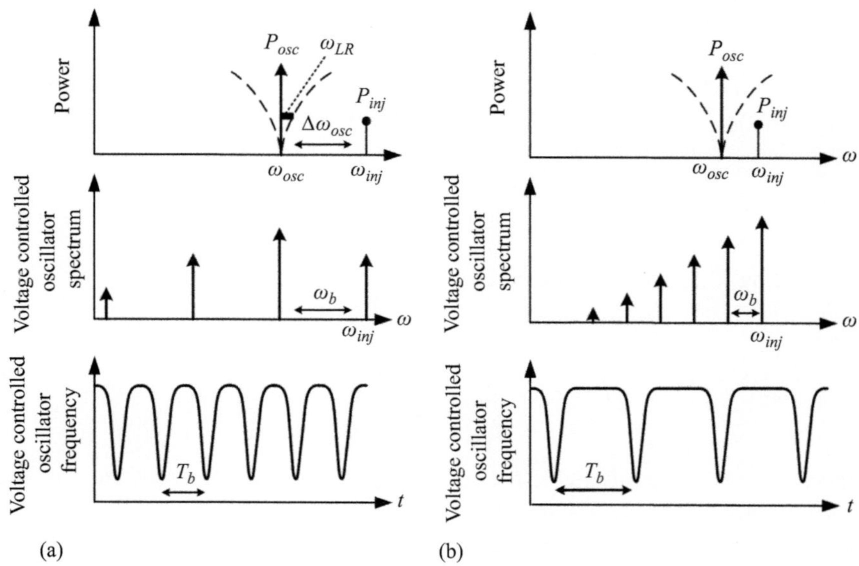

Figure 3.3 Injection-pulling behavior of an oscillator. (a) Fast-beat. (b) Quasi-lock (from [35])

power versus frequency for the inherent oscillation signal $S_{osc}(t)$ and the injection signal $S_{inj}(t)$. The broken curve is the injection-locking characteristic curve of the oscillator, and ω_{LR} is the locking range with respect to an injection signal power P_{inj}. The middle graph shows the spectrum of the oscillator that is pulled by the injection signal. The oscillator output spectral lines are equally separated, with a frequency spacing of ω_b given by (3.18). The bottom graph plots the oscillator output frequency $\omega_{out}(t)$, which is time-varying with a period of $T_b = 2\pi/\omega_b$ during the injection-pulling process.

Figure 3.3(a) shows the fast-beat condition with ω_{inj} far beyond the locking range, $|\omega_{inj} - \omega_{osc}| > \omega_{LR}$, whereas Figure 3.3(b) shows the quasi-lock condition with ω_{inj} approaching the locking range, $|\omega_{inj} - \omega_{osc}| \rightarrow \omega_{LR}$. Under the fast-beat condition, the oscillator output spectral lines are separated by ω_b and have approximately equal power levels at ω_{inj} and $\omega_{inj} - 2\omega_b$, as shown in the middle graph of Figure 3.3(a). Under the quasi-lock condition, the oscillator output spectrum has a peak of significant magnitude at ω_{inj}, as shown in the middle graph of Figure 3.3(b). The bottom graphs of Figure 3.3(a) and (b) reveal that the oscillator output frequency $\omega_{out}(t)$ varies with an average beat cycle of T_b, which is longer under the quasi-lock condition than under the fast-beat condition.

3.2.4 Synchronization

Since an oscillator can synchronize with a weak injection signal, two of the popular uses of an injection-locked oscillator (ILO) are as a high-gain amplifier [28,29,31,39] and as a frequency demodulator [40–42]. For amplifier applications, the technique can amplify an injected angle-modulated signal such as a phase-modulated (PM) or a frequency-modulated (FM) signal under the conditions $E_{inj}(t) = E_{inj}$ and $d\theta_i(t)/dt \neq 0$. This subsection uses an approach that is based on (3.7) to investigate the synchronization of an oscillator with a general angle-modulated signal. When the oscillator is tuned to the injection frequency for amplifier applications and then synchronized with the injected angle-modulated signal, $\Delta\omega_{osc} = 0$ and $d\alpha(t)/dt$ equals $d\theta_i(t)/dt$. From (3.7):

$$\frac{d\alpha(t)}{dt} - \frac{d\theta_i(t)}{dt} = -B(t)\sin\alpha(t) - 2\frac{d\theta_i(t)}{dt} = 0 \tag{3.19}$$

Substituting (3.8) into (3.19) and rearranging yields:

$$\frac{d\theta_i(t)}{dt} = -\frac{1}{2}\left[\frac{\omega_{osc}}{2Q}\frac{E_{inj}}{E_{osc}}\frac{\sin\alpha(t)}{1+\dfrac{E_{inj}}{E_{osc}}\cos\alpha(t)}\right] \tag{3.20}$$

where E_{inj} is the amplitude of the modulated signal with the constant envelope. Since in the steady state, the maximum value of the bracket term on the right-hand side of (3.20) equals the locking range in (3.14). Therefore:

$$\left|\frac{d\theta_i(t)}{dt}\right| < \frac{\omega_{LR}}{2} \tag{3.21}$$

Inequality (3.21) states that the instantaneous modulation frequency cannot exceed half of the locking range if the synchronization condition is satisfied. It motivates consideration of practical PM or FM signals. First, an injection is assumed to be a PM signal of the form:

$$S_{PM}(t) = \cos\left(\omega_{inj}t + \theta_i(t)\right) \tag{3.22}$$

where

$$\theta_i(t) = \beta \cos(\omega_m t) \tag{3.23}$$

is the instantaneous phase variation, ω_m is the modulation rate, and β is the modulation index of the PM signal with a center frequency of ω_{inj}. Notably, β is also treated as a peak phase deviation $\Delta\phi_{peak}$ of the PM signal. Substituting (3.23) into (3.21) yields:

$$|\Delta\phi_{peak}| < \frac{1}{2}\frac{\omega_{LR}}{\omega_m} \tag{3.24}$$

Next, considering an injected FM signal by substituting modulation index β in terms of $\Delta\omega_{peak}/\omega_m$ into (3.21) and (3.23) gives:

$$|\Delta\omega_{peak}| < \frac{\omega_{LR}}{2} \tag{3.25}$$

where $\Delta\omega_{peak}$ represents the peak frequency deviation of the FM signal. This result is consistent with the claim [28] that the distortion will be negligible if the peak frequency deviation of an FM signal is less than half of the locking range when the ILO is used as an FM amplifier. Finally, a digital angle-modulated signal with a modulation index h and a bit rate f_{br} has a peak frequency deviation given by

$$\Delta\omega_{peak} = \pi f_{br} h \tag{3.26}$$

Substituting (3.26) into (3.25) and rearranging yields:

$$f_{br} < \frac{1}{2\pi}\frac{\omega_{LR}}{h} \tag{3.27}$$

Inequalities (3.24), (3.25), and (3.27) constrain the synchronization conditions that are related to the locking range for PM, FM, and digital angle-modulated signals, respectively. The above derivations provide the design criteria for an injection-locked receiver for detecting various constant-envelope-modulated signals.

Two experiments were conducted to detect Gaussian minimum shift keying (GMSK) signals with an injection-locked receiver. Example A is a GMSK signal with a power of −66.56 dBm and a data rate of 100 kbps, which is less than the upper bound on the data rate (satisfying (3.27)). In contrast, example B has a power of −56.56 dBm and a data rate of 2 Mbps, which exceeds the upper bound on the data rate. Figure 3.4(a) and (b) shows demodulated signal constellation and eye diagrams of examples A and B, respectively. The error vector magnitude of example A is 1.39% while that of example B is 22.59%. These two examples

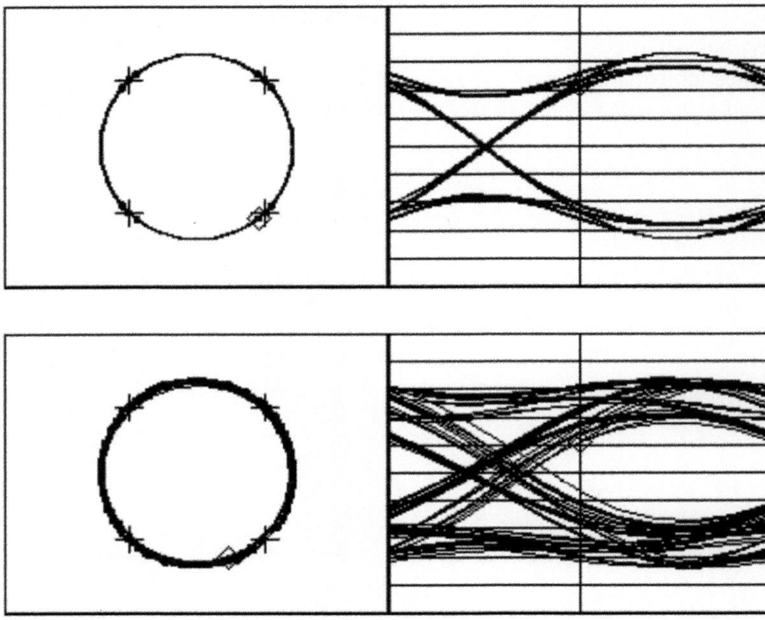

(a)

(b)

Figure 3.4 Output constellation and eye diagrams of the injection-locked receiver. (a) Detection of a GMSK signal with −66.56 dBm power and 100 kbps data rate. (b) Detection of a GMSK signal with −56.56 dBm power and 2 Mbps data rate (from [35])

provide evidence that the injection-locked receiver can perform signal detection well only below the upper bound, as indicated by (3.27).

3.2.5 Discrete-time analysis

To predict the behavior of the LO of an ILQR with an injection signal from a nonconstant envelope modulation system, a discrete-time dynamic model was established, based on (3.7). First, (3.7) was discretized with respect to time to obtain:

$$\frac{\alpha(t_i) - \alpha(t_{i-1})}{\Delta t} = \Delta\omega_{osc} - B(t_i)\sin\,\alpha(t_i) - \frac{\theta(t_i) - \theta(t_{i-1})}{\Delta t} \tag{3.28}$$

where $\Delta t = t_i - t_{i-1}$ is the period for calculation, and

$$B(t_i) = \frac{\omega_{osc}}{2Q}\frac{E_{inj}(t_i)}{E_{osc}}\frac{1}{1 + \dfrac{E_{inj}(t_i)}{E_{osc}}\cos\,\alpha(t_i)} \tag{3.29}$$

From (3.1) and (3.2), the oscillator output can be viewed as a signal that is centered at a frequency ω_{inj} and phase-modulated by a function $\phi_m(t)$, satisfying:

$$\frac{d\phi_m(t)}{dt} = \frac{d\alpha(t)}{dt} + \frac{d\theta(t)}{dt} \tag{3.30}$$

In the discrete-time domain, this oscillator phase modulation function can be represented as

$$\phi_m(t_i) = \phi_m(t_{i-1}) + \alpha(t_i) - \alpha(t_{i-1}) + \theta(t_i) - \theta(t_{i-1}) \tag{3.31}$$

Substituting (3.28) into (3.31) yields:

$$\phi_m(t_i) = \phi_m(t_{i-1}) + [\Delta\omega_{osc} - B(t_i)\sin\,\alpha(t_i)]\Delta t \tag{3.32}$$

When Δt is small, $\alpha(t_i)$ is approximately $\alpha(t_{i-1})$. Replacing $\alpha(t_i)$ with $\alpha(t_{i-1})$ in the right-hand side of (3.28), (3.29), and (3.32) and rearranging gives:

$$\alpha(t_i) = \alpha(t_{i-1}) + [\Delta\omega_{osc} - B(t_i)\sin\,\alpha(t_{i-1})]\Delta t - \theta(t_i) + \theta(t_{i-1}) \tag{3.33}$$

$$B(t_i) = \frac{\omega_{osc}(t_i)}{2Q}\frac{E_{inj}(t_i)}{E_{osc}}\frac{1}{1 + \dfrac{E_{inj}(t_i)}{E_{osc}}\cos\,\alpha(t_{i-1})} \tag{3.34}$$

$$\phi_m(t_i) = \phi_m(t_{i-1}) + [\Delta\omega_{osc} - B(t_i)\sin\,\alpha(t_{i-1})]\Delta t \tag{3.35}$$

Using (3.33) and (3.34), $\alpha(t_i)$ can recursively be calculated. To do so, (3.33) and (3.35) are rewritten with an index n that represents $\alpha(t_i)$ and $\phi_m(t_i)$ as a combination of data sequences in the discrete-time domain [43]:

$$\alpha[n] = \alpha[n-1] + (\Delta\omega_{osc} - B[n]\cdot\,\sin\,\alpha[n-1])\Delta t - \theta[n] + \theta[n-1] \tag{3.36}$$

$$\phi_m[n] = \phi_m[n-1] + (\Delta\omega_{osc} - B[n]\cdot\,\sin\,\alpha[n-1])\Delta t \tag{3.37}$$

where

$$B[n] = \frac{\omega_{osc}}{2Q}\frac{E_{inj}[n]}{E_{osc}}\frac{1}{1 + \dfrac{E_{inj}[n]}{E_{osc}}\cos\,\alpha[n-1]} \tag{3.38}$$

The phase modulation function is finally expressed as a pulse train that is weighted by $\phi_m[n]$, given by (3.37). The oscillator output waveform can be reconstructed as

$$S_{out}(t) = E_{out}\cos\left(\omega_{inj}t + \sum_n \phi_m[n]\cdot h_r(t - n\Delta t)\right) \tag{3.39}$$

where

$$h_r(t) = \frac{\sin(\pi t/\Delta t)}{(\pi t/\Delta t)} \tag{3.40}$$

is the impulse response of the reconstruction filter. The oscillator output spectrum can then be predicted by taking the Fourier transform of (3.39).

3.3 Passive radar using Wi-Fi/LTE signals

This section presents a passive radar that receives the reference and echo signals of Wi-Fi/LTE for processing to measure Doppler shifts for moving targets. Figure 3.5 shows a simplified model of the radar in which an input reference signal of Wi-Fi/LTE $S_{REF}(t)$ is delivered to an antenna for transmission and an oscillator for injection locking. Part of the reference signal $S_{REF}(t)$ is reflected from a moving target back to the antenna and sent to a mixer-based detector as an echo signal of Wi-Fi/LTE, $S_{ECHO}(t)$. The ILO that is obtained by injecting the reference signal $S_{REF}(t)$ into the oscillator is equivalent to a cascade of an extremely narrow band-pass filter with a center frequency at the oscillation frequency ω_{osc} and an amplifier-limiter with a fixed output magnitude that equals the oscillation amplitude. As inferred from (3.21), the ILO output signal $S_{ILO}(t)$ is partially synchronized to the reference signal $S_{REF}(t)$ with the same frequency components as the latter signal within half of the locking range. The mixing of the echo and ILO output signals in the detector largely cancels the noise due to the Wi-Fi/LTE modulation within half of the locking range, so the Doppler shift $\phi_d(t)$ can be extracted from the echo signal after low-pass filtering.

Wi-Fi/LTE signals are ubiquitous so they can be used to sense moving targets anytime and anywhere. Furthermore, Wi-Fi/LTE-based radar can be operated with existing Wi-Fi/LTE communication devices that adopt time division duplex (TDD). This is attributed to that the reception processing of echo Wi-Fi/LTE signals does not interfere with that of downlink Wi-Fi/LTE signals because the former occurs only in the absence of the latter.

3.3.1 System architecture

As shown in the block diagram of Figure 3.6, the passive radar has a core ILQR, a coupler, a splitter, an echo signal receive (RXe) antenna, and a reference port that is connected to a Wi-Fi/LTE transmitter or a reference signal receive (RXr) antenna. The ILQR comprises an ILO, a quadrature demodulator, and two low-pass

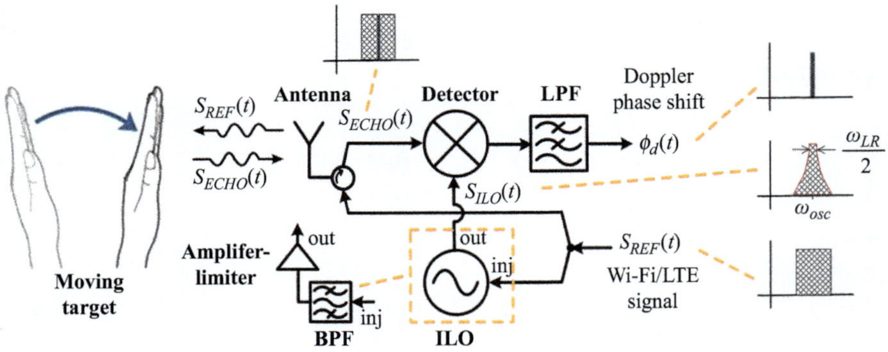

Figure 3.5 A simplified model of the passive radar using Wi-Fi/LTE signals. BPF: band-pass filter

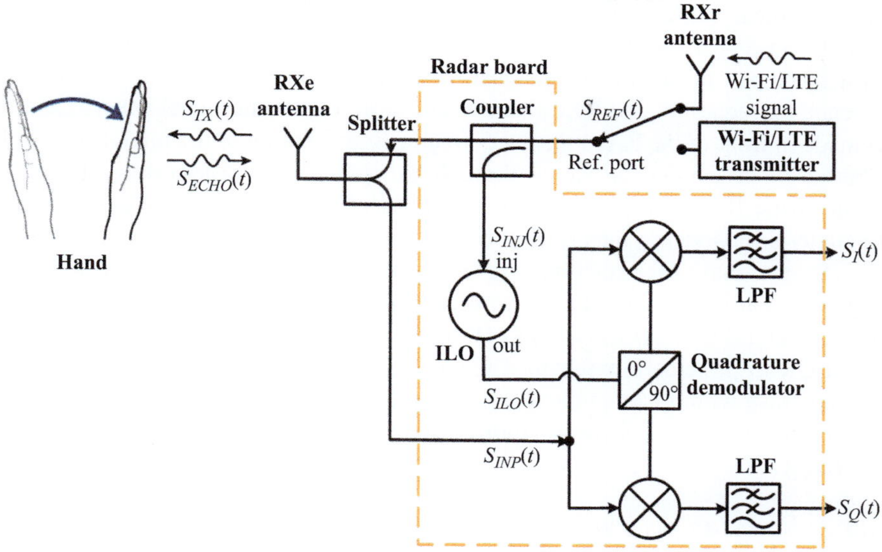

Figure 3.6 Block diagram of the ILQR-based passive radar

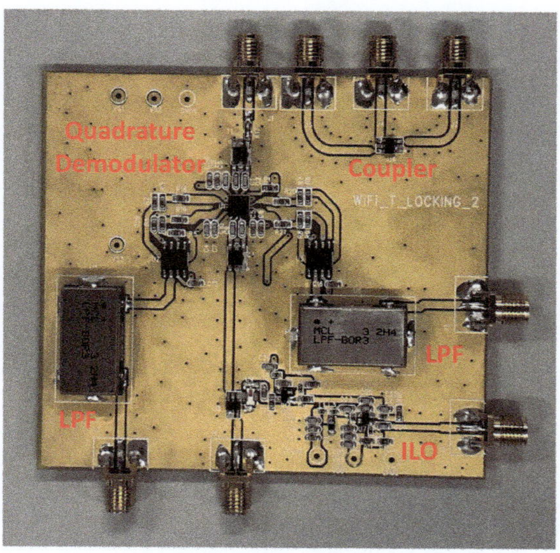

Figure 3.7 A prototype board of the ILQR-based passive radar

filters (LPFs). The output baseband in-phase (I) and quadrature (Q) signals of the ILQR are sampled by a data acquisition (DAQ) unit and then processed by a LabVIEW program to obtain Doppler information. Figure 3.7 shows a prototype board of the radar that operates in the 2.3–2.5 GHz frequency range to cover the 2.4 GHz Wi-Fi band and LTE band 40.

The radar operates as follows. It receives a Wi-Fi/LTE signal as a reference signal from either the input of the Wi-Fi/LTE transmitter or the RXr antenna and then sends part of the reference signal through the coupler and splitter to the RXe antenna for radiation. The RXe antenna then receives an echo signal when a gesture is made in front of the RXe antenna during the transmission of the reference signal. This received echo signal passes through the splitter to the radio frequency (RF) input of the ILQR. Notably, the polarizations of the RXr and RXe antennas are mutually orthogonal to maximize their mutual isolation.

Another part of the reference signal is coupled through the coupler to the ILO, achieving partial synchronization with the reference signal as a result of injection locking. The ILO, serving as an LO of the ILQR, mixes with the input RF signal of the ILQR to produce a baseband Doppler signal with canceled Wi-Fi/LTE modulation noise within half of the locking range. The combination of the coupler and the splitter ensures excellent isolation between the echo and the reference signal in the ILQR.

3.3.2 Sensing principle

In the analysis of the system in Figure 3.6, the Wi-Fi/LTE signal at the reference port is mathematically described by

$$S_{REF}(t) = A_m(t)\cos(\omega_c t + \phi_m(t)) \tag{3.41}$$

where $A_m(t)$ and $\phi_m(t)$ are, respectively, the amplitude and phase modulations of the carrier wave at frequency ω_c. Assume that the path attenuation between the reference port and RXe antenna is A_t. The signal that is transmitted by the RXe antenna is written as

$$S_{TX}(t) = A_t \cdot A_m(t)\cos(\omega_c t + \phi_m(t)) \tag{3.42}$$

When a gesture is made, the transmitted signal $S_{TX}(t)$ is reflected from the hand and then received by the RXe antenna as an echo signal $S_{ECHO}(t)$. This echo signal is further coupled via the splitter into the ILQR to become an input signal of the ILQR $S_{INP}(t)$, given by

$$S_{INP}(t) = A_s A_t \rho(\tau_p(t)) A_m(t - \tau_p(t))\cos(\omega_c t + \phi_m(t - \tau_p(t)) + \phi_d(t)) \tag{3.43}$$

where A_s is the insertion attenuation of the splitter; $\tau_p(t)$ is the instantaneous round-trip time between the hand and the RXe antenna; $\rho(\tau_p(t))$ is the echo coefficient of the moving hand, which depends on the distance or round-trip time between the hand and the RXe antenna; and $\phi_d(t) = \omega_c \tau_p(t)$ is the Doppler phase shift that is caused by the hand motion associated with the gesture. Part of the reference signal $S_{REF}(t)$ is coupled to the ILQR as an injection signal, as represented by (3.44):

$$S_{INJ}(t) = C \cdot A_m(t)\cos(\omega_c t + \phi_m(t)) \tag{3.44}$$

where C is the coupling coefficient of the coupler.

In the ILQR, the input signal $S_{INP}(t)$ is split into two in-phase signals of equal amplitude to provide the RF signals of the quadrature demodulator. The injection signal $S_{INJ}(t)$ is injected into the ILO, which functions as a high-gain and high-selectivity band-pass amplifier at an offset frequency ω_s that is less than half of the locking range ω_{LR} of the ILO, yielding an ILO output signal that can be expressed as

$$S_{ILO}(t) = A_{ILO}\cos(\omega_c \cdot (t - \tau_l) + \phi_m(t - \tau_l)|_{\omega_s<\omega_{LR}/2} + \phi_p(t)) \tag{3.45}$$

where A_{ILO} is the amplitude of oscillation and ω_{LR} is as given by (3.14). Equation (3.45) indicates that the output signal of the ILO preserves the low-frequency phase modulation of the reference signal within half of the locking range of the ILO. In (3.45), $\phi_p(t)$ represents the injection-pulled part of the phase modulation that happens when the offset frequency ω_s exceeds $\omega_{LR}/2$ or the reference signal is near zero crossings of its envelope. The term τ_l is the ILO's lock-in time, which is given by (3.17).

The ILO output signal, given by (3.45), is split into two quadrature signals of equal amplitude as the LO signals of the quadrature demodulator. Low-pass filtering with a cutoff frequency ω_{LPF} set below $\omega_{LR}/2$ yields the baseband I and Q signals of the ILQR as

$$\begin{aligned} S_I(t) = {} & G_r A_s A_t \rho(\tau_p(t)) A_m\big(t - \tau_p(t) - \tau_r\big) \\ & \times \cos\big(\omega_c \cdot (\tau_l - \tau_r) + \Delta\phi_m(t) + \Delta\phi_p(t) + \phi_d(t)\big) \end{aligned} \tag{3.46}$$

$$\begin{aligned} S_Q(t) = {} & G_r A_s A_t \rho(\tau_p(t)) A_m\big(t - \tau_p(t) - \tau_r\big) \\ & \times \sin\big(\omega_c \cdot (\tau_l - \tau_r) + \Delta\phi_m(t) + \Delta\phi_p(t) + \phi_d(t)\big) \end{aligned} \tag{3.47}$$

where G_r is the gain of the quadrature demodulator; τ_r is the relative delay of the RF path to the LO path in the ILQR; and $\Delta\phi_m(t) = \phi_m(t - \tau_p(t) - \tau_r) - \phi_m(t - \tau_l)|\omega_s < \omega_{LPF}$ and $\Delta\phi_p(t) = -\phi_p(t)|\omega_s < \omega_{LPF}$ represent the residual phase modulations that arise from the injection locking and injection pulling of the ILO, respectively. Generally, G_r depends on the ILO output signal level and $\tau_p(t)$ is negligible compared to τ_r or τ_l for gesture detection at short range. Finally, the magnitude and phase of the in-phase and quadrature (IQ) signal vector are obtained as

$$A_{BB}(t) = \sqrt{S_I(t)^2 + S_Q(t)^2} = G_r A_s A_t \rho(\tau_p(t)) A_m\big(t - \tau_p(t) - \tau_r\big) \tag{3.48}$$

$$\begin{aligned} \phi_{BB}(t) = \tan^{-1}\frac{S_Q(t)}{S_I(t)} = {} & \mathrm{MOD}(\omega_c(\tau_l - \tau_r), \pi) \\ & + \Delta\phi_m(t) + \Delta\phi_p(t) + \phi_d(t) \end{aligned} \tag{3.49}$$

where MOD stands for a modulus function. The first term and the combination of the second and third terms on the right-hand side of (3.49) are the dc offset and modulation noise, respectively, that are added to the Doppler phase shift $\phi_d(t)$. Notably, the first two terms can be minimized by satisfying the delay matching condition $\tau_r = \tau_l$, while the third term can be suppressed by reducing the modulation or zero-crossing rates of the reference signal.

Equation (3.49) reveals that under the delay matching condition, the phase modulation of the baseband signal is canceled out within half of the locking range of the ILO and then the low-frequency phase of the baseband signal is mostly the Doppler phase shift $\phi_d(t)$. To evaluate more rigorously the quality of the Doppler signal in terms of the reference signal parameters, such as modulation type, symbol rate, received power, and RF path delay, a system simulation platform that is based on a discrete-time model of the ILO is utilized and discussed in the following subsection.

3.3.3 System performance simulations and verification

A system simulation platform, which was established in the Keysight Advanced Design System Ptolemy environment, is used to evaluate the performance of the passive radar with various modulated signals. As shown in Figure 3.8, the platform is composed of a wireless communication signal generator, a two-way power splitter, a Doppler phase modulator, a variable time-delay unit, a discrete-time ILO model, and a quadrature demodulator. In the simulation process, the generator output signal is split into two paths—an RF path and an LO path. The signal along the RF path mimics the received RF signal that has been phase-modulated by a Doppler signal $V_d(t)$ and then delayed for a variable time τ_r. The signal that travels along the LO path is injected into the ILO to generate an LO signal for use in the quadrature demodulator. The output baseband I and Q signals from the quadrature demodulator are used to recover the Doppler signal for comparison with the actual signal. A Chebyshev digital finite impulse response LPF with a cutoff frequency of 50 Hz is applied to these baseband signals.

In the RF path, the Doppler signal is specified as a periodic triangular function $\Lambda(t)$ to modulate the phase of the signal that is input to the phase modulator. Mathematically, this modulated phase is written as

$$\phi_d(t) = m_p V_d(t) = \frac{2\omega_c}{c}\Lambda(t) \tag{3.50}$$

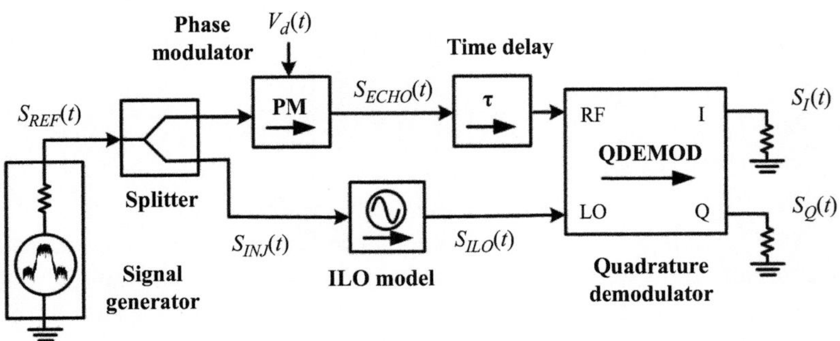

Figure 3.8 A simulation platform built with Advanced Design System (ADS) Ptolemy for the ILQR-based passive radar (from [24])

where $m_p = 2\omega_c/c$ is a phase modulation index and c is the speed of light. The ILO model that is used in the LO path is based on a discrete-time analysis that was derived in Section 3.2.5.

Figure 3.9 presents a measurement setup to verify the results of simulations of using the passive radar to detect the motion of a 15 cm × 10 cm metal plate that is about size of a palm. The plate was controlled, using an actuator, to move in a periodic back-and-forth motion with a frequency of 0.1 Hz and a peak-to-peak displacement of 5 cm. The minimum distance between this plate and the RXe antenna of the radar was 40 cm. A Keysight E4433B digital RF signal generator was used to generate continuous-wave (CW) and various wireless communication signals.

Figure 3.10(a) shows, as an example, the simulated spectrum of a generated 2.4 GHz LTE TDD signal that uses orthogonal frequency-division multiplexing (OFDM) 64-quadrature amplitude modulation (64-QAM) with a symbol rate of 20 Msps and an output power of −20 dBm. After it has been injected with this signal that has been passed through the coupler, the ILO yields an output spectrum which is also displayed in Figure 3.10(a) and determined by the combination of

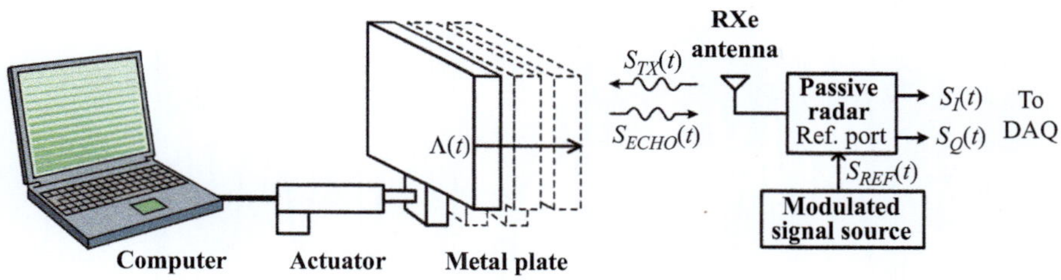

Figure 3.9 Setup sketch for actuator-controlled motion measurements

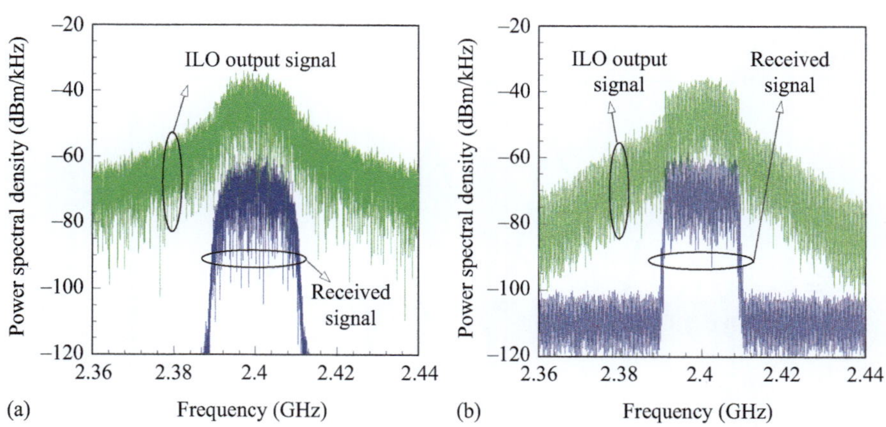

Figure 3.10 Reference LTE TDD signal spectrum at a power level of −20 dBm and resultant ILO output signal spectrum at a power level of 7 dBm. (a) Simulation. (b) Measurement (from [24])

injection-locking and injection-pulling effects, as described by (3.45). These simulated spectra in Figure 3.10(a) are in close agreement with the corresponding measured spectra that are shown in Figure 3.10(b). In this case, the locking range of the ILO is estimated from the simulation results to be 2.65 MHz.

Equalizing the frequencies of the free-running oscillator signal and the received reference modulation signal, $\Delta\omega_{osc} = 0$, yields a delay τ_l of approximately 60 ns in the locking of the ILO by the injected modulation signal, as determined by (3.17). Therefore, for delay matching, the RF path delay τ_r is set to the same value in both the simulation and measurement. The actual Doppler phase shift $\phi_d(t)$ that is caused by the metal plate with a 5-cm periodic back-and-forth motion has a phase modulation index of 57.6°/cm and a peak-to-peak phase change of 288°, according to (3.50).

The simulation with the aforementioned settings yields the quadrature Doppler signals that are shown in Figure 3.11(a). Since the applied Doppler phase-modulation frequency is only 0.1 Hz, which is much less than half of the ILO's locking range, the dc offset and modulation noise that are added to the Doppler phase shift, given by the first two terms on the right-hand side of (3.49), are

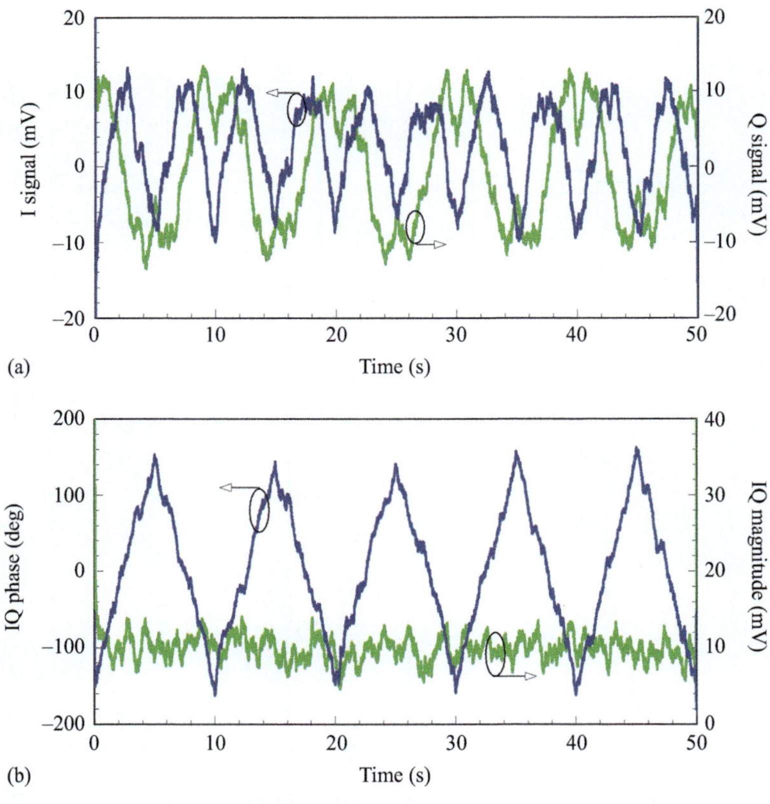

Figure 3.11 Demodulated quadrature Doppler signals. (a) I and Q signals. (b) IQ magnitude and phase (from [24])

eliminated if the delay matching condition is satisfied, as it is in this case. Figure 3.11(b) plots the magnitude and phase of the IQ signal vector, showing an rms magnitude of 13 mV and an rms percentage phase error (rms difference between the actual and detected phases, divided by the actual peak-to-peak phase change) of 3.3%. The glitches in the quadrature Doppler signals in Figure 3.11(a) and (b) arise mainly from the effect of injection pulling on the ILO output signal, particularly during periods close to the zero crossings of the injected modulation signal.

The measurement of the motion of the actuator-controlled metal plate with the radar yields the baseband I and Q signals. Further processing of the I and Q signals to obtain the desired Doppler signals requires an arctangent demodulation procedure that involves dc calibration and phase unwrapping [44,45]. Figures 3.12–3.15 compare the simulated and measured results, which are discussed in the following.

Figure 3.12(a) and (b) compares the simulated and measured rms magnitudes and percent phase errors, respectively, of the IQ signal vectors when the radar receives 2.4 GHz OFDM 64-QAM signals with the same settings as above, except for the change in symbol rate. Close agreement is observed. These figures reveal that as the symbol rate increases, the magnitude of the IQ signal vector decreases while the percentage phase error of the vector increases, so the Doppler signal is degraded. This effect arises because a larger fraction of the signal bandwidth falls beyond the half locking range of the ILO when a modulation signal with a higher symbol rate is injected. This phenomenon generally weakens injection locking but strengthens injection pulling. Unfortunately, for the radar system, the former effect contributes mostly to the signal while the latter contributes mostly to the noise in the Doppler band.

Figure 3.13(a) and (b) compares the simulated and measured results when the radar receives the modulation signal in Figure 3.10 with various delays. These delays are realized by inserting one to several coaxial cables with a 14-ns delay into

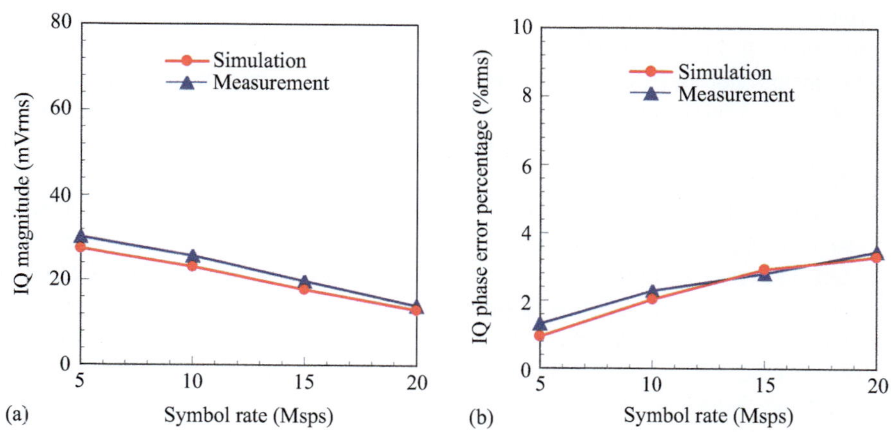

Figure 3.12 Effects of the symbol rate of the reference LTE TDD signal on quadrature Doppler demodulation. (a) IQ magnitude. (b) IQ phase error percentage

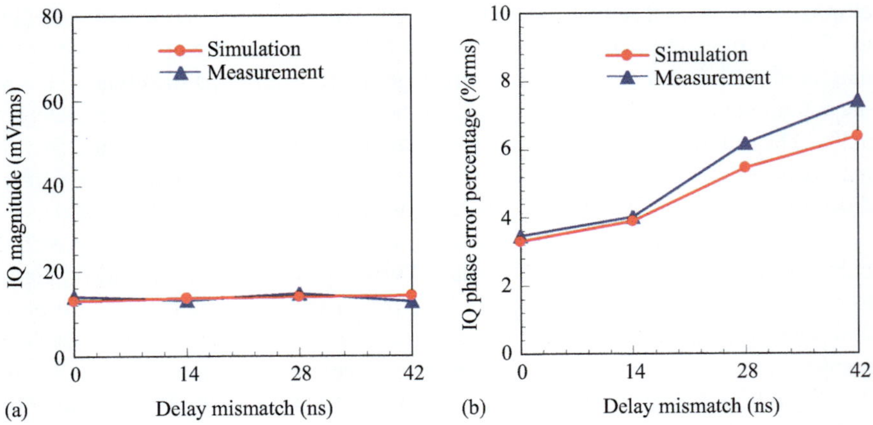

Figure 3.13 Effects of the delay mismatch of the reference LTE TDD signal on quadrature Doppler demodulation. (a) IQ magnitude. (b) IQ phase error percentage

the RF path whose original delay matches that of the LO path. The simulated and measured results reveal similar trends. Figure 3.13(a) demonstrates that the delay mismatch has no observable effect on the magnitude of the IQ signal vector because the RF path delay does not affect the locking range of the ILO. However, Figure 3.13(b) indicates that the associated percentage phase error increases with the delay mismatch, primarily because the modulation noise term $\Delta\phi_m(t)$ in (3.49) becomes proportionally larger.

Figure 3.14(a) and (b) compares the simulated and measured results for the radar with the modulation signal in Figure 3.10 but with variable received power. The simulated magnitude and percentage phase error of the IQ signal vector are strongly consistent with the measured data. In this radar, the amplitude of the echo and injection signals increases with the square root of the received power. This relationship not only causes the magnitude of the output IQ signal vector to increase with the received power, according to (3.48), but also improves the locking range of the ILO, according to (3.14), and thereby lowers the associated percentage phase error.

Figure 3.15(a) and (b) compares the simulated and measured results when the radar receives variously modulated 2.4 GHz signals. The symbol rate, delay mismatch, and received power of these signals are set to 20 Msps, 0 ns, and −20 dBm, respectively. The comparisons again reveal a close agreement between simulated and measured data. Since the ILO produces an output signal that is completely synchronized with the received unmodulated CW signal, feeding this signal into the radar leads to the largest magnitude and lowest percentage phase error of the output IQ signal vector, and hence achieves the highest-quality Doppler signal among all of the received signals. Moreover, the GMSK signal concentrates more power density around the carrier frequency than does the quadrature phase shift keying (QPSK) or OFDM 64-QAM signal. The ILO can therefore capture more phase

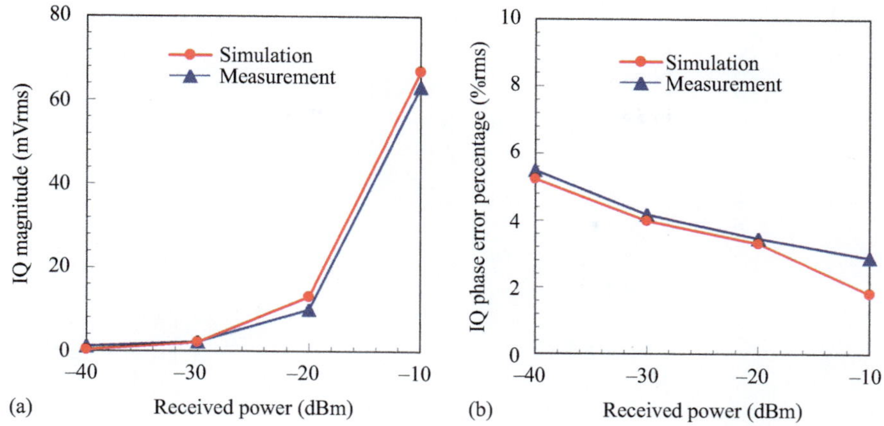

Figure 3.14 Effects of the power of the reference LTE TDD signal on quadrature Doppler demodulation. (a) IQ magnitude. (b) IQ phase error percentage

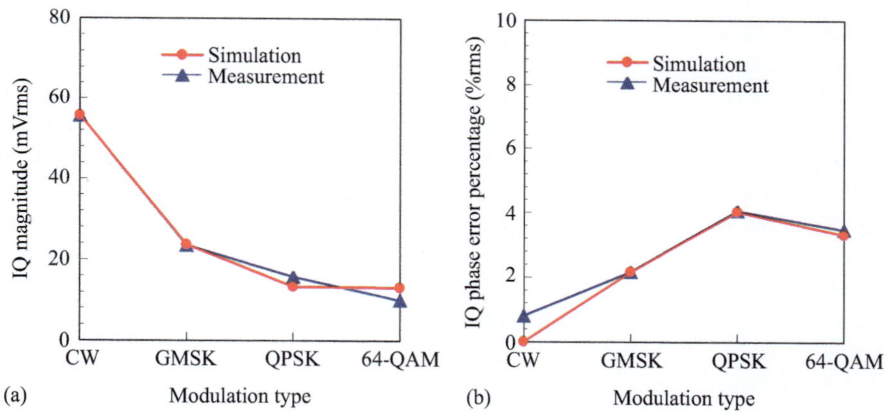

Figure 3.15 Effects of the modulation type of the reference signal on quadrature Doppler demodulation. (a) IQ magnitude. (b) IQ phase error percentage

information from the former signal within half of the locking range, causing the radar to produce a Doppler signal of higher quality than the latter two signals. The QPSK and 64-QAM OFDM signals are modulation signals with time-varying envelopes. The radar yields similar Doppler signal magnitudes in demodulating these two signals. However, the QPSK signal appears to have a higher zero-crossing rate than the OFDM 64-QAM signal, yielding the highest percentage phase error among all of the signals that were received by the radar.

The limitation of the radar is that the locking range of the ILO must greatly exceed the maximum Doppler frequency shift that is caused by the motion to be

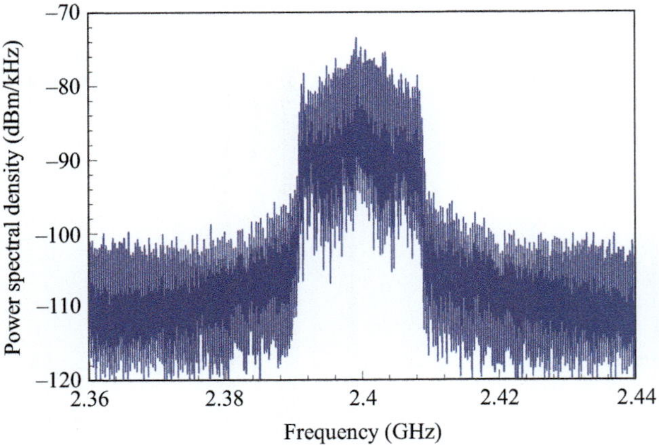

Figure 3.16 Spectrum of the reference LTE TDD signal with the ILO leakage (from [24])

measured. This limitation restricts the minimum received reference signal power but can be substantially loosened by adding an low noise amplifier (LNA) in front of the ILO. Nonetheless, this minimum received reference signal power may be of interest as it is a measure of system sensitivity. In this study, the lowest received reference signal power was set to −40 dBm owing to the limitation on ILO leakage. As depicted in Figure 3.16, when the radar receives the OFDM 64-QAM signal with a power of −50 dBm, the ILO leakage signal clearly distorts the signal spectrum that is measured in the air near the RXe antenna. Hence, the ILO leakage signal becomes an important part of the radar source signal if the radar receives weak communication signals with power below −40 dBm. Operating in this situation, the radar still functions but may no longer be a passive radar.

3.3.4 Experimental results

For the purposes of demonstration, the radar is placed in a laboratory for use in gesture sensing with two experimental setups. In the first setup, the radar used the RXr antenna that was connected to the input terminal of the coupler to capture the through-wall Wi-Fi signals that were generated by an IEEE 802.11b/g/n AP in an adjacent office room. This AP ran in mixed b/g/n mode with a maximum bandwidth of 20 MHz. The Wi-Fi signal that was received at the RXr antenna of the radar had a center frequency of 2.412 GHz and a power of about −40 dBm. Figure 3.17(a) and (b) shows the measured spectra of the received Wi-Fi signal and the resultant ILO output signal, respectively. Figure 3.17(b) reveals that the weak Wi-Fi injection signal caused the ILO to significantly reduce the locking range below that in the previously discussed case, shown in Figure 3.10(b). In the second setup, the radar used a signal generator, instead of the RXr antenna, that was connected to the input terminal of the coupler to input a 2.412 GHz LTE TDD signal with a power of −20 dBm and a symbol rate of 20 Msps. The operating frequency of the

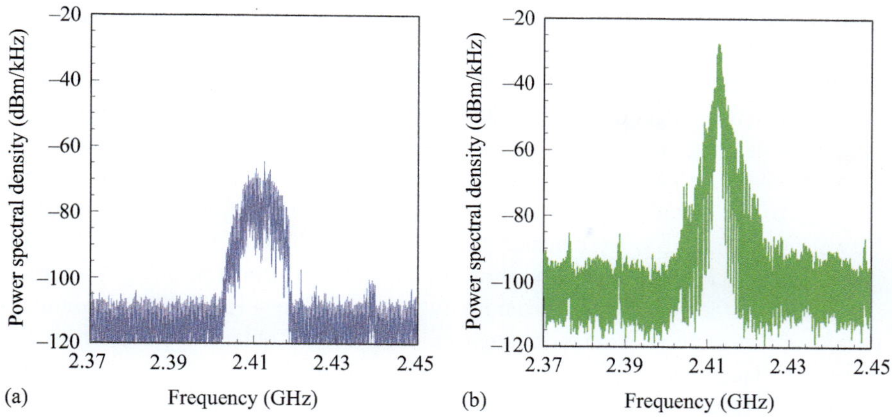

Figure 3.17 Ambient Wi-Fi signal spectrum at a power level of −40 dBm and resultant ILO output signal spectrum at a power level of 7 dBm. (a) Simulation. (b) Measurement (from [24])

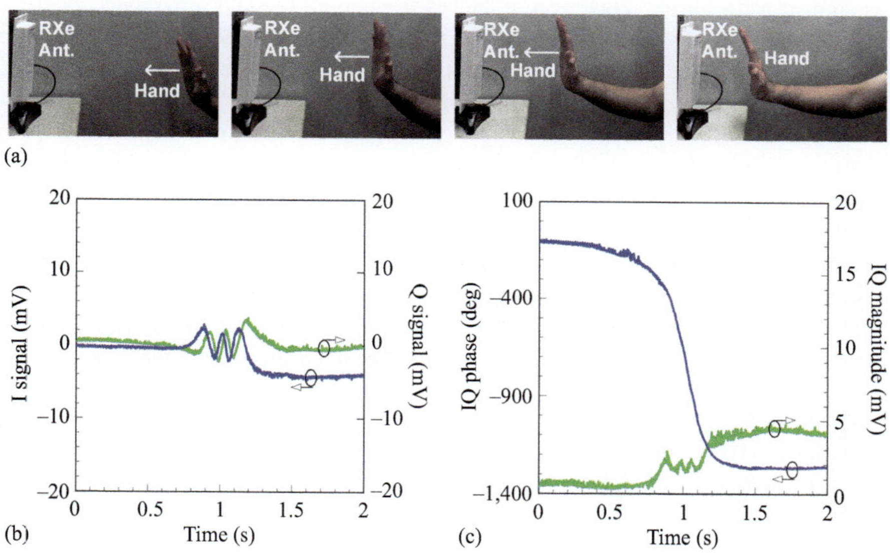

Figure 3.18 Experimental results of the ILQR-based passive radar in detecting a "push" gesture with the ambient Wi-Fi signal. (a) Continuous shooting of the gesture. (b) Output I and Q signals. (c) Magnitude and phase of the IQ signal vector

ILO in these two setups was set to 2.412 GHz to make $\Delta\omega_{osc} = 0$ in (3.17) to minimize the delay during injection locking. When a gesture is made, the shortest distance from the hand to the RXe antenna of the radar is at least 5 cm.

Figure 3.18(a) presents the detection of a "push" gesture by the radar that uses the captured Wi-Fi signals from the AP. The subject moves his palm toward the

RXe antenna with a displacement of about 20 cm. Figure 3.18(b) plots the quadrature Doppler signals that are output by the radar. The I signal leads the Q signal. In Figure 3.18(c), the magnitude of the IQ signal vector, represented by the green curve, rises from 0.3 to 4.7 mV during the period of the gesture. This rise in magnitude is attributed to an increase of the echo signal power as the palm moves closer to the RXe antenna. Nevertheless, due to multipath reflection interference, the plot of magnitude against time exhibits irregular fluctuations in the period between 0.8 and 1.1 s. By contrast, the unwrapped phase of the IQ signal vector, represented by the blue curve, provides more useful information on the gesture. In this example, the phase changes smoothly from −94° to −1,271.4° throughout the period of the gesture. A phase change of −1,177.4° is obtained, yielding an estimated displacement of 20.3 cm of the palm toward the RXe antenna. This estimated displacement is close to the actual one, as mentioned above.

Figure 3.19(a) and (b) shows the results concerning the demodulated quadrature Doppler signals when the radar detects the "push" gesture using the LTE TDD signal from the signal generator. The noise on the data curves is clearly smaller than in Figure 3.18(b) and (c) because the received signal power in this case is significantly larger. This gesture causes the magnitude and unwrapped phase to vary from 2.9 to 18.9 mV and from −65.9° to −1,235°, respectively, as plotted in Figure 3.19(b). The latter result provides an estimate of the displacement that is associated with the gesture, which is 20.2 cm.

Figure 3.20(a) presents the detection of a "pull" gesture by the radar using ambient Wi-Fi AP signals. The displacement that is associated with the gesture is again around 20 cm. The results of the detection of this gesture, shown in Figure 3.20(b) and (c), reveal an opposite trend in the quadrature Doppler signals to that revealed by those for a "push" gesture. In Figure 3.20(b), the Q signal leads the I signal. In Figure 3.20(c), the magnitude of the IQ signal vector decreases from 4.88 to 3.68 mV and the unwrapped phase of the vector increases from −1,300.6° to −127.3° during the period of the gesture. The magnitude curve indicates that the

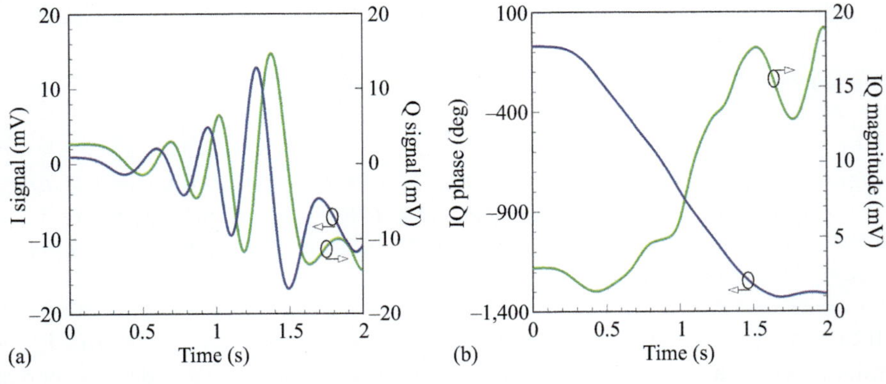

Figure 3.19 Experimental results of the ILQR-based passive radar in detecting a "push" gesture with the reference LTE TDD signal. (a) Output I and Q signals. (b) Magnitude and phase of the IQ signal vector

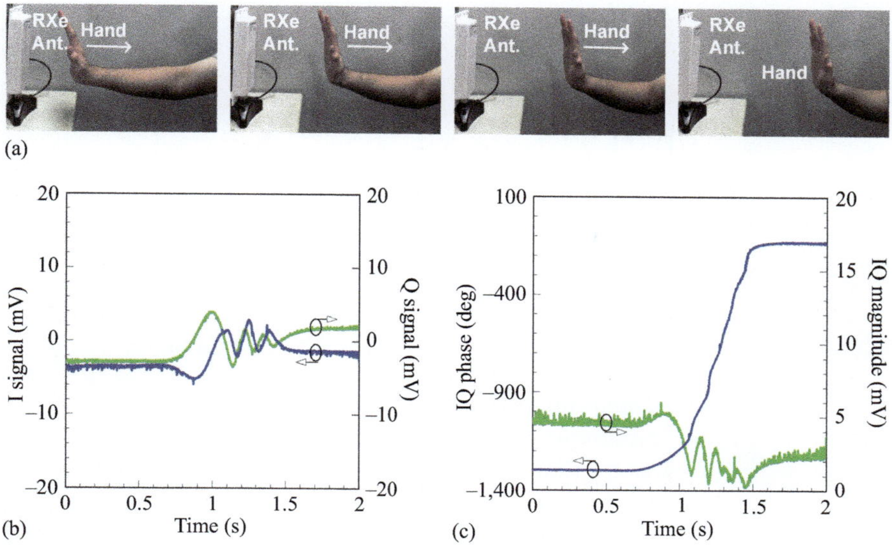

Figure 3.20 Experimental results of the ILQR-based passive radar in detecting a "pull" gesture with the ambient Wi-Fi signal. (a) Continuous shooting of the gesture. (b) Output I and Q signals. (c) Magnitude and phase of the IQ signal vector

echo signal becomes weaker as the palm moves away from the RXe antenna. Irregular fluctuations in magnitude occur throughout most of the period of the gesture. The phase curve yields an estimated displacement of 20.3 cm, which is close to the actual displacement as the palm moves away from the RXe antenna.

Figure 3.21(a) and (b) shows the results of detection of the "pull" gesture by the radar using the LTE TDD signal. The magnitude and unwrapped phase of the IQ signal vector in Figure 3.21(b) change from 19.852 to 3.058 mV and from −1,287.6° to −123.9°, respectively, and the latter result yields an estimated displacement of 20.1 cm associated with the gesture.

Figure 3.22(a) presents the last gesture, "push–pull," with a one-way displacement of approximately 20 cm. Figure 3.22(b) and (c) shows the outputs of detection by the radar using ambient Wi-Fi AP signals. The Doppler characteristics of a "push–pull" gesture can be approximated as those of a "push" gesture followed by those of a "pull" gesture. In Figure 3.22(b), the I signal leads the Q signal in the interval 0–0.9 s while the Q signal leads the I signal in the interval 1.1–2 s, with a phase reversal period between 0.9 and 1.1 s. In Figure 3.22(c), the magnitude of the IQ signal vector clearly drops in the middle of the phase reversal period as a result of multipath fading. The unwrapped phase of the IQ signal vector, with a maximum value of 38.3° and a minimum value of −1,151.3°, reveals an estimated one-way displacement of 20.6 cm.

The "push–pull" gesture detection is repeated using the LTE TDD signal, yielding the results in Figure 3.23(a) and (b). In Figure 3.23(a), a phase reversal

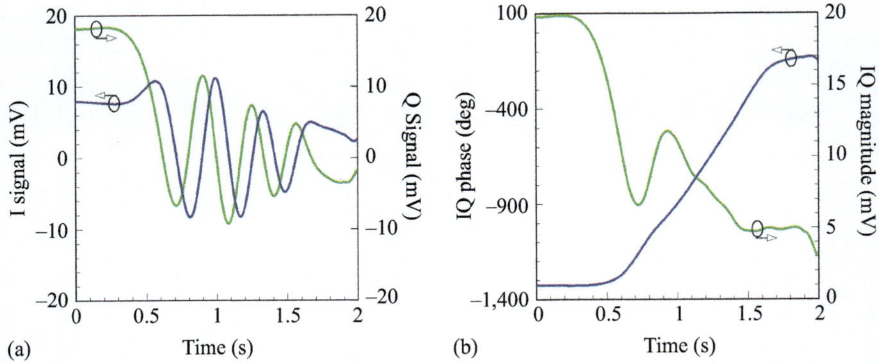

Figure 3.21 Experimental results of the ILQR-based passive radar in detecting a "pull" gesture with the reference LTE TDD signal. (a) Output I and Q signals. (b) Magnitude and phase of the IQ signal vector

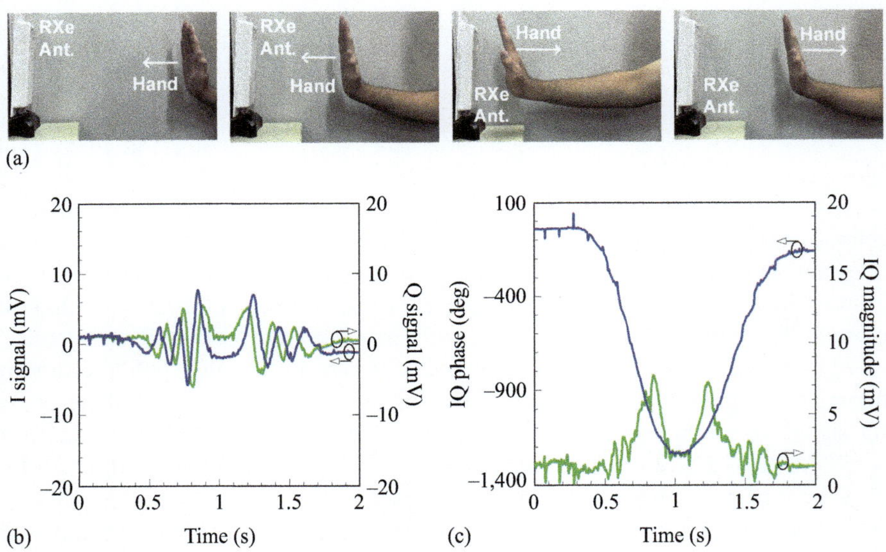

Figure 3.22 Experimental results of the ILQR-based passive radar in detecting a "push–pull" gesture with the ambient Wi-Fi signal. (a) Continuous shooting of the gesture. (b) Output I and Q signals. (c) Magnitude and phase of the IQ signal vector

period is identified from 0.9 to 1.2 s. The curve of the unwrapped phase in Figure 3.23(b) reaches a maximum value of 7.8° and a minimum value of −1,167°. The one-way displacement is therefore calculated to be 20.3 cm.

Based on Figure 3.18–3.23, the radar uses ambient Wi-Fi AP or input LTE TDD signals to detect three gestures: "push," "pull," and "push–pull." Regardless of the received signal that is used, the locking ranges of the ILO in the radar are

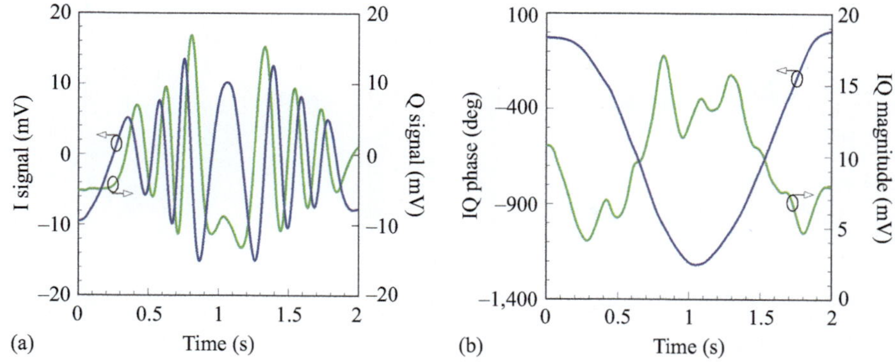

Figure 3.23 Experimental results of the ILQR-based passive radar in detecting a "push–pull" gesture with the reference LTE TDD signal. (a) Output I and Q signals. (b) Magnitude and phase of the IQ signal vector

much larger than the frequency range of the Doppler signals that are caused by the gestures, so the additive modulation noise that arises from the wireless communication signals can be effectively canceled, so the Doppler signals can be detected. Therefore, the actual and detected displacements that are associated with the gestures agree very closely. The worst detected displacement error in the gesture sensing experiments is estimated to be below 1 cm. Undoubtedly, the detected phase information is much more useful than the detected magnitude information for determining the displacement that is associated with the gesture.

3.4 Applications of Wi-Fi-based gesture sensing

This section demonstrates applications of gesture-sensing technology that is based on a passive radar that uses ambient Wi-Fi signals or those of an AP. These applications include gesture control of a tablet and a video game. Moreover, with the help of machine learning, gesture recognition can be carried out to distinguish accurately various hand gestures. To extend the technology to the detection of 3D gesture motion, the radar is combined with a 2D RGB camera to construct the 3D trajectory of a hand gesture that was made within the FOV of the camera.

3.4.1 Gesture control

Gesture control is a natural, easy, and dynamic way to interact with electronic devices when touchpads and voice control are not convenient or feasible. It has a wide range of applications in the electronics, automotive, and medical industries and is used, for example, in mobile devices and automotive control panels and in combination with virtual reality and augmented reality (AR) technologies for entertainment or medical purposes. Gesture control systems use sensors that are based on either optical, ultrasonic, or radar techniques. Radar, and particularly Wi-Fi-based passive radar, is the most suitable method for gesture detection over a long

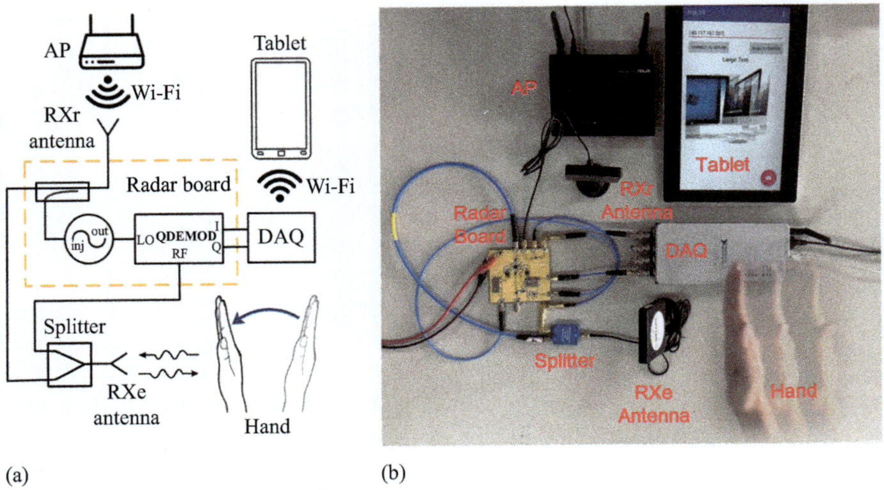

Figure 3.24 A gesture control system using the Wi-Fi-based passive radar to swipe pages on a tablet. (a) Block diagram. (b) Photograph

and wide range. When something moves within a space that is covered by a Wi-Fi signal, the Wi-Fi signal frequency slightly changes as a result of the Doppler frequency shift, which may be detected by sensors. These sensors can operate anytime and anywhere because Wi-Fi signals are ubiquitous and always present in today's internet age.

Figure 3.24(a) and (b) shows a schematic diagram and a photograph of a gesture control system that uses the radar board shown in Figure 3.7 to swipe pages on a tablet. The radar board receives a Wi-Fi signal from an AP and routes it through a splitter to an antenna for transmission to a moving hand. Simultaneously, the board receives an echo signal via the same antenna and splitter to be processed by an ILQR, and then outputs the quadrature Doppler signals to a DAQ unit. The DAQ unit delivers digital I and Q signals over Wi-Fi to the tablet in which is installed a gesture control APP for swiping pages back or forward as determined by the direction of the hand motion. To ensure that insignificant hand motions have no effect, the control action is enabled only when the speed or range of the hand motion exceeds a set threshold.

3.4.2 Gesture games

In gesture games, a player makes gestures instead of using a conventional joystick or keyboard. A 3D camera is the mainstream interface for gesture games. A larger range of motion in gesture games is generally associated with greater player satisfaction and immersion [46]. However, since cameras have a limited FOV, gesture games that use a 3D camera interface often require a large space to accommodate the players. A Doppler radar system serves as a better gesture interface than a 3D camera because it allows for an almost full range of hand motions with a low gain antenna. Moreover, a Doppler radar system responds more quickly to a gesture than does a 3D camera. Therefore, a Doppler radar-based gesture game can provide players with an intense exercise challenge in a small space. Modern life involves a wide range of electronic devices such as smartphones, tablets, and various multimedia products. Many of these devices have a display and a Wi-Fi module. The aforementioned Wi-Fi-based passive radar has the potential to be the most popular interface for gesture games because it can be easily integrated into these devices.

For demonstration purposes, a gesture gaming system with two-hand motion control was developed based on Wi-Fi passive radar technology, as shown in the schematic diagram and photograph in Figure 3.25(a) and (b), respectively. This system consists of two antennas, three splitters, two of the radar boards that are

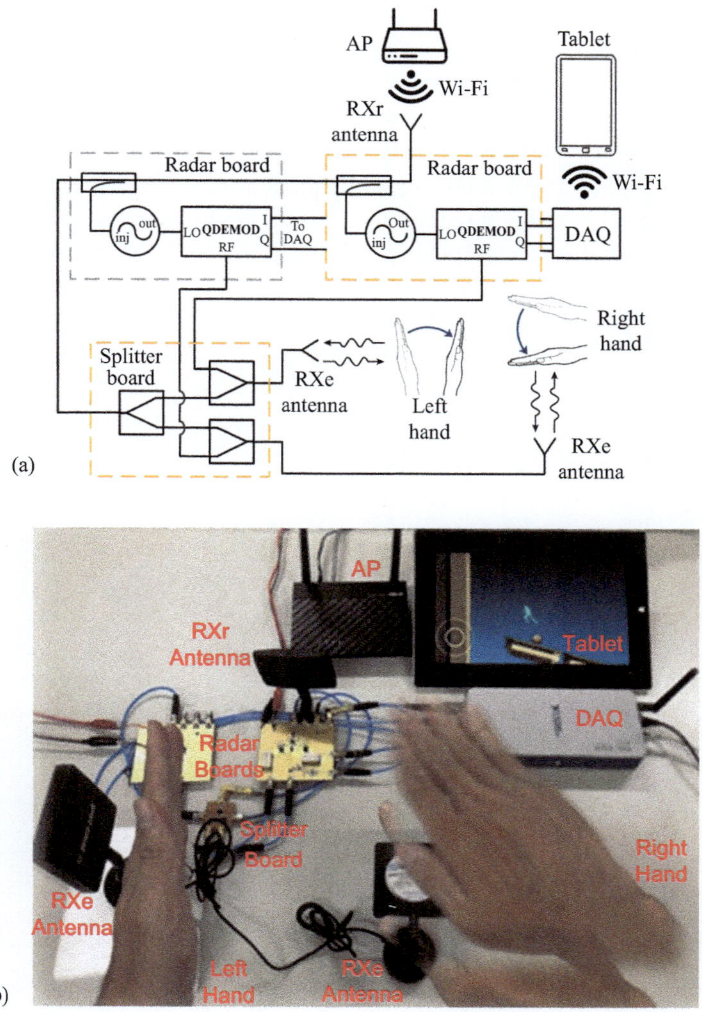

Figure 3.25 A gesture gaming system using the Wi-Fi-based passive radar to realize two-hand motion control. (a) Block diagram. (b) Photograph

shown in Figure 3.7, and one DAQ unit. The two antennas transmit Wi-Fi signals from the AP to, and simultaneously receive echo signals from, the two hands of a player for detection by the two radar boards. The two antennas are effectively isolated because they are orthogonally polarized. The three splitters provide not only equal transmission power to the two antennas but also mutual isolation of the echo signals that are received by the two antennas. The two radar boards output two pairs of quadrature Doppler signals that are associated with the two hand motions to the DAQ unit. After analog-to-digital conversion, the DAQ unit sends the digital I and Q signals via Wi-Fi to a tablet for further processing, enabling the video game to be controlled. A player of the game makes a "pull right" gesture and a "push left" gesture with his or her left hand to cause a character to run right and left, respectively, and a "pull up" gesture and a "push down" gesture to cause the character to jump and stop, respectively. All of these control gestures work only when the speed or range of motion of the hand exceeds a threshold.

3.4.3 Gesture recognition with machine learning

Gesture recognition is inherently a classification problem, which can be effectively solved by a machine learning algorithm. First, a Wi-Fi-based passive radar is used to detect four hand gestures: "push–pull," "push," "pull," and "rotating fingers," as shown in Figure 3.26. Each gesture lasts for 2 s and is carried out 5,000 times, yielding a collection of 20,000 samples, each of I and Q data points. Figure 3.27 shows four samples that correspond to the four-hand gestures.

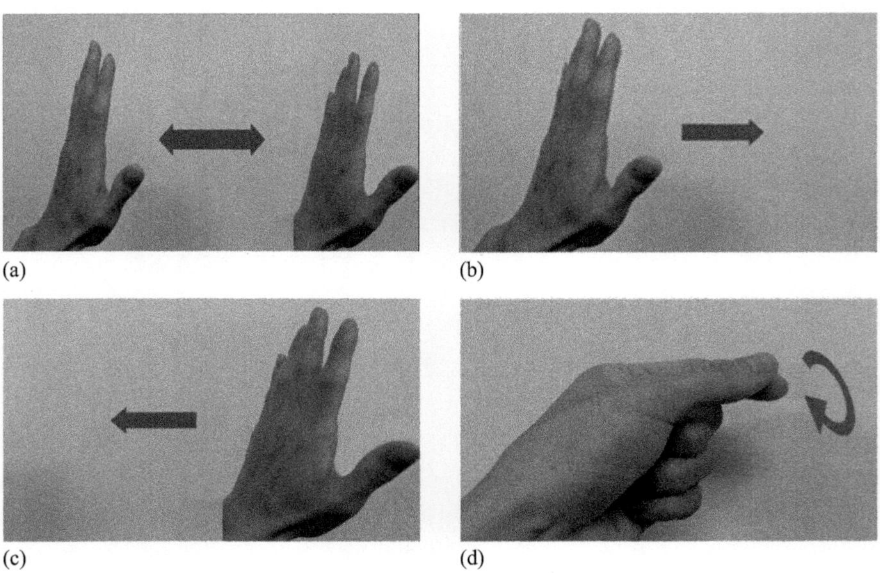

Figure 3.26 Four different hand gestures used in a machine learning-based recognition application: (a) push–pull, (b) push, (c) pull, and (d) rotating fingers

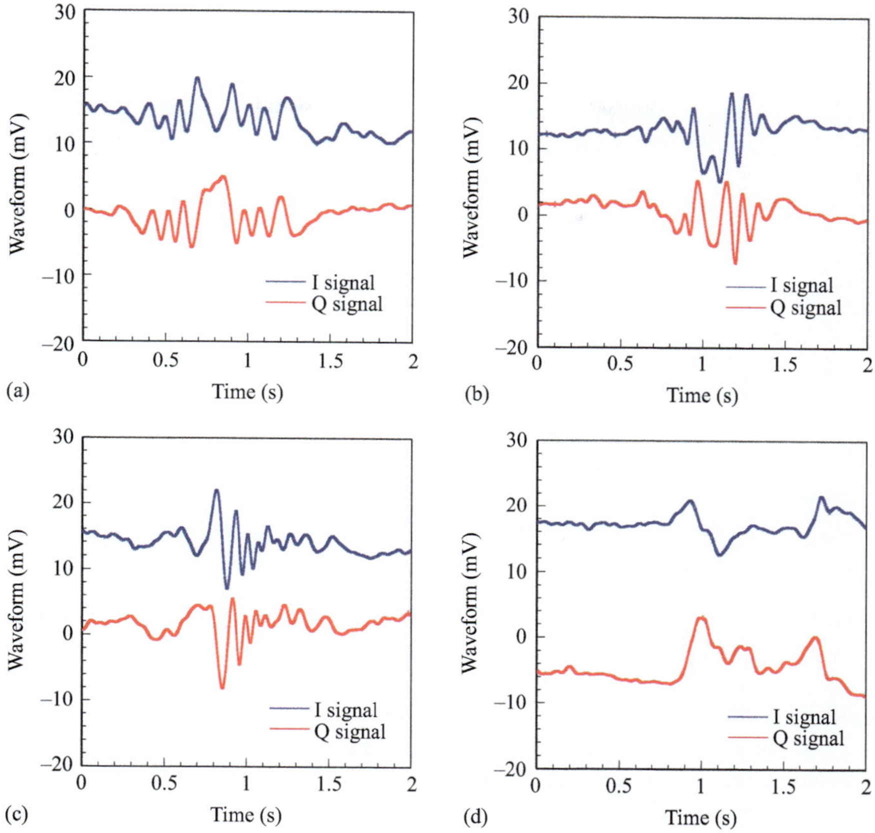

Figure 3.27 Radar detection outputs corresponding to the four different hand gestures: (a) push–pull, (b) push, (c) pull, and (d) rotating fingers

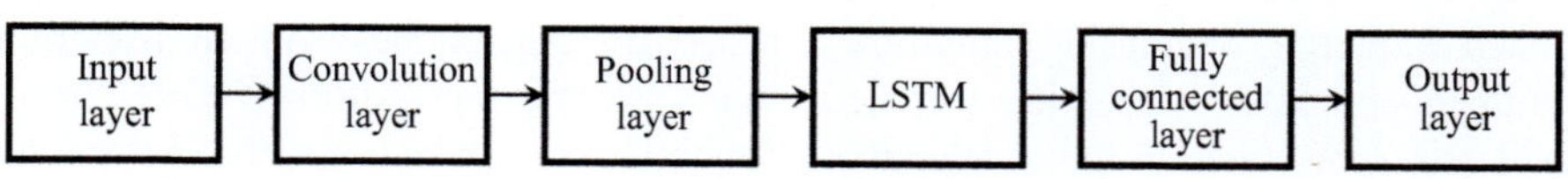

Figure 3.28 Combined CNN and LSTM neural network model

In this study, a neural network model that combines a convolutional neural network (CNN) [47] with a long short-term memory (LSTM) network [48], as depicted in Figure 3.28, is used to recognize hand gestures [13]. The CNN is used to extract a compact feature vector from an input normalized IQ data vector, while the LSTM network further captures sequential information about the feature vector for classification.

As the input data to the neural network model, the 20,000 samples are normalized and randomly divided into groups of 14,000, 3,000, and 3,000 samples for training, validation, and testing, respectively. Keras is then used with the CuDNN

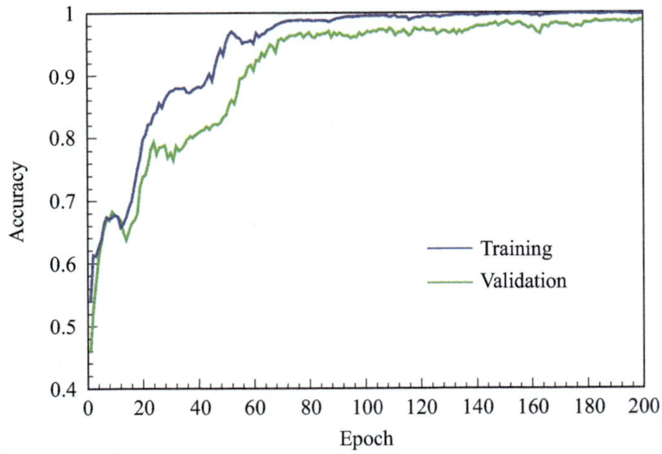

Figure 3.29 Training and validation accuracy of the combined CNN and LSTM neural network model

accelerated TensorFlow deep learning framework to build and train the gesture recognition model rapidly without overfitting. In the training process, the conventional rectified linear unit activation function [49] and the Adam optimizer [50] are used to update the network weights iteratively. The batch size is set to 64 so the model completes one epoch in 219 iterations.

Training was conducted for 200 epochs. Figure 3.29 displays a training and validation accuracy graph. The training converges well after 100 epochs with a training accuracy of 99.67% and a validation accuracy of 98.92% at 200 epochs. The trained model results in a test accuracy of 98.21%.

3.4.4 Sensor fusion with camera

Recently, since the game Pokemon Go became hugely popular worldwide, AR has been attracting increasing attention for use in smartphones. As an important human–computer interaction technique, hand gesture recognition (HGR) is considered to be a step in the acceleration of mobile AR applications; the major issue concerns its application to smartphones as inexpensively as possible. Mainstream technologies for HGR involve 3D cameras such as Microsoft Kinect, Intel RealSense, and Leap Motion. However, equipping a smartphone with a 3D camera requires the use of an IR camera, which consumes considerable computational resources and power. Additionally, IR cameras often function poorly outdoors or in bright light.

This subsection presents a vision and Wi-Fi sensing fusion system for capturing the trajectory of 3D hand motion with a smartphone. Figure 3.30 presents the use of a smartphone that is Wi-Fi-connected to an AP to recognize the user's hand gestures. The technique involves the computer vision processing of image data from a camera and depth displacement measurement with a Wi-Fi-based passive radar to obtain the transverse (x- and y-directed) and longitudinal (z-directed)

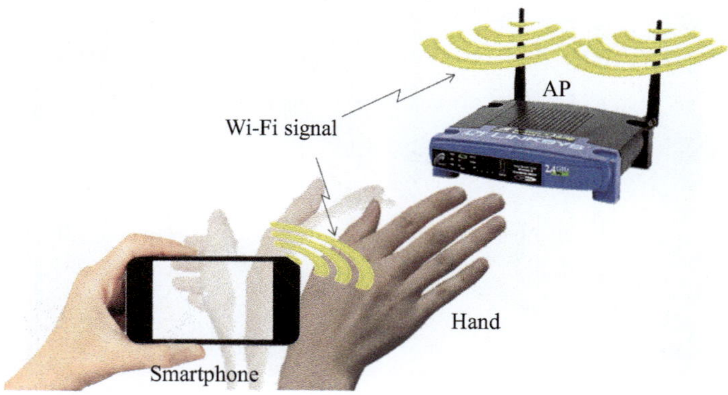

Figure 3.30 3D hand gesture detection using a vision and Wi-Fi sensing fusion system (from [26])

Figure 3.31 ROI detection using the sampled color profile

components of hand motion, respectively, and the combination of these components to construct a 3D hand trajectory.

3.4.4.1 Vision-based hand gesture detection

Hand gesture detection is one of the most important research topics in the field of computer vision because hand gestures are a simple and natural way to communicate with electronic devices. For this purpose, a vision-based sensing system is established herein; it performs two main functions, which are color segmentation and hand detection. Owing to the need for a low computational burden, color segmentation is often used to distinguish regions of interest (ROIs), such as may contain a hand, from the background. In related applications, fixed threshold values, i.e. hue–saturation–value (HSV), in color space [51] for skin are commonly used. However, different skin tones and objects like gloves may cause failure in determining the ROIs. A sampled color profile is used to solve this problem [52]. In Figure 3.31, seven 40×40 sampling grids are placed over an object to produce median color profiles by averaging the colors of pixels in each sampling grid cell. These profiles are used to generate binary images, which are summed so that the ROIs can be identified from the white pixels.

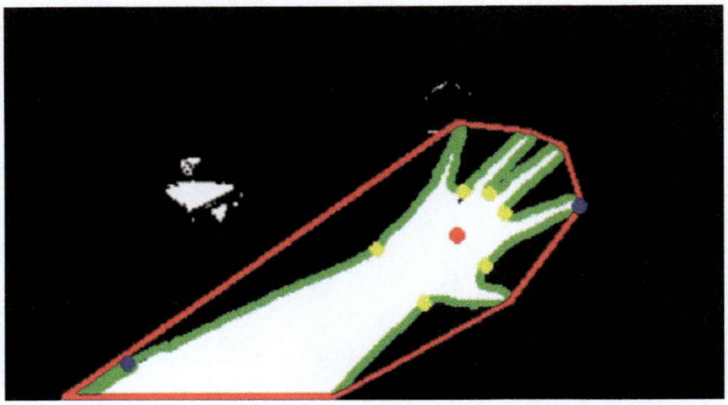

Figure 3.32 Hand detection using the structural analysis

The presence or absence of a hand region as an ROI can be determined from structural information. The structural analysis of Andresen [52] assumed that the hand is still in front of the camera, so the forearm cannot be seen by the camera. However, a serious problems of noise arises in many situations because the hand moves fast and the forearm is often captured. Therefore, the system uses methods that involve contour detection [53], a convex hull [54], convexity defects, constraints on convexity defects, and Kalman filtering to reduce the effect of noise. The first of these methods detects a contour that delineates the most probable hand region, as indicated by the green boundary in Figure 3.32. The second method yields a corresponding convex hull that is represented by the red contour in Figure 3.32. The points in yellow and blue in Figure 3.32 are the convexity defects that are farthest from the convex hull. Generally, convexity defects in the junctions of fingers are far from the convex hull. Therefore, if the shortest distance between the convexity defects and the convex hull is less than a predefined threshold value, then the convexity defects are regarded as noise that is generated in the finger and forearm regions and are excluded for consideration in identifying the hand region. Moreover, the hand region is considered to be successfully detected if it contains at least three confirmed convexity defects from which a centroid can be calculated and treated as its center. Finally, the Kalman filter is used to track the center of the hand region in consecutive frames to provide a more accurate estimate of the 2D hand gesture trajectory than could otherwise be obtained, as shown in Figure 3.33.

3.4.4.2 Sensor-fusion calibration scheme

To provide the depth trajectory of a hand, the presented system uses Wi-Fi-based passive radar technology, introduced in Section 3.3, to measure the instantaneous longitudinal displacement of the hand. For example, Figure 3.34(a) and (b) shows the results of using the radar to capture a "push–pull" hand gesture. In Figure 3.34(a), the Q signal leads the I signal in the "push" stage and vice versa in the "pull" stage, causing a phase reversal of the signals between these two stages. Following arctangent demodulation and phase unwrapping, the instantaneous displacement of the

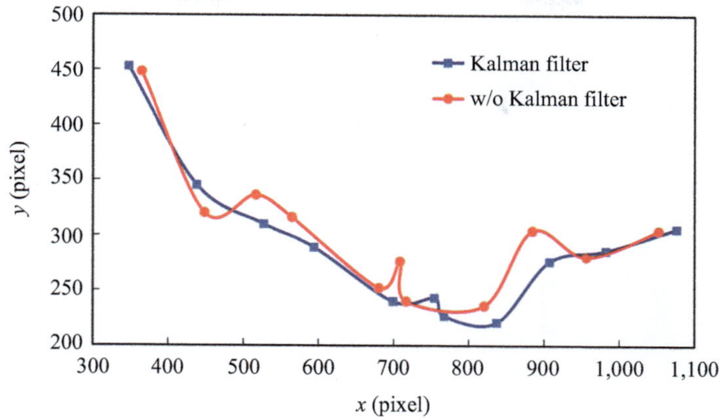

Figure 3.33 Comparison between trajectory estimation with and without Kalman filter

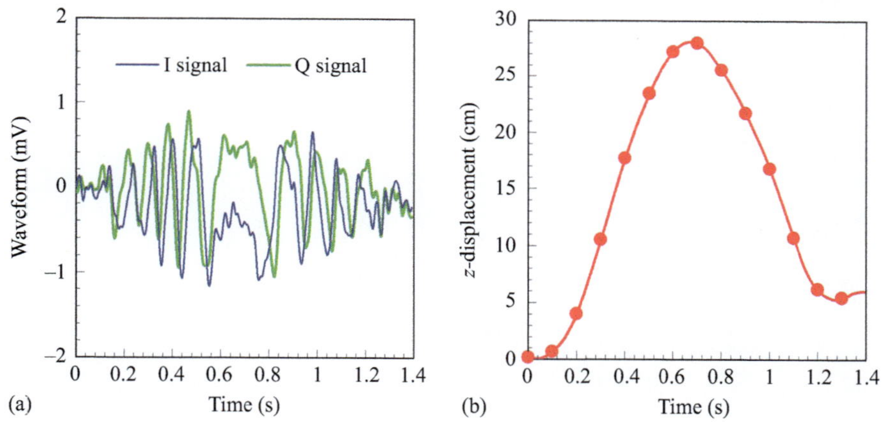

Figure 3.34 Capturing the longitudinal displacement of a "push–pull" hand gesture with the Wi-Fi-based passive radar. (a) Quadrature Doppler signals. (b) Longitudinal displacement

hand gesture is found; from Figure 3.34(b), the maximum displacement of the hand is 28.1 cm and the gesture takes about 1.4 s.

A problem arises when the transverse and longitudinal trajectories that were captured by the camera and radar, respectively, are combined to form a 3D trajectory because the transverse trajectory comprises data points in pixel coordinates rather than in spatial coordinates. Therefore, the transverse trajectory must be transformed into spatial coordinates. This transformation can be performed by using the system to detect hand displacements along straight lines on the xz and yz planes at predetermined angles of θ_x and θ_y relative to x and y axes as depicted in Figure 3.35(a) and (b), respectively. From the longitudinal displacements (Δz) that

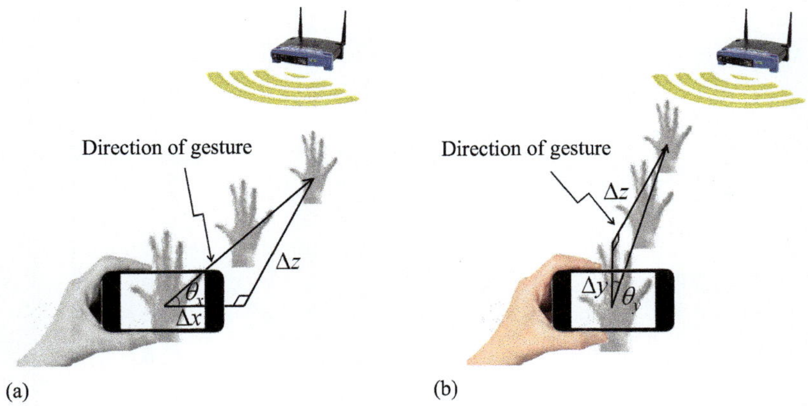

Figure 3.35 System calibration setup. (a) Calibration of x-displacement with $\theta_x = 75°$. (b) Calibration of y-displacement with $\theta_y = 75°$ (from [26])

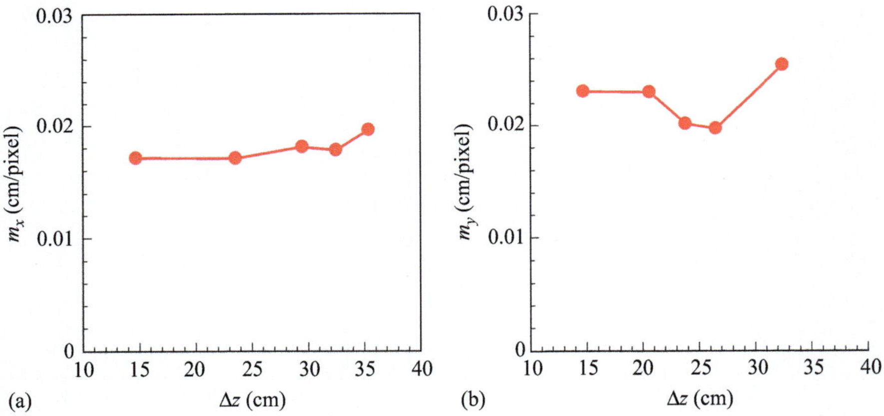

Figure 3.36 Results of conversion coefficients from pixel to centimeter in the calibration scheme. (a) Conversion coefficient of the x-displacement. (b) Conversion coefficient of the y-displacement

were obtained by radar detection, the conversion coefficients of the transverse displacements (Δx and Δy) from pixel to centimeter can be found as

$$m_x = \frac{\Delta z\ (\text{cm})}{\Delta x\ (\text{pixel}) \cdot \tan\theta_x};\ m_y = \frac{\Delta z\ (\text{cm})}{\Delta y\ (\text{pixel}) \cdot \tan\theta_y} \tag{3.51}$$

Figure 3.36(a) and (b) shows the conversion coefficients that were calculated using (3.51) for five samples in the x- and y-displacement calibrations. The coefficients m_x and m_y have mean values of 0.0179 and 0.0222, respectively.

3.4.4.3 3D Hand trajectories

Figure 3.37 shows an experimental setup of the gesture detection system fusing vision and radar data. In this figure, a user holds a smartphone in his left hand to capture the gestures of his right hand with an in-phone camera and an external Wi-Fi-based passive radar. The smartphone is Wi-Fi-connected to an AP that runs in mixed b/g/n mode with a maximum bandwidth of 20 MHz. The radar is connected via a cable to one of the antenna ports of the AP and uses the Wi-Fi signals from the AP to detect the longitudinal displacement of the hand. A directional antenna with a gain of 12 dBi is used to transmit and receive signals in this radar, and its distance from the smartphone is 50 cm. The digital signal processing unit of this radar is a laptop, which outputs the radar detection results via the AP to the smartphone. Therefore, the smartphone can construct the 3D trajectory of the hand by combining the transverse and longitudinal displacement data from the camera and the radar, respectively. Notably, in this system, all data sets are synchronized to coordinated universal time (UTC) and can be made to correspond to a single time axis. The laptop downloads a large file of about 10 GB during the experiment to ensure continuous transmission of Wi-Fi signals from the AP over a long period.

Figure 3.38(a) presents the capture by the system of a hand gesture with an arc-shaped trajectory. To make this hand gesture, the phone user moves his right hand up, forward and to the right, and then down, forward and to the right. Figure 3.38(b) shows the constructed 3D trajectory of the hand. The figure demonstrates that this hand gesture takes 0.56 s to complete and its maximum x, y, and z displacements

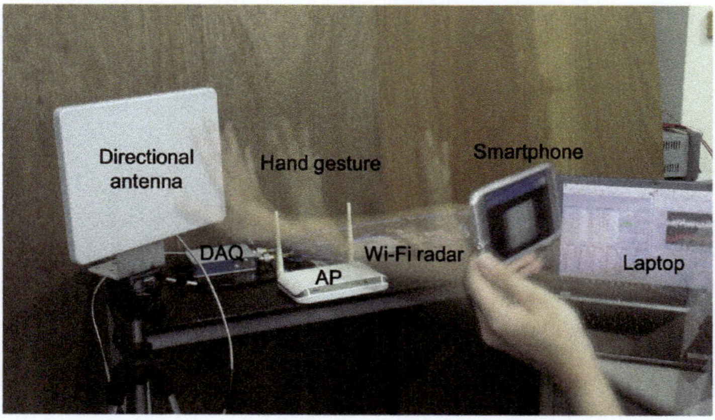

Figure 3.37 Experimental setup of the vision and Wi-Fi sensing fusion system (from [26])

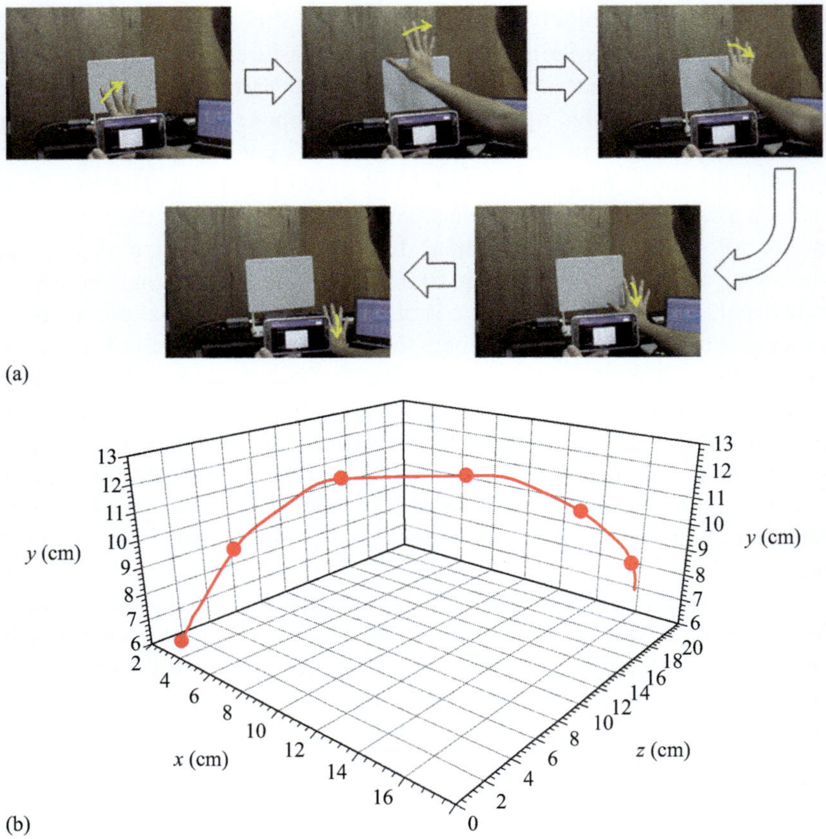

Figure 3.38 Capturing an arc-shaped hand trajectory with the vision and Wi-Fi sensing fusion system. (a) Continuous shooting of the hand gesture. (b) Constructed 3D hand trajectory (from [26])

are 12.8, 6, and 20 cm, respectively. Figure 3.39(a) presents the capture by the system of another hand gesture with an "awareness ribbon"-shaped trajectory. In this hand gesture, the right hand of the phone user moves forward and to the right, reverses, and then moves backward to the left. Figure 3.39(b) presents the constructed 3D trajectory of the hand; the time taken is 1.2 s, and the maximum x, y, and z displacements are 14.8, 5.3, and 17.3 cm, respectively. The processing time of the system to construct a 3D hand trajectory is very short—less than 0.1 s. Figure 3.40 compares the construction of a 1.5-s-long C-shaped hand trajectory using this system with that obtained using a Leap Motion device. The comparison shows excellent agreement.

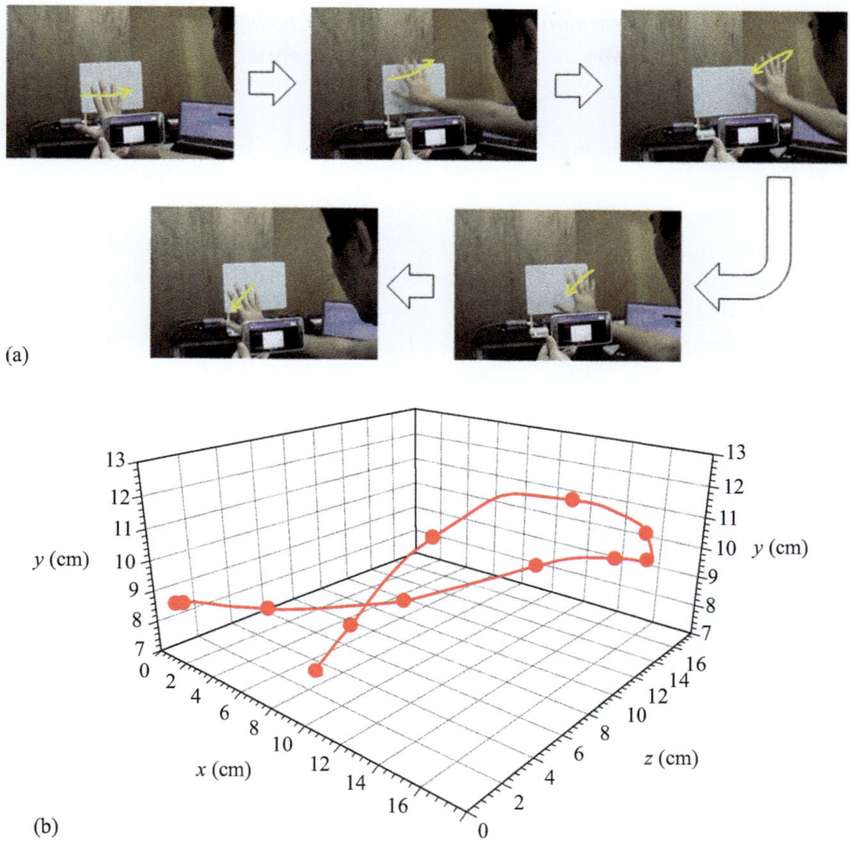

Figure 3.39 Capturing an "awareness ribbon"-shaped hand trajectory with the vision and Wi-Fi sensing fusion system. (a) Continuous shooting of the hand gesture. (b) Constructed 3D hand trajectory (from [26])

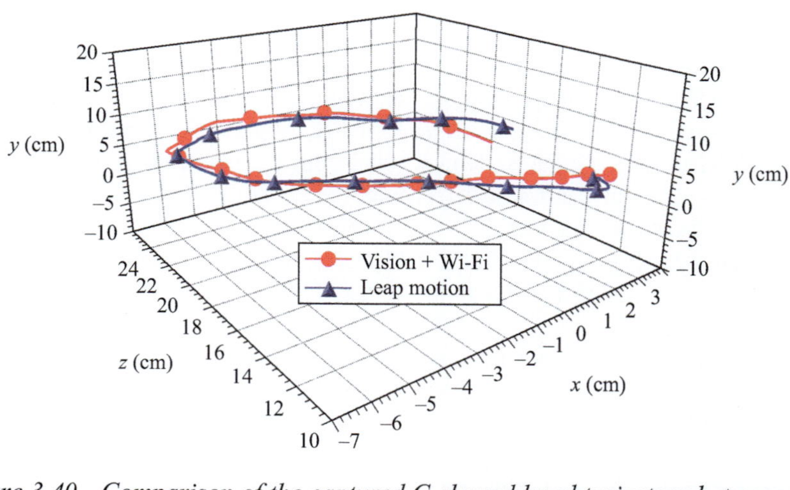

Figure 3.40 Comparison of the captured C-shaped hand trajectory between Leap Motion and the developed sensing system

References

[1] Schlömer T., Poppinga B., Henze N., and Boll S. "Gesture recognition with a Wii controller." *Proc. Int. Conf. Tangible Embedded Interaction*; Bonn, Germany, Feb 2008. pp. 11–14.

[2] Dipietro L., Sabatini A., and Dario P. "A survey of glove-based systems and their applications." *IEEE Trans. Syst. Man Cybern. C, Appl. Rev.* 2008, vol. 38(4), pp. 461–482.

[3] Attenberger A., and Buchenrieder K. "RemoteHand: a wireless myoelectric interface." *Proc. Human-Computer Interaction Int. Conf., Part II*; Crete, Greece, Jun 2014. pp. 3–11.

[4] Zhang Z. "Microsoft kinect sensor and its effect." *IEEE Multimedia.* 2012, vol. 19(2), pp. 4–10.

[5] Kim Y.S., and Baek K.-H. "A motion gesture sensor using photodiodes with limited field-of-view." *Optics Express.* 2013, vol. 21(8), pp. 9206–9214.

[6] Valtonen M., Kaila L., Mäentausta J., and Vanhala J. "Unobtrusive human height and posture recognition with a capacitive sensor." *J. Ambient Intelligence Smart Environments*. 2011, vol. 3(4), pp. 305–332.

[7] Raj B., Kalgaonkar K., Harrison C., and Dietz P. "Ultrasonic Doppler sensing in HCI." *IEEE Pervasive Computing*. 2012, vol. 11(2), pp. 24–29.

[8] Gupta S., Morris D., Patel S., and Tan D. "Soundwave: using the Doppler effect to sense gestures." *Proc. SIGCHI Conf. Human Factors Computing Systems*; Austin, USA, May 2012. pp. 1911–1914.

[9] Przybyla R.J., Tang H., Shelton S.E., Horsley D.A., and Boser B.E. "3D ultrasonic gesture recognition." *Int. Solid-State Circuits Conf. Dig. Tech. Papers*; San Francisco, USA, Feb 2014. pp. 210–211.

[10] Adib F., Kabelac Z., Katabi D., and Miller R.C. "3D tracking via body radio reflections." *Proc. USENIX Conf. Networked Systems Design Implementation*; Seattle, USA, Apr 2014. pp. 1–13.

[11] Wan Q., Li Y., Li C., and Pal R. "Gesture recognition for smart home applications using portable radar sensors." *Proc. Int. Conf. IEEE Eng. Medicine Biology Society*; Chicago, USA, Aug 2014. pp. 6414–6417.

[12] Molchanov P., Gupta S., Kim K., and Pulli K. "Short-range FMCW monopulse radar for hand-gesture sensing." *Proc. IEEE Radar Conf.*; Johannesburg, South Africa, Oct 2015. pp. 1491–1496.

[13] Zhang Z., Tian Z., and Zhou M. "Latern: dynamic continuous hand gesture recognition using FMCW radar sensor." *IEEE Sensors J.* 2018, vol. 18(8), pp. 3278–3289.

[14] Howland P.E. "Target tracking using television-based bistatic radar." *IEE Proc. Radar Sonar Navig*. 1999, vol. 146(3), pp. 166–174.

[15] Griffiths H.D., Baker C.J., Baubert J., Kitchen N., and Treagust M. "Bistatic radar using satellite-borne illuminators." *Proc. IEE Int. Radar Conf.*; Edinburgh, UK, Oct 2002. pp. 1–5.

[16] Howland P.E., Maksimiuk D., and Reitsma G. "FM radio based bistatic radar." *IEE Proc. Radar Sonar Navig*. 2005, vol. 152(3), pp. 107–115.

[17] Tan D.K.P., Sun H., Lu Y., Lesturgie M., and Chan H.L. "Passive radar using global system for mobile communication signal: theory, implementation and measurements." *IEE Proc. Radar Sonar Navig.* 2005, vol. 152(3), pp. 116–123.

[18] Wang Q., Hou C., and Lu Y. "An experimental study of WiMAX-based passive radar." *IEEE Trans. Microw. Theory Techn.* 2010, vol. 58(12), pp. 3502–3510.

[19] Berger C.R., Demissie B., Heckenbach J., Willett P., and Zhou S. "Signal processing for passive radar using OFDM waveforms." *IEEE J. Sel. Topics Signal Process.* 2010, vol. 4(1), pp. 226–238.

[20] Colone F., Woodbridge K., Guo H., Mason D., and Baker C.J. "Ambiguity function analysis of wireless LAN transmissions for passive radar." *IEEE Trans. Aerosp. Electron. Syst.* 2011, vol. 47(1), pp. 240–264.

[21] Palmer J.E., Harms H.A., Searle S.J., and Davis L.M. "DVB-T passive radar signal processing." *IEEE Trans. Signal Process.* 2013, vol. 61(8), pp. 2116–2126.

[22] Pu Q., Gupta S., Gollakota S., and Patel S. "Whole-home gesture recognition using wireless signals." *Proc. 19th Annual Int. Conf. Mobile Computing Networking*; Miami, USA, Sep 2013. pp. 27–38.

[23] Kellogg B., Talla V., and Gollakota S. "Bringing gesture recognition to all devices." *Proc. USENIX Conf. Networked Systems Design Implementation*; Seattle, USA, Apr 2014. pp. 1–14.

[24] Wang F.-K., Tang M.-C., Chiu Y.-C., and Horng T.-S. "Gesture sensing using retransmitted wireless communication signals based on Doppler radar technology." *IEEE Trans. Microw. Theory Techn.* 2015, vol. 63(12), pp. 4592–4602.

[25] Tang M.-C., Wang F.-K., and Horng T.-S. "Human gesture sensor using ambient wireless signals based on passive radar technology." *IEEE MTT-S Int. Microwave Symp. Dig.*; Phoenix, USA, May 2015. pp. 1–4.

[26] Tang M.-C., Chen C.-L., Lin M.-H., Wang F.-K., Yeh C.-H., and Horng T.-S. "A hybrid computer vision and Wi-Fi Doppler radar system for capturing the 3-D hand gesture trajectory with a smartphone." *IEEE MTT-S Int. Microwave Symp. Dig.*; Honolulu, USA, Jun 2017. pp. 1–4.

[27] Adler R. "A study of locking phenomena in oscillators." *Proc. IRE.* 1946, vol. 34(6), pp. 351–357.

[28] Huntoon R.D., and Weiss A. "Synchronization of oscillators." *Proc. IRE.* 1947, vol. 35(12), pp. 1415–1423.

[29] Mackey R.C. "Injection locking of klystron oscillators." *IRE Trans. Microw. Theory Techn.* 1962, vol. 10(4), pp. 228–235.

[30] Paciorek L.J. "Injection locking of oscillators." *Proc. IEEE.* 1965, vol. 53(11), pp. 1723–1727.

[31] Kurokawa K. "Injection locking of microwave solid-state oscillators." *Proc. IEEE.* 1973, vol. 61(10), pp. 1386–1410.

[32] Calandra E.F., and Sommariva A.M. "Stability analysis of injection-locked oscillators in their fundamental mode of operation." *IEEE Trans. Microw. Theory Techn.* 1981, vol. 29(11), pp. 1137–1143.

[33] Razavi B. "A study of injection locking and pulling in oscillators." *IEEE J. Solid-State Circuits.* 2004, vol. 39(9), pp. 1415–1424.

[34] Mirzaei A., Heidari M.E., and Abidi A.A. "Analysis of oscillators locked by large injection signals: generalized Adler's equation and geometrical interpretation." *Proc. IEEE Custom Integrated Circuits Conf.*; San Jose, USA, Sep 2006. pp. 737–740.

[35] Li C.-J., Wang F.-K., Horng T.-S., and Peng K.-C. "A novel RF sensing circuit using injection locking and frequency demodulation for cognitive radio applications." *IEEE Trans. Microw. Theory Techn.* 2009, vol. 57(12), pp. 3143–3152.

[36] Li C.-J., Hsiao C.-H., Wang F.-K., Horng T.-S., and Peng K.-C. "A rigorous analysis of a phased-locked oscillator under injection." *IEEE Trans. Microw. Theory Techn.* 2010, vol. 58(5), pp. 1391–1400.

[37] Chen C.-T., Hsiao C.-H., Horng T.-S., Peng K.-C., and Li C.-J. "Cognitive polar receiver using two injection-locked oscillator stages." *IEEE Trans. Microw. Theory Techn.* 2011, vol. 59(12), pp. 3484–3493.

[38] Pozar D.M. *Microwave Engineering*. 4th edn. Singapore: Wiley; 2011. pp. 272–278.

[39] Stover H.L., and Shaw R.C. "Injection-locked oscillators as amplifiers for angle-modulated signals." *IEEE G-MTT Int. Symp. Dig.*; Palo Alto, USA, May 1966. pp. 60–66.

[40] Main E., and Coffing D. "An FSK demodulator for bluetooth applications having no external components." *IEEE Trans. Circuits and Systems II: Analog and Digital Processing*. 2002, vol. 49(6), pp. 373–378.

[41] Ramírez F., Araña V.A., and Suárez A. "Frequency demodulator using an injection-locked oscillator: analysis and design." *IEEE Microwave and Wireless Component Lett.* 2008, vol. 18(1), pp. 43–45.

[42] Ye R.-F., Horng, T.-S., and Wu J.-M. "Ultralow power injection-locked GFSK receiver for short-range wireless systems." *IEEE Trans. Circuits Syst. II, Express Briefs*. 2012, vol. 59(11), pp. 706–710.

[43] Oppenheim A.V., Schafer R.W., and Buck J.R. *Discrete-Time Signal Processing*. 2nd edn. Upper Saddle River: Prentice Hall; 1999.

[44] Park B.-K., Boric-Lubecke O., and Lubecke V.M. "Arctangent demodulation with DC offset compensation in quadrature Doppler radar receiver systems." *IEEE Trans. Microw. Theory Techn.* 2007, vol. 55(5), pp. 1073–1079.

[45] Lv Q., Ye D., Qiao S., *et al.* "High dynamic-range motion imaging based on linearizer Doppler radar sensor." *IEEE Trans. Microw. Theory Techn.* 2014, vol. 62(9), pp. 1837–1846.

[46] Lee J.Y., and Woo T. "A study on gesture game interface." *Proc. 2nd International Conference on Electronics, Electrical Engineering, Computer Science: Innovation and Convergence*; Qingdao, China, Aug 2016. pp. 176–180.

[47] Donahue J., Hendricks L.A., Guadarrama S. *et al.* "Long-term recurrent convolutional networks for visual recognition and description." *Proc. IEEE Conference on Computer Vision and Pattern Recognition*; Boston, USA, Jun 2015. pp. 2625–2634.

[48] Hochreiter S., and Schmidhuber J. "Long short-term memory." *Neural Computation.* 1997, vol. 9(8), pp. 1735–1780.

[49] Agarap A.F. "Deep learning using rectified linear units (ReLU)." *Neural and Evolutionary Computing*. 2018, vol. arXiv:1803.08375.

[50] Kingma D.P., and Ba J. "Adam: a method for stochastic optimization." *Machine Learning*. 2014, vol. arXiv:1412.6980.

[51] Oliveira V.A., and Conci A. "Skin detection using HSV color space." *Proc. XXII Brazilian Symposium on Computer Graphics and Image Processing*; Rio de Janeiro, Brazil, Oct 2009. pp. 1–2.

[52] Andresen S. *Hand tracking and recognition with OpenCV* [Online]. 2013. Available from http://simena86.github.io/blog/2013/08/12/hand-tracking-and-recognition-with-opencv/ [Accessed 30 Aug 2015].

[53] Suzuki S. "Topological structural analysis of digitized binary images by border following." *Computer Vision, Graphics, and Image Processing*. 1985, vol. 30(1), pp. 32–46.

[54] Graham R.L., and Yao F. F. "Finding the convex hull of a simple polygon." *Journal of Algorithms*. 1983, vol. 4(4), pp. 324–331.

Chapter 4

Hand gesture recognition based on SIMO Doppler radar sensors

Yi Zhang[1], Shuqing Dong[1], Chengkai Zhu[1], Anjie Zhu[1], Zhitao Gu[1], Tenlong Fan[1], Qinyi Lv[1], Jingyu Wang[1], Changzhan Gu[1], and Lixin Ran[1]

Humans communicate with each other with acoustic and visual signals from the very beginning. Typical examples are voices and hand gestures. In the computer age, it is naturally desired that the communication between humans and machines, i.e., the human–machine interface (HMI), can also use voices and gestures. With the rapid development of computer science and artificial intelligence (AI), the computer-based natural language understanding has achieved significant progress. However, the computer-based hand gesture recognition (HGR) remains a technical challenge due to complicated image processing algorithms and excessive occupation of computational resources.

Recently, wireless sensing and detection based on continuous-wave (CW) Doppler radar sensors (DRSs) have been intensively investigated, based on which experimental HGRs implemented with cost-effective, miniaturized hardware and linearized, highly efficient algorithms have been experimentally demonstrated. This DRS-based solution is able to recognize definitive, meaningful signatures of human gestures retrieved from low-sampling-rate data, exhibiting promising potential for practical applications.

This chapter aims to introduce this progress. The content mainly includes an introduction to the basis of Doppler radar sensing, the architecture, and algorithms suitable for HGRs, and experimental HGR demonstrations based on such architecture and algorithms.

4.1 Doppler radar sensing

The frequency, or equivalently the phase, of electromagnetic, optic, and acoustic waves scattered off a moving object will be shifted. This physical phenomenon is

[1]Laboratory of Applied Research on Electromagnetics (ARE), Zhejiang University, China
All authors contributed equally to this chapter.

known as the Doppler effect, named after the physicist Christian Andreas Doppler. As described by the Doppler theory [1], the frequency of a wave will be increased (red shift) when the object is moving toward the source generating this wave, or, alternatively, will be decreased (blue shift) when the object is moving away from the source. The resultant frequency shift Δf, i.e., Doppler frequency or Doppler signal, can be calculated by

$$\Delta f = \frac{2f_{inc} \cdot |V_r|}{c} \cos(\theta) \tag{4.1}$$

where f_{inc} is the frequency of the incident wave, c is the speed of light, V_r is the relative velocity between the source and the moving object, and θ is the angle included by the two vectors, i.e., the relative velocity and the k vector of the incident wave.

In the past decades, the Doppler effect has been widely used in CW and pulsed radar systems for multipurpose remote detections. Typical applications include detecting the displacement and velocity of moving targets and improving the ability to identify moving objects from strong background scatterings caused by stationary clutters [1]. With the rapid progress of radio frequency integrated circuits (RFICs), miniaturized, cost-effective, low-power-consumption CW DRSs have been developed so that wireless detections to various motions can be realized in short-range and indoor environments [2–4]. Working with simple structure, sharp sensitivity, and high spectrum efficiency, DRSs have been used in a wide range of applications. Examples include noncontact human vital signs detection (VSD) [5–8], mechanical vibration measurement and malfunction localization [9], structural health monitoring for buildings and bridges [10–16], and indoor objects' tracking and localization [17]. Recently, single-chip integrated DRS front ends have been reported [18–20], implying that low-cost, mass-produced micro-sized DRSs will be available for a variety of new applications in the near future.

Figure 4.1 shows the system schematic of a single-input single-output (SISO) CW DRS that can be used to detect a one-dimensional (1-D) motion $x(t)$. As a SISO system, it only has one transmitting antenna T and one receiving antenna R, respectively. The transmitting antenna radiates a CW radio frequency (RF) signal to illuminate a moving target at a distance d_0 from the DRS. The backscattered RF signal carrying the Doppler phase information is received by the receiving antenna and is further demodulated by the zero-intermediate frequency (zero-IF) receiver, outputting the baseband in-phase (I), and quadrature-phase (Q) signals:

$$I(t) = A_I(t)\cos\left[\theta_0 + \frac{4\pi x(t)}{\lambda} + \varphi_0(t)\right] + DC_I(t) \tag{4.2}$$

and

$$Q(t) = A_Q(t)\sin\left[\theta_0 + \frac{4\pi x(t)}{\lambda} + \varphi_0(t)\right] + DC_Q(t) \tag{4.3}$$

respectively. Here $DC_I(t)/DC_Q(t)$ and $A_I(t)/A_Q(t)$ denote the direct current (DC) offsets and the alternate current (AC) amplitudes of the I/Q signals, respectively. In the arguments of the trigonometric functions, $\varphi_0(t)$ represents the residual phase noise of the receiver, λ is the wavelength of the CW signal, and θ_0 represents the accumulated phase shift due to the initial distance d_0 between the moving object and the DRS and the phase delay inside the transceiver system. Mathematically, the motion $x(t)$ can be retrieved from the above *I/Q* signals if an appropriate algorithm can be applied.

DRSs implemented based on the architecture described in Figure 4.1 have been intensively investigated in the detection of human vital signs, mainly human respirations and heartbeats. In a sense, human gestures, as well as human gaits, can also be considered as another kind of vital signs. However, there are essential differences. First, human respirations and heartbeats are small-scale motions. The range of such motions is normally shorter than the wavelength of the frequently used Industrial Scientific Medical (ISM) frequencies, normally 2.4 and 5.8 GHz. Second, from the application point of view, detecting the 1-D component of a Doppler vital sign signal has been enough, and therefore a SISO-structured DRS can be simply used. Human gestures, however, can be 2-D or even 3-D motions. 1-D detection to human gestures is obviously insufficient for the further recognition. In the meantime, the gesture motions can be both small scale and large scale compared with the used wavelengths. Therefore, conventional zero-IF architecture and algorithms used in VSDs may be unsuitable for human gesture detections. These differences imply that for the HGR applications, the architecture of DRS transceivers, and the algorithms used to retrieve the Doppler signals caused by human gestures should be specifically considered so that highly dynamic motions including both small- and larger-scale ones can be linearly reconstructed using the same DRS.

4.2 Architecture

4.2.1 Optimal architecture for HGR application

The RF receiver shown in Figure 4.1 is known as the zero-IF receiver. Without the requirement of external filters, zero-IF transceivers can be easily integrated into a single chip. Owning to this unique advantage, zero-IF receivers have been widely used in modern portable electric devices. However, it has been recognized that this kind of receiver would suffer from the inherent flicker noise and DC offset issues. Unfortunately, Doppler signals are caused by human behaviors and actions such as gestures are hertz-range signals. The spectra of most of such Doppler signals would

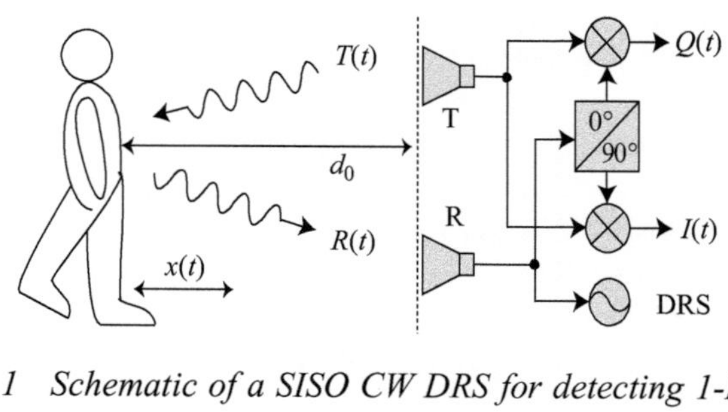

Figure 4.1 Schematic of a SISO CW DRS for detecting 1-D motions

include DC components and extend from DC to a frequency about several hertz. In this case, the DC offset and the flicker noise normally distributed close to the DC frequency will potentially interfere the Doppler signal, especially in small signal-to-noise ratio (SNR) cases. Especially, if the DC offset output by the zero-IF receiver cannot be properly separated from the useful DC component of the Doppler signal, achieving a satisfactory linear retrieval of the Doppler signal would become impossible. Therefore, a DRS architecture that can eliminate the DC offset and flicker noise from its baseband output signals would be highly desired for HGR applications.

On the other hand, the algorithm for the linear Doppler signal retrieval would also have its own requirement for the transceiver architecture. Mathematically, the motion $x(t)$ shown in Figure 4.1 can be solved from (4.2) and (4.3) by computing an arctangent function, i.e.:

$$x(t) = \frac{\lambda}{4\pi}\arctan\left\{\frac{Q(t) - DC_Q(t)/A_Q(t)}{I(t) - DC_I(t)/A_I(t)} - \theta_0 - \varphi_0(t)\right\} \tag{4.4}$$

To investigate the impacts of the DC offsets $DC_I(t)$ and $DC_Q(t)$ and the imbalance between the amplitudes $A_I(t)$ and $A_Q(t)$ to the solution $x(t)$, the sine and cosine functions in (4.2) and (4.3) can be eliminated so that:

$$\left[\frac{I(t) - DC_I(t)}{A_I(t)}\right]^2 + \left[\frac{Q(t) - DC_Q(t)}{A_Q(t)}\right]^2 = 1 \tag{4.5}$$

It is seen that the $I(t)$ and $Q(t)$ data are located on an ellipse defined by the DC offsets $DC_I(t)$, $DC_Q(t)$ and the amplitudes $A_I(t)$, $A_Q(t)$. It implies that with the $I(t)$ and $Q(t)$ detected by the DRS, simultaneously solving the DC offsets and the amplitudes from (4.5) is difficult due to the lack of sufficient equations. In order to linearly retrieve $x(t)$ from the $I(t)$ and $Q(t)$ signals, the amplitude imbalance between $A_I(t)$ and $A_Q(t)$ has to be eliminated so that (4.5) can be simplified to

$$[I(t) - DC_I(t)]^2 + [Q(t) - DC_Q(t)] = A_R^2(t) \tag{4.6}$$

in which $A_I = A_Q = A_R$. It is seen that after eliminating the imbalance between $A_I(t)$ and $A_Q(t)$, the ellipse described by (4.5) reduces to a circle with the point ($DC_I(t)$, $DC_Q(t)$) and the balanced amplitude A_R being the circle center and the radius, respectively. In this case, as long as the circle center can be determined based on $I(t)$ and $Q(t)$, the motion $x(t)$ can be mathematically calculated by (4.4).

According to the above discussion, the architecture of a CW DRS that is optimal for the HGR application should be able to eliminate the DC offset and flicker noise from its baseband output and ensure the balance of the amplitudes of the orthogonal I/Q signals. Among the available state-of-the-art architectures of RF transceiver, the digital intermediate frequency (digital-IF) architecture shown in Figure 4.2 would satisfy the above requirements [7]. By sampling the IF signal and performing a band-pass filtering and a quadrature zero-IF downconversion in digital domain, optimal elimination of the DC offset, the flicker noise, and the amplitude imbalance can be simultaneously achieved.

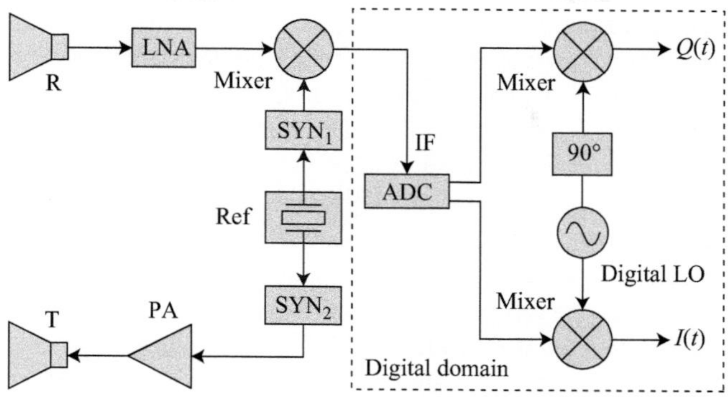

Figure 4.2 Schematic of a SISO digital-IF CW DRS

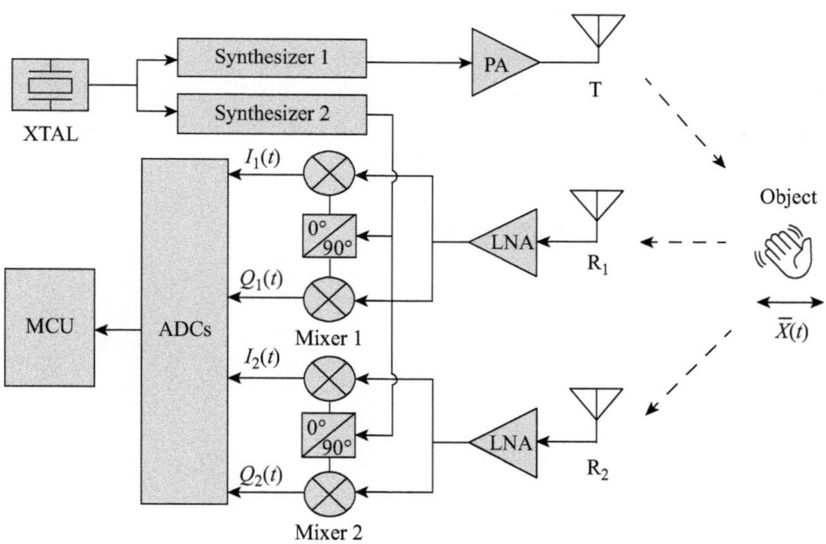

Figure 4.3 Schematic of a SIMO digital-IF CW DRS with one transmitting and two receiving channels

4.2.2 SIMO-structured CW DRS

As discussed, the SISO-structured DRS can only be used in the detection of 1-D motions. For 2-D and 3-D hand gestures, single-input multiple-output (SIMO) DRS can be used. As an example, Figure 4.3 shows the schematic of such a DRS with one transmitting channel and two identical receiving channels, each implemented with a digital-IF architecture.

The key to implement a SIMO CW DRS is to ensure the coherence of the reference signals between all the transmitting and receiving channels. In Figure 4.3, this is realized by sharing the same reference signal generated by a highly stable, temperature compensated crystal oscillator (TCXO). In order to realize the digital-IF structure, two frequency synthesizers sharing the same reference signal can be used to generate two CW signals with different frequencies. One is amplified by an RF power amplifier (PA) and then radiated by the transmitting antenna. Another is used as the local oscillation (LO) signal of the two identical receivers. The frequency difference between these two CW signals will become the IF.

Since the Doppler signals modulated in the phase of the IF signal are hertz-range narrowband ones, the passband sampling technique [21] can be used to sample the IF signal. The sampling rate can be as low as 10–100 Hz, meaning that the built-in analog-to-digital convertors (ADCs) of ordinary microcontroller units (MCUs) can be used to finish the digital-IF sampling and the digital-domain zero-IF downconversion. Note that the band-pass sampling is simultaneously equivalent to a zero-IF downconversion [19].

4.2.3 Experimental implementation of a digital-IF DRS

As an example, Figure 4.4 shows the schematic and the picture of a 24-GHz DRS originally designed for highly sensitive VSDs [8]. It was implemented based on the monolithic millimeter-wave-integrated circuit (MMIC), Infineon BGT24MTR11.

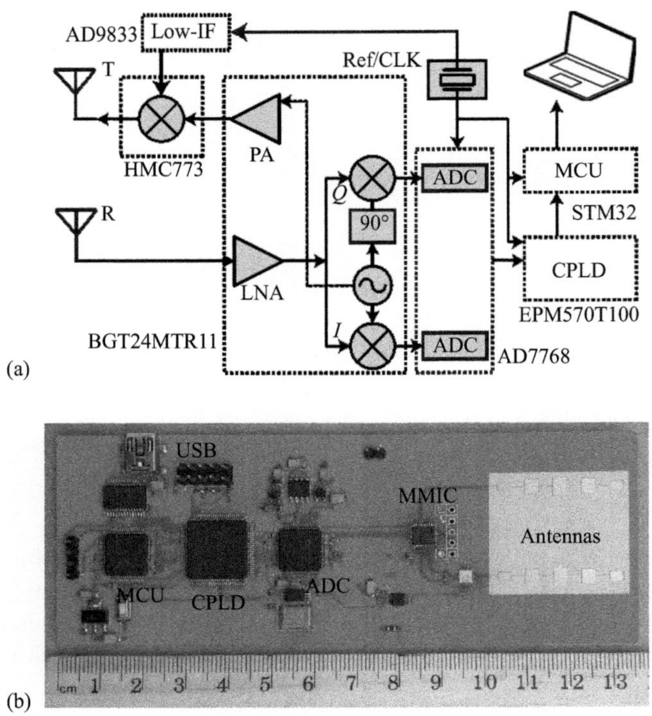

Figure 4.4 Experimental implementation of a 24-GHz digital-low-IF DRS. (a) Schematic. (b) Implemented DRS

Table 4.1 Main devices used in the 24-GHz DRS

Number	Device name	Manufacturer	Function
1	BGT24MTR11	Infineon	24-GHz MMIC
2	HMC773	Hittite	RF mixer
3	AD9833	Analog Devices	Programmable waveform generator
4	AD7768	Analog Devices	ADC
5	EPM570T100C5	Altera	Complex programmable logic device
6	STM32F106	ST Electronics	MCU

The main devices as well as their functions and vendor information are listed in Table 4.1.

A special feature of this DRS is the use of a 1-MHz subcarrier generated by the signal source AD9833, to implement a double-sideband digital-low-IF modulation. By doing so, the DC offset and flicker noise generated by the MMIC can be completely filtered by a band-pass sampling to the low-IF *I/Q* signals. For this DRS, the used band-pass sampling rate is only 90 Hz. This frequency is much lower than the 1-MHz IF frequency, but much higher than the frequency of human activities. Also, a USB interface is designed to transmit the digitized data to and receive the control information from a computer, where the algorithms for the Doppler signal retrieval will be run to recover the time-domain Doppler signal. For this DRS, the conversion gain and the transmitting power of the receiving and the transmitting channels are 26 dB and 6 dBm, respectively.

Compared with this SISO DRS, a SIMO DRS would have more receiving channels. As mentioned, the key to implement a SIMO DRS is to ensure the coherence between all the transmitting and receiving channels. An experimental SIMO CW DRS will be introduced in the experiment section in the following.

4.3 Algorithms

Apart from an optimal DRS, different algorithms are needed in the implementation of HGRs. These algorithms will be discussed in the following.

4.3.1 Algorithms for the linear retrieval of Doppler signals

In the early stage of the studies on noncontact VSDs, the small-angle approximation (SAA) was used to retrieve the Doppler vital sign signals from the single-ended output of zero-IF receivers [5,6]. It is equivalent to demodulating $x(t)$ from (4.2) or (4.3) only. SAA works only when the phase item $\theta_0+\varphi_0(t)$ in (4.4) is close to 0 so that the function sin(x) can be approximated as x. The limitation of the SAA is that the motion $x(t)$ should be small scale and the distance d_0 should be adjusted in advance so that the DRS is able to work at an “optimal distance” [5]. Obviously, SAA is not suitable for HGR applications due to the fact that hand gestures can be large-scale motions and the distance between the hand and the DRS will not be fixed.

According to (4.6), after the imbalance between $A_I(t)$ and $A_Q(t)$ have been eliminated by a digital-domain baseband downconversion, if the circle center ($DC_I(t)$, $DC_Q(t)$) can be further located, the motion $x(t)$ can be linearly recovered from (4.4). As indicated before, since $DC_I(t)$ and $DC_Q(t)$ also include the DC component of the useful Doppler signal, they are essentially time variant, especially for larger-scale motions. In order to linearly recover large-scale, highly dynamic motions from the arctangent function, a dynamic tracking to the time-variant DC offsets should be carried out before the Doppler phase can be retrieved.

4.3.1.1 Dynamic DC offset tracking

This algorithm aims to track the DC offsets based on the $I(t)$ and $Q(t)$ data sampled by the DRS for large-scale motions. The n sampled I/Q data can be denoted as a discrete time series ($I[1]$, $Q[1]$), ($I[2]$, $Q[2]$), ..., ($I[n]$, $Q[n]$). For balanced amplitudes of the I/Q signals of a digital-IF DRS, we denote $A_I[k] = A_Q[k] = A_R[k]$. For the k-th sample, ($DC_I[k]$, $DC_Q[k]$, $A_R[k]$) together form an array of unknown parameters, where the index k represents the time evolution. In [6], a highly efficient first-order optimization based on the gradient descent algorithm was proposed. If a large-scale motion can be segmented to many consecutive small-scale, subwavelength motions, for each subwavelength segment, $DC_I[k]$, $DC_Q[k]$, $A_R[k]$ can be approximated to constants DC_I, DC_Q, A_R, resulting in a fixed circle center and a fix radium. In this case, (4.6) can be simplified to

$$\sqrt{(I[k] - DC_I)^2 + (Q[k] - DC_Q)^2} - A_R = 0 \tag{4.7}$$

The dynamic DC offset tracking aims to minimize the left side of (4.7) so that it can approach zero. Therefore, for each segment, the target function of the optimization can be defined by

$$F\left(DC_I, DC_Q, A_R\right) = \min \sum_{k=1}^{n} \left| \left(\sqrt{(I[k] - DC_I)^2 + (Q[k] - DC_Q)^2} - A_R \right) \right| \tag{4.8}$$

When $F(DC_I, DC_Q, A_R)$ reaches its minimum, the DC offsets for the current segment can be obtained. This optimization can be continued until all the sampled I/Q data have been processed.

4.3.1.2 Linearized arctangent algorithm

After eliminating the amplitude imbalance and tracking the time-variant DC offset, mathematically, a direct arctangent computation can be used to obtain the large-scale motion $x(t)$ in (4.4). However, due to the inherent multiple-value nature, the arctangent function would only return the principal values limited in a codomain range from $-\pi/2$ to $\pi/2$. This may potentially result in strongly nonlinear discontinuities in the computed motion $x(t)$. Although the conventional phase unwrapping process can be used to correct such discontinuities, this issue is difficult to be reliably avoided, especially in practical scenarios when noises exist [7].

In order to effectively solve this nonlinearity issue, the differentiate and cross-multiply (DACM) algorithm can be used. Instead of the direct computation of the arctangent function, the DACM algorithm would calculate a derivative to (4.4) first, obtaining:

$$\omega(t) = \frac{d}{dt}\left[\arctan\frac{Q(t)}{I(t)}\right] = \frac{I(t)\dot{Q}(t) - \dot{I}(t)Q(t)}{I^2(t) + Q^2(t)} \tag{4.9}$$

where $\omega(t)$ represents the Doppler frequency shift caused by the motion $x(t)$, while $\dot{I}(t)$ and $\dot{Q}(t)$ represent the time derivative to $Q(t)$ and $I(t)$, respectively. In order to obtain the Doppler signal $x(t)$, and in the meantime to suppress noises potentially caused by the derivatives, an integration can be performed to (4.9), resulting in an extended DACM algorithm [9].

In digital domain, this extended DACM algorithm can be expressed in a discrete form:

$$x[n] = \frac{\lambda}{4\pi}\sum_{k=2}^{n}\frac{I[k]\Delta Q[k] - \Delta I[k]Q[k]}{I^2[k] + Q^2[k]} \tag{4.10}$$

where the derivatives are approximated by forward differences, and the integration is replaced by an accumulation, which would help to effectively suppress the noises with zero means. The block diagram of the digital-domain extended DACM algorithm is illustrated in Figure 4.5, in which the module "D" represents a unit delay element. With the assistance of the extended DACM algorithm, the linear retrieval of the discrete motion $x[n]$, either small-small or large-scale, can be accomplished. It is seen that the extended DACM algorithm is simple and straightforward, and the obtained discrete $x[n]$ will be free from the phase unwrapping problem.

In [7], the extended DACM algorithm was further improved for low-data-rate sampling. This is important for the HGR hardware because of the use of SIMO architectures. With increased receiving channels, allowing lower-data-rate sampling implies the cost-effective baseband implemented based on MCUs can still be used.

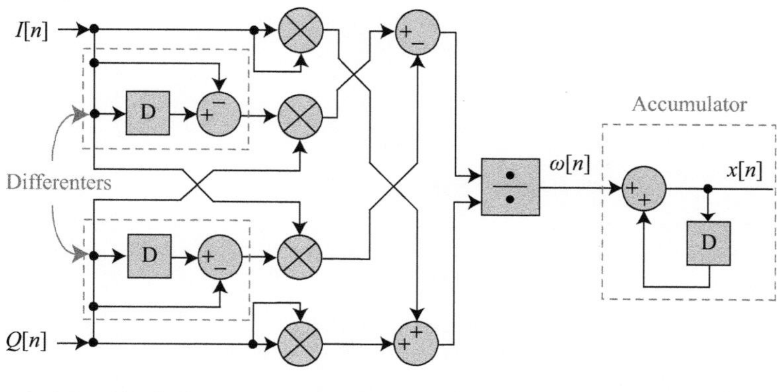

Figure 4.5 Block diagram of the digital-domain extended DACM algorithm

For the *I/Q* signals sampled with a low sampling rate, the error due to the discrete integration in the extended DACM algorithm could be potentially accumulated. To solve this problem, as shown in Figure 4.6, *I*(*t*) and *Q*(*t*) can be treated as a phasor denoted as $\bar{\Gamma} = [I(t), Q(t)]$. After applying the dynamic DC offset tracking, the neighboring sampling data would locate close to the circle perimeter shown in Figure 4.6. Between the *k*th and the (*k* − 1)th sampling data, the incremental phase $\Delta\Phi_k$ can be obtained by

$$\begin{aligned}\Delta\Phi_k &= \arcsin\frac{\hat{n}\cdot(\bar{\Gamma}_{k-1}\times\bar{\Gamma}_k)}{|\bar{\Gamma}_{k-1}|\cdot|\bar{\Gamma}_k|}\\ &= \arcsin\frac{I[k-1]Q[k]-I[k]Q[k-1]}{\sqrt{I[k-1]^2+Q[k-1]^2}\cdot\sqrt{I[k]^2+Q[k]^2}}\end{aligned} \tag{4.11}$$

where $\hat{n}$ is the unit vector pointing to the normal direction of the complex *I*(*t*)-*Q*(*t*) plane. Then $x[n]$ in (4.10) changes to

$$\begin{aligned}\Phi[n] &= \Phi[0]+\sum_{k=2}^{n}\Delta\Phi_k\\ &= \Phi[0]+\sum_{k=2}^{n}\arcsin\frac{I[k-1]Q[k]-I[k]Q[k-1]}{\sqrt{I[k-1]^2+Q[k-1]^2}\cdot\sqrt{I[k]^2+Q[k]^2}}\end{aligned} \tag{4.12}$$

It can be seen that when the phase difference between the *k*th and the (*k* − 1)th samplings approaches zero, (4.12) will reduce to

$$\Phi'[n] = \Phi[0]+\sum_{k=2}^{n}\frac{I[k-1]Q[k]-I[k]Q[k-1]}{\sqrt{I[k-1]^2+Q[k-1]^2}\cdot\sqrt{I[k]^2+Q[k]^2}} \tag{4.13}$$

which is equivalent to the extended DACM equation (4.10). Finally, according to the proportional relationship between the displacement and phase shift, the discrete form of the motion $x(t)$ can be obtained by

$$x[n] = \frac{\lambda}{4\pi}\Phi[n] \tag{4.14}$$

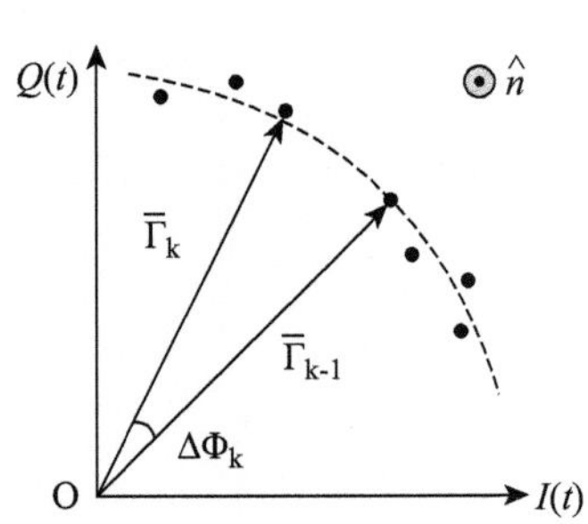

Figure 4.6 Derivation of the arcsine algorithm for low-date-rate samplings

Compared with the extended DACM method, the arcsine algorithm is no longer relevant to the arctangent function. Without accumulating sampling errors, it is especially suitable for the SIMO DRSs implemented with low-cost baseband with low-sampling-rate ADCs.

4.3.2 Algorithms for HGRs based on a SIMO DRS

In order to recognize human gestures, the 2-D or 3-D trajectories of hand motions should be tracked and, if necessary, localized by a SIMO DRS in a predefined coordinate system. In cases when the gesture motions are interfered by other motions such as human respirations and random body movements (RBMs), an additional algorithm would be needed to separate the gesture motions from the interfering motions. Such algorithms relevant to HGRs will be discussed in the following.

4.3.2.1 2-D motion tracking algorithm

Figure 4.7(a) shows an *x-y-z* coordinate system defined for deriving the 2-D motion tracking algorithm. Here the moving object such as a human hand would only move in the *x-y* plane (or equivalently $z = 0$ plane), and thus the time-variant location of this object can be denoted as $(x(t), y(t), 0)$. A realistic example of this *x-y* plane can be the desktop of a table.

To detect a 2-D motion, a SIMO DRS with one transmitting antenna T and two receiving antennas R_1 and R_2 would suffice. Assume such antennas are located at (x_T, y_T, z_T), (x_1, y_1, z_1), and (x_2, y_2, z_2), respectively. All the antennas can be placed in or out of the $z = 0$ plane, as long as the signals scattered by the moving object can be properly received by receiving antennas R_1 and R_2. The distances between antennas T, R_1, and R_2 and the moving object are denoted as d_T, d_{R1}, and d_{R2}. In this case, the round-trip distance from the transmitting antenna to the object, and then to the two receiving antennas will be $d_1 = d_T + d_{R1}$, $d_2 = d_T + d_{R2}$, respectively.

According to the geometry, when the object moves in the *x–y* plane, it must locate at a point intersected by the *x-y* plane and the surfaces of two ellipsoids whose focuses are at the locations (T, R_1) and (T, R_2), respectively. The ellipsoid functions can be derived as

$$\begin{cases} \sqrt{(x_T - x(t))^2 + (y_T - y(t))^2 + z_T{}^2} + \\ \sqrt{(x_1 - x(t))^2 + (y_1 - y(t))^2 + z_1{}^2} = d_1 \\ \sqrt{(x_T - x(t))^2 + (y_T - y(t))^2 + z_T{}^2} + \\ \sqrt{(x_2 - x(t))^2 + (y_2 - y(t))^2 + z_2{}^2} = d_2 \end{cases} \tag{4.15}$$

respectively.

The point intersected by the ellipsoid surfaces and the *x-y* plane is shown in Figure 4.7(b). By solving this time-variant intersected point, the location of the object at time t_0, i.e., $(x(t_0), y(t_0), 0)$ can be obtained. Similarly, at the next time t_1,

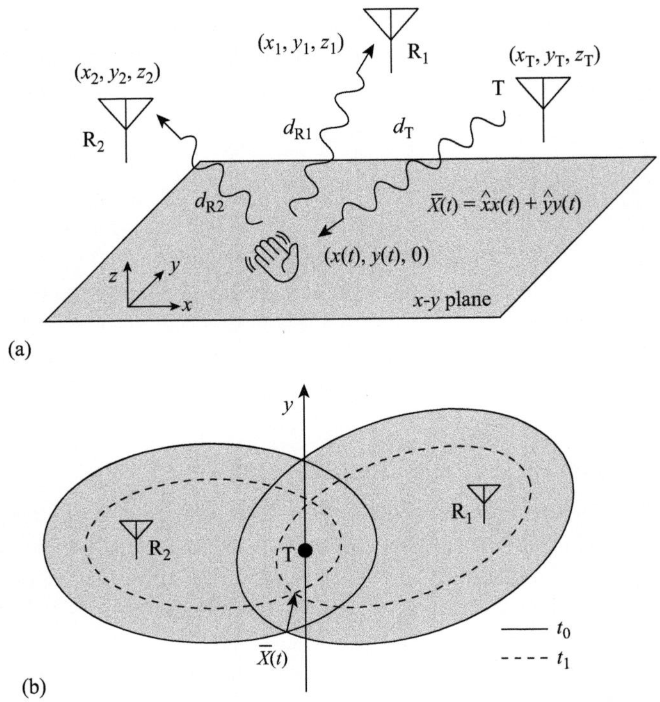

Figure 4.7 Derivation of the 2-D motion tracking algorithm. (a) Coordinates and the location of antennas. (b) Mathematical model

$(x(t_1), y(t_1), 0)$ can be obtained, based on which the displacement of the moving object between time t_0 and time t_1, i.e., $\hat{x}(x(t_1) - x(t_0)) + \hat{y}(y(t_1) - y(t_0))$, can be directly calculated by a subtraction between these coordinates. This displacement is denoted as a black arrow in Figure 4.7(b). By continuing this process, the real-time displacement, or the trajectory, of the moving object can be obtained.

Mathematically, the general solution of the above displacement can always be obtained by solving (4.13) and (4.14), although this solution can be very complicated and thus inconvenient for practical applications. On the other hand, if antennas T, R_1, and R_2 can be placed at some special locations, such as on a straight line parallel with the $z = 0$ plane and the projection of this straight line on the x-y plane can be redefined as the x-axis, (4.15) can be significantly simplified to

$$\begin{cases} \sqrt{x^2(t) + y^2(t) + z_T{}^2} + \\ \sqrt{(x_1 - x(t))^2 + y^2(t) + z_T{}^2} = d_1 \\ \\ \sqrt{x^2(t) + y^2(t) + z_T{}^2} + \\ \sqrt{(x_2 - x(t))^2 + y^2(t) + z_T{}^2} = d_2 \end{cases} \tag{4.16}$$

respectively. In this case, the coordinates of T, R_1, and R_2 are changed to $(0, 0, z_T)$, $(x_1, 0, z_T)$, and $(x_2, 0, z_T)$, respectively. In this case, the displacement in the *x*-*y* plane can be calculated by

$$\begin{cases} x(t) = \dfrac{d_1 d_2 (d_1 - d_2) + d_1 {x_2}^2 - d_2 {x_1}^2}{2(d_1 x_2 - d_2 x_1)} \\ y(t) = \dfrac{\sqrt{\begin{array}{l}(d_1 + x_1)(d_1 - x_1)(d_2 + x_2)(d_2 - x_2) \cdot \\ (d_1 - d_2 + x_1 - x_2)(d_1 - d_2 - x_1 + x_2)\end{array}}}{2(d_1 x_2 - d_2 x_1)} - z_T \end{cases} \quad (4.17)$$

respectively, note that d_1 and d_2 are also time variant.

Finally, when the transmitting antenna T and the receiving antennas R_1 and R_2 are located at (0, 0, 0), $(-a, 0, 0)$, and $(a, 0, 0)$, the ellipsoid equations can be further reduced to

$$\begin{cases} \sqrt{x^2 + y^2} + \sqrt{(x - (-a))^2 + y^2} = d_1 \\ \sqrt{x^2 + y^2} + \sqrt{(x - a)^2 + y^2} = d_2 \end{cases} \quad (4.18)$$

and the 2-D motion tracking equations can be simplified accordingly to

$$\begin{cases} x = \dfrac{(d_1 d_2 + a^2)(d_1 - d_2)}{2a(d_1 + d_2)} \\ y = \dfrac{\sqrt{(d_1^2 - a^2)(d_2^2 - a^2)(4a^2 - (d_1 - d_2)^2)}}{2a(d_1 + d_2)} \end{cases} \quad (4.19)$$

4.3.2.2 3-D motion tracking algorithm

Clearly, a 3-D motion tracking algorithm can be derived by adding the third receiving antenna R_3 at the location (x_3, y_3, z_3). In this case, the number of the ellipsoid functions will be increased to three, i.e.:

$$\begin{cases} \sqrt{x^2 + y^2 + z^2} + \sqrt{(x - x_1)^2 + (y - y_1)^2 + (z - z_1)^2} = d_1 \\ \sqrt{x^2 + y^2 + z^2} + \sqrt{(x - x_2)^2 + (y - y_2)^2 + (z - z_2)^2} = d_2 \\ \sqrt{x^2 + y^2 + z^2} + \sqrt{(x - x_3)^2 + (y - y_3)^2 + (z - z_3)^2} = d_3 \end{cases} \quad (4.20)$$

Similarly, the general solution of (4.20) can be very complicated. Again, for specially configured antennas, the solution can be significantly simplified. If the locations of antennas T, R_1, R_2, and R_3 are (0, 0, 0), $(0, -a, 0)$, $(0, a, 0)$, and $(a, 0, 0)$, the corresponding ellipsoid functions become:

$$\begin{cases} \sqrt{x^2 + y^2 + z^2} + \sqrt{x^2 + (y - (-a))^2 + z^2} = d_1 \\ \sqrt{x^2 + y^2 + z^2} + \sqrt{x^2 + (y - a)^2 + z^2} = d_2 \\ \sqrt{x^2 + y^2 + z^2} + \sqrt{(x - a)^2 + y^2 + z^2} = d_3 \end{cases} \quad (4.21)$$

and its analytical solutions are

$$\begin{cases} x = \dfrac{(d_1 d_3 + a^2)(d_1 - d_3) + (d_2 d_3 + a^2)(d_2 - d_3)}{2a(d_1 + d_2)} \\ y = \dfrac{(d_1 d_2 + a^2)(d_1 - d_2)}{2a(d_1 + d_2)} \\ z = \pm \dfrac{\sqrt{(a^2 - d_3^2)((a - 2x)^2 - d_3^2) - 4y^2 d_3^2}}{2d_3} \end{cases} \tag{4.22}$$

respectively. If the moving object is constrained in the $z > 0$ space, z would take the positive value.

4.3.2.3 Motion separation algorithm

The same SIMO DRS can also be used to separate interfering motions from the motion of interest [22]. For HGR applications, the typical interfering motions can be the human respirations and the RBMs.

In theory, separating multiple human actions is similar to separating multiple human voices. The later one has been widely investigated in the past decade [23]. An important difference is that, unlike multiple voices that can be modeled as a summation of sinusoidal signals, Doppler signals are hidden in the phase of the quadrature I/Q signals. For two independent motions $x_1(t)$ and $x_2(t)$ detected by a SIMO DRS, the I/Q signals received by the two receiving antennas are

$$\begin{cases} I_i(t) = A_{i,1}(t)\cos[4\pi x_1(t)/\lambda] + A_{i,2}(t)\cos[4\pi x_2(t)/\lambda],\ i = 1,2 \\ Q_i(t) = A_{i,1}(t)\sin[4\pi x_1(t)/\lambda] + A_{i,2}(t)\sin[4\pi x_2(t)/\lambda],\ i = 1,2 \end{cases} \tag{4.23}$$

where λ is the wavelength of the transmitted signal, and $A_{i,1}(t)$ and $A_{i,2}(t)$ represent the amplitudes of the signals received by the ith receiver.

Since human actions such as inspiration and hand gestures are independent, the independent component analysis (ICA) theory can be used to separate the scattered signals described by (4.2) and (4.3) [24]. To apply the ICA, signals $I_1(t)$, $I_2(t)$ and $Q_1(t)$, $Q_2(t)$ can be denoted in vector forms as $\boldsymbol{I}(t) = [I_1(t), I_2(t)]^{\mathrm{T}}$ and $\boldsymbol{Q}(t) = [Q_1(t), Q_2(t)]^{\mathrm{T}}$. To avoid the permutation issue [24], the I/Q signals can be further synthesized into a complex form $\boldsymbol{x}(t) = \boldsymbol{I}(t) + i\boldsymbol{Q}(t)$. Then, letting:

$$\mathbf{x} = \mathbf{A}\mathbf{s} \tag{4.24}$$

where the 2×2 matrix $\mathbf{A}$ is the "mixed coefficient matrix" [24] and $\mathbf{s} = [\exp(j4\pi m_1(t)/\lambda), \exp(j4\pi m_1(t)/\lambda)]^{\mathrm{T}}$ represents the column vector containing the Doppler signals.

To recover $\mathbf{s}$ from $\mathbf{x}$, an optimal 2×2 "unmixing coefficient matrix" $\mathbf{W_{opt}}$ can be optimized so that

$$\mathbf{y} = \mathbf{W_{opt}}\mathbf{x} \tag{4.25}$$

where $\mathbf{y}$ is the estimated matrix of $\mathbf{s}$. Then, the two elements in $\mathbf{y}$ can be considered as the separated action signals. Solving (4.5) is essentially an inverse problem. To solve the complex-valued problem described by (4.5), the complex-valued maximum pseudo SNR (PSNR) as the target function in the optimization of $\mathbf{W_{opt}}$ [22], and the unmixing coefficient matrix can be defined as

$$\mathbf{W_{opt}} = \left\{eig\left[((\tilde{\boldsymbol{x}} - \boldsymbol{x})(\tilde{\boldsymbol{x}} - \boldsymbol{x})^*)^{-1}(\tilde{\boldsymbol{x}} \cdot \tilde{\boldsymbol{x}}^*)\right]\right\}^* \tag{4.26}$$

where $\tilde{\boldsymbol{x}}$ is the moving average of the received signals, and $eig(\cdot)$represents the calculation of the matrix eigenvectors. Note that (4.6) can be directly calculated without any iteration, implying that the above blind motion separation based on (4.6) can be implemented in real time.

Once the composite signals containing the Doppler signals of two moving objects have been separated, the extended DACM or the arcsine algorithms can be used to retrieve the motions from the phases of the separated signals. For the separated signals denoted as

$$y_k(t) = w_k \exp(j4\pi x_k(t)/\lambda) = I_k'(t) + jQ_k'(t),\ k = 1,2 \tag{4.27}$$

where w_k is an element in $\mathbf{W}_{\text{opt}}$ corresponding to the kth separated motion $I'_k(t) + jQ'_k(t)$, and the digital-domain-extended DACM algorithm can be used to recover motions $x_k(t)$, i.e.:

$$x_k[n] = \frac{\lambda}{4\pi}\sum_{i=2}^{M}\frac{I_k'[i]\{Q_k'[i] - Q_k'[i-1]\} - \{I_k'[i] - I_k'[i-1]\}Q_k'[i]}{I_k'^2[i] + Q_k'^2[i]} \tag{4.28}$$

where $x_k\ [n]$, $I'_k\ [i]$, and $Q'_k\ [i]$ are the discrete forms of $x_k(t)$, $I'_k(t)$, and $Q'_k(t)$, respectively.

4.3.2.4 Localization with a redundant SIMO DRS

Working with the principle of interferometry, CW DRSs are able to detect the displacement and velocity with very high sensitivity and accuracy. However, working with a CW with a single frequency, they are unable to determine the absolute location of a moving object. It means that in Doppler radar sensing applications when the objects under detection are moving in an arbitrary, undefined space, an additional localization algorithm may be needed for the CW DRS so that an initial location of the displacement can be determined. Otherwise, the trajectory tracked by the SIMO DRS can be severely distorted.

Fortunately, a SIMO DRS with a redundant receiving antenna can be used to solve this problem. Figure 4.8(a) and (b) compares the original SIMO DRS and the redundant SIMO DRS. As discussed before, with one transmitting and two receiving channels, the DRS shown in Figure 4.8(a) are able to detect 2-D motions. In comparison, with an additional receiving channel, the redundant SIMO DRS shown in Figure 4.8(b) is able to provide an ability to determine the absolute location of the displacement tracked by the DRS. Note that this redundant SIMO DRS can be considered to consist of two "sub-DRSs" with different antenna groups, and each can be used to detect the same 2-D motion based on the previous motion tracking algorithm. Since the 2-D motion tracked by the two sub-DRSs are the same motion, it provides a prior knowledge for solving the ambiguity of the Doppler phase.

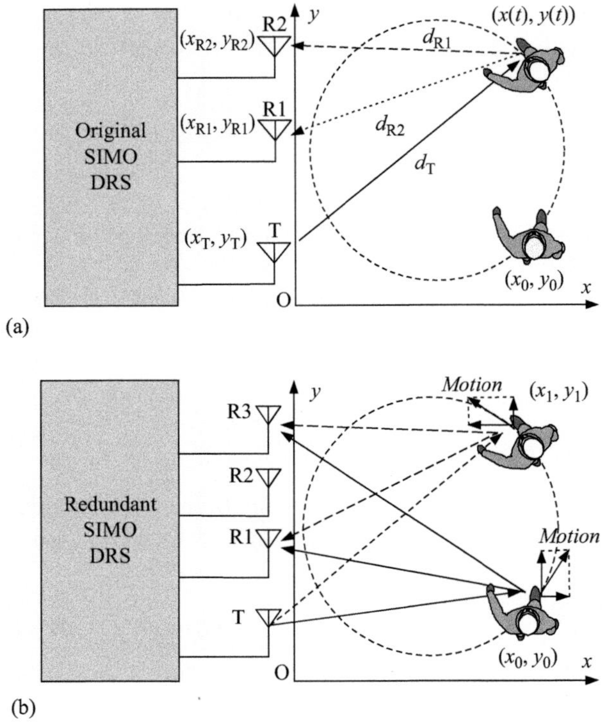

Figure 4.8 Redundant SIMO DRS for localizing 2-D motions without any prior knowledge of the initial location. (a) Original SIMO DRS. (b) Redundant SIMO DRS

Assume all the antennas are in the x-y plane. If antenna T is located at origin (0, 0), receiving antennas R_1, R_2, and R_3 are located at (x_{Ri}, y_{Ri}), $i = 1, 2, 3$, the ellipsoid equations are

$$\sqrt{x^2(t)+y^2(t)}+\sqrt{(x(t)-x_{Ri})^2+(y(t)-y_{Ri})^2}=d_i(t) \tag{4.29}$$

where $d_i(t)$ are the round-trip distances. For a given initial location of the moving object at (x_0, y_0), the detected motions using the two sub-DRSs with the antenna groups (T, R_1, R_2) and (T, R_1, R_3) are

$$\begin{cases} x_1(t) \triangleq f_1(d_1(t), d_2(t)) \\ \quad = \dfrac{d_1(t)d_2(t)(d_1(t)-d_2(t)) + x_{R2}^2 d_1(t) - x_{R1}^2 d_2(t)}{2(x_{R2}d_1(t) - x_{R1}d_2(t))} \\ y_1(t) \triangleq g_1(d_1(t), d_2(t)) \\ \quad = \dfrac{\sqrt{\begin{array}{c}(d_1(t)+x_{R1})(d_1(t)-x_{R1})(d_2(t)+x_{R2})(d_2(t)-x_{R2}) \\ (d_1(t)-d_2(t)+x_{R1}-x_{R2})(d_1(t)-d_2(t)-x_{R1}+x_{R2})\end{array}}}{2(x_{R2}d_1(t) - x_{R1}d_2(t))} \end{cases} \tag{4.30}$$

and

$$\begin{cases} x_2(t) \triangleq f_2(d_1(t), d_3(t)) \\ \quad = \dfrac{d_1(t)d_3(t)(d_1(t) - d_3(t)) + x_{R3}^2 d_1(t) - x_{R1}^2 d_3(t)}{2(x_{R3}d_1(t) - x_{R1}d_3(t))} \\ y_2(t) \triangleq g_2(d_1(t), d_3(t)) \\ \quad = \dfrac{\sqrt{\begin{array}{c}(d_1(t) + x_{R1})(d_1(t) - x_{R1})(d_3(t) + x_{R3})(d_3(t) - x_{R3}) \\ (d_1(t) - d_3(t) + x_{R1} - x_{R3})(d_1(t) - d_3(t) - x_{R1} + x_{R3})\end{array}}}{2(x_{R3}d_1(t) - x_{R1}d_3(t))} \end{cases} \tag{4.31}$$

respectively. When the initial location (x_0, y_0) is unknown, $d_1(t)$, $d_2(t)$, and $d_3(t)$ are not only functions of time, but also functions of (x_0, y_0). In this case:

$$d_i(t, x_0, y_0) = d_{0i}(x_0, y_0) + \sum_{t_s=1}^{t} \Delta d_i(t_s), \ i = 1, 2, 3 \tag{4.32}$$

where $\Delta d_1(t_s)$, $\Delta d_2(t_s)$, and $\Delta d_3(t_s)$ are the differences between round-trip distances d_1, d_2, and d_3 at adjacent sampling time t_s and $t_s - 1$, respectively, and

$$d_{0i}(x_0, y_0) = \sqrt{x_0^2 + y_0^2} + \sqrt{(x_0 - x_{Ri})^2 + (y_0 - y_{Ri})^2}, \ i = 1, 2, 3 \tag{4.33}$$

represent the round-trip distances from antenna T to the initial location (x_0, y_0) and back to antennas R_i, with $i = 1, 2, 3$, which is the unknown to be solved.

Based on the priori knowledge that the two sub-DRSs are measuring the same motion, the relationship of

$$x_1(t) = x_2(t), \ y_1(t) = y_2(t) \tag{4.34}$$

should be strictly satisfied not only for the first sampling, but also for all the following samplings. Due to the nonlinear characteristic of (4.7)–(4.12), it is mathematically difficult to confirm that for each sample, (4.13) would have a unique solution. However, with increased samplings, the number of functions is proportionally increased, and all the obtained equations would satisfy (4.13). Therefore, if a cooperative tracking-localization algorithm can be performed along with the continuous sampling, even if (4.13) does not have a unique solution for each sample, inverse optimization can always be used to determine the optimal solution.

In order to derive the inverse optimization equation, error functions can be defined for each measurement at time t based on (4.7)–(4.10), i.e.:

$$\begin{cases} e_1(t, x_0, y_0) = f_1(d_1(t, x_0, y_0), d_2(t, x_0, y_0)) \\ \qquad\qquad - f_2(d_1(t, x_0, y_0), d_3(t, x_0, y_0)) \\ e_2(t, x_0, y_0) = g_1(d_1(t, x_0, y_0), d_2(t, x_0, y_0)) \\ \qquad\qquad - g_2(d_1(t, x_0, y_0), d_3(t, x_0, y_0)) \end{cases} \tag{4.35}$$

where functions g_1, g_2 and f_1, f_2 have been defined in (4.7)–(4.10), respectively. Then, the optimization function can be further defined as

$$F(x_0, y_0) = \min \sum_{t=1}^{N} \sqrt{e_1^2(t, x_0, y_0) + e_2^2(t, x_0, y_0)} \tag{4.36}$$

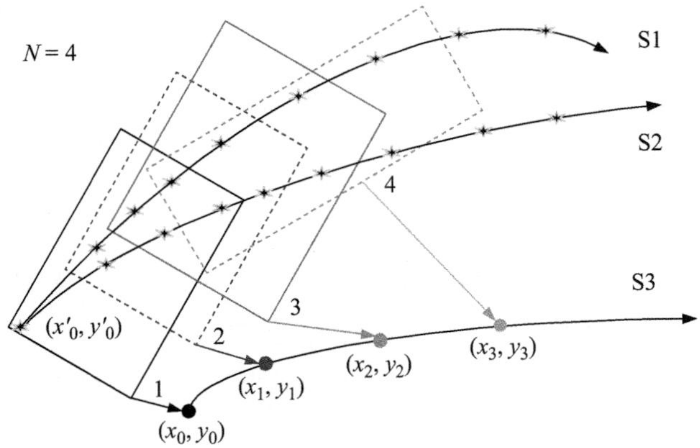

Figure 4.9 Flowchart for the tracking-localization algorithm for $N = 4$

where (x_0, y_0) is the unknown initial location, and N is the number of points used in the optimization at time t.

In order to obtain an effective optimization, the motion trajectories obtained by the two sub-DRSs can be divided into different segments using a time window with a fixed length T, and the points separating such segments can be used to count number N. The above dividing process can be considered as a piece-wise linear approximation to the motion trajectory.

For illustration, Figure 4.9 shows the process how the optimization can be implemented with $N = 4$. Since the initial location (x_0, y_0) is unknown, an assumed initial location (x'_0, y'_0) can be randomly chosen to start the tracking algorithms. Trajectories S1 and S2 detected by different sub-DRSs can be obtained with increased samplings. With the same window length T, segments for the localization optimization can be truncated on both trajectories. The blocks marked as 1, 2, 3, and 4 include four pairs of segments with $N = 4$. The error between the two trajectory segments for S1 and S2 in the first block can be calculated and optimized, and the optimized initial location (x_0, y_0) (or its approximation) can be obtained.

With the increasing time, this process can be iterated using the segments included in the following blocks 2, 3, 4, etc., until the tracking and localization of this moving object are finished. Finally, the resulted S3 can be considered as the localized trajectory.

4.3.2.5 Micro-Doppler analysis

In previous discussions, the algorithms aiming to linearly track, localize, and separate different motions have been introduced. Such algorithms are based on the ability to linearly retrieve Doppler signals provided by the extended DACM or arcsine algorithms. Instead, micro-Doppler analysis can potentially provide an alternative approach to process the baseband signals in frequency domain without needing to linearly retrieve the Doppler information.

In the above mathematical derivations, the moving objects have been modeled as an electrically small scatterer by default. In practice, the signal scattered by an electrically large object can be considered as a summation of all the signals scattered from each point on the reflective surface of the object. In this case, the "micro-motions" will contribute to the Doppler signal, resulting in the so-called micro-Doppler effect [25].

The micro-motion parasitic on the "main motion" can be frequently observed in many combined motions. Examples of such motions include the swinging limbs of a human in walking, the flapping wings of a bird in flying, the rotating wheels of a vehicle in driving, and so on. In 2000, the first study on the micro-Doppler phenomenon in radar systems was published by Chen [25]. It was indicated that the micro-Doppler effect would depend on the geometrical shape and the movement form of the moving object. If the scattered signal can be properly processed, the "micro-Doppler signature" can be extracted for the recognition and classification purposes. Fundamental analyses to the micro-Doppler effects in radars can be found in [26–29].

Technically, time–frequency spectrum analysis can be effectively used to convert a 1-D time-domain signal into a 2-D representation of signal amplitude versus time and frequency with the assistance of a moving window. The "spectrogram" defined as the square modulus of the short-time Fourier transform (STFT) is frequently used in time-frequency analysis. For the previously defined motion $x(t)$, its STFT can be defined as the Fourier transform of $x(t)$ multiplied by a window function $w(t)$, normally a Hanning or Gaussian window centered around zero, i.e.:

$$STFT\{x(t)\}(\tau,\omega) \equiv X(\tau,\omega) = \int_{-\infty}^{\infty} x(t)w(t-\tau)e^{-jwt}dt \tag{4.37}$$

where $X(\tau,\omega)$ is essentially the Fourier Transform of $x(t)w(t-\tau)$, a complex function representing the phase and magnitude of the signal over time and frequency.

In practical HGR applications, micro-Doppler analysis can be effective since the motion of fingers and hands can be considered as parasitic motions on the "main motion" of human arms and bodies. The extracted "signature" from a micro-Doppler "spectrogram" can potentially provide an angle of view for HGRs from a new dimension. Compared with the motion separation based on a customized ICA algorithm, micro-Doppler analysis does not require additional receiving channels. This unique advantage is benefit for practical HGR applications.

4.4 Experimental demonstration

This section aims to demonstrate how the algorithms discussed in Section 4.3 can be experimentally realized. The SIMO DRS used in the demonstration is designed to work at an ISM frequency of 5.8 GHz. A picture of this SIM DRS consisting of one transmitting and two receiving channels is shown in Figure 4.10(a). The key components used in this DRS are marked in the figure, and more details of them can be found in Table 4.2.

The transmitting and receiving channels are designed as separate modules based on a pair of commercial chipsets originally designed for Wi-Fi applications.

(a) (b)

Figure 4.10 A SIMO DRS designed for 2-D HGRs. (a) Implemented SIMO DRS with an additional receiving channel. (b) Detection of "click" and "moving" hand gestures

Table 4.2 Main devices used in the 5.8-GHz DRS

Number	Device name	Manufacturer	Function
1	MAX2829	Maxim Integrated	RF Transceiver
2	AWL6951	Anadigics	RF PA
3	AD7357	Analog Devices	ADC
4	EPM570T100C5	Altera	CPLD
5	STM32F106	ST Electronics	MCU

The IF frequency is chosen as 6 MHz. Two ADC chips (Analog Devices AD7357) are used for the band-pass sampling. The used band-pass sampling frequency is 100 Hz, much lower than that of the IF frequency at 6 MHz. Finally, A USB port is designed to transmit sampled data to a computer, in which the algorithms discussed before can be run.

The experimental setup used in the demonstration is shown in Figure 4.10(b). In the demonstration, the gestures of "finger clicks" and "hand moving" are chosen to simulate the basic HMI functions of a mouse. In the following, the linear retrieval of large-scale 2-D motions, the generation of the 2-D Doppler signal patterns, and the separation of interfering respiration motions will be demonstrated. Since the motions of finger clicks and hand moving are limited in a specified small area, the cooperative tracking-localization algorithm is not necessary to be applied. The details of the application of this algorithm can be found in [7].

4.4.1 Linear retrieval of large-scale 2-D motions

To examine the accuracy of the linearized arctangent algorithm developed for large-scale 2-D motions discussed in Section 4.3, large-scale "test motions" that

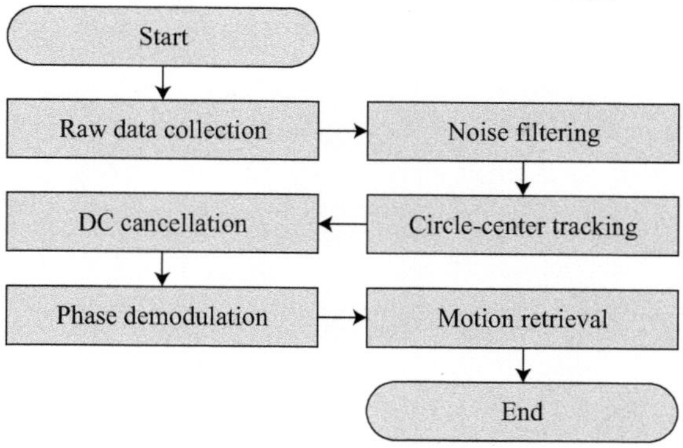

Figure 4.11 Flowchart for the linearized arctangent algorithm

have accurate, predefined displacements can be used. In [7], this signal was generated by driving a 200-mm-long aluminum cylinder by a 2-axis actuator, ZABER's X-LSM200A. The cylinder was driven to precisely move in the horizontal plane with predefined patterns, such as a perfect circle with a diameter around 10 cm, much larger than the wavelength at 5.8 GHz. In the meantime, the SIMO DRS was used to measure this circular movement at a distance of about 20 cm from the DRS, and the sampled *I/Q* data can be processed following the flowchart shown in Figure 4.11.

Corresponding to the main steps in the flowchart, the 2-D constellations of the raw *I/Q* signals and the corresponding constellations after applying the low-pass filtering and DC offset calibration are shown in Figure 4.12, and the retrieved motions by the arcsine algorithm and the reconstructed large-scale 2-D circular motion as well as the original circle are shown in Figure 4.13(a). It is seen that the reconstructed motions comply well with the original one. The small derivation mainly occurs when the cylinder is moved to the left and right edges of the moving region. At these places, the signal received by the far-side antennas would have larger noises, resulting in a distortion of the measured movement trajectory. Similarly, the measurement to a Z-shaped motion resulted in a similar result, as seen in Figure 4.13(b). These results verified the performance of the DC-offset tracking and the linearized arctangent algorithm.

4.4.2 Reconstruction of 2-D gesture patterns

Using the same signal processing, the displacements and trajectories of the finger clicks and the hand's moving on the surface of the table can be reconstructed. Figure 4.14(a) shows the retrieved Doppler signals (left) and the reconstructed trajectory of a double-click action. It is seen that both the recovered displacements and the reconstructed motion patterns can be used to identify the double-click action. Figure 4.14(b) shows an example of a complex motion including a single click and a hand moving. It is seen that the single click and the moving actions can also be clearly identified.

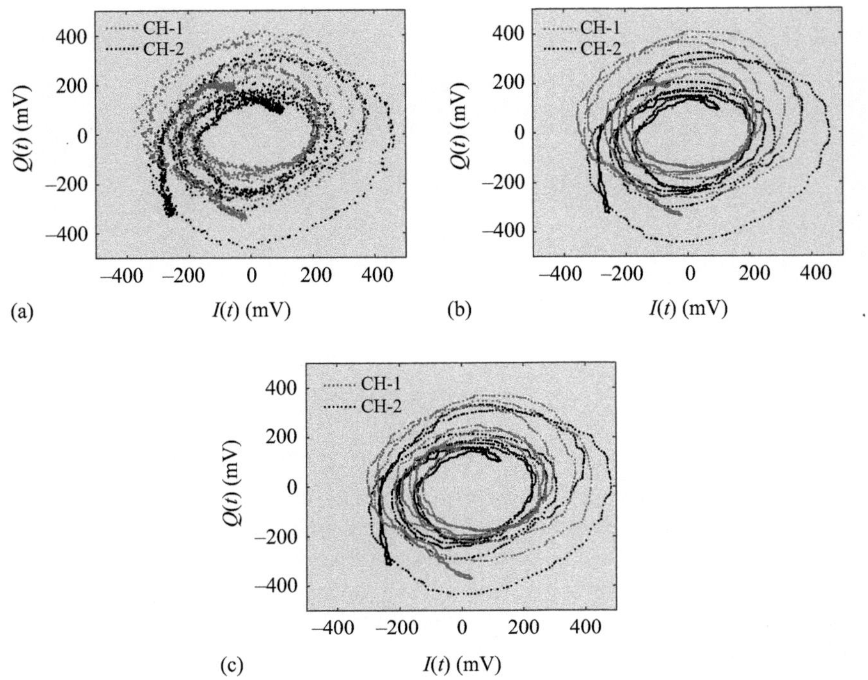

Figure 4.12 Constellations of the I/Q signals after the different processing steps. (a) Raw data. (b) Filtered data. (c) Constellations after DC offset compensation

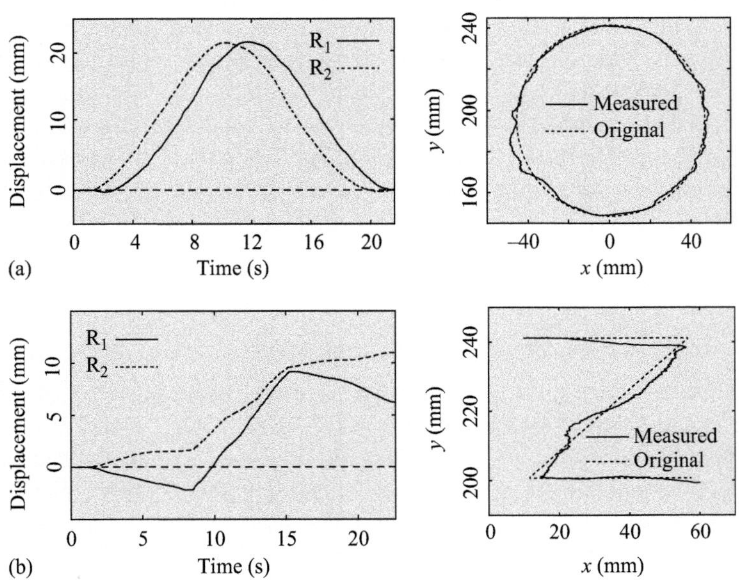

Figure 4.13 Experimental reconstruction for the test motions. (a) Circular motion. (b) Z-shaped motion

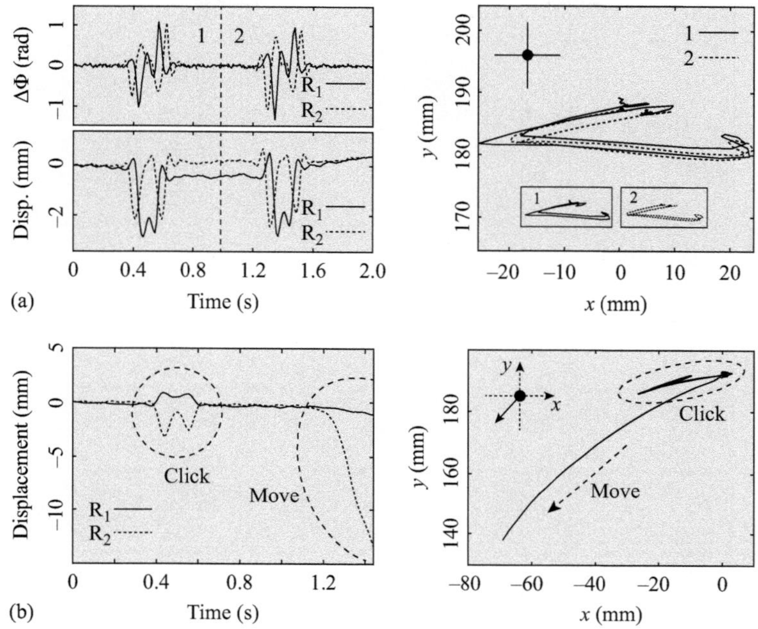

Figure 4.14 Reconstructed patterns of "double click" (a) and "click + moving" (b) hand gestures

4.4.3 Separation of interfering Doppler signal

In order to verify the effectiveness of the blind motion separation algorithm, an experiment was designed to separate click actions from the interfering human respirations. The experimental setup is similar to Figure 4.10(b). In the experiments, a subject sat before the DRS at a distance around 0.8 meter. First, while keeping a normal respiration, the object periodically clicked the table using the fingers of his hand. The clicking was about three times per second. The sampled data were processed using the proposed two-step blind motion separation algorithm, to separate the clicking actions from the interfering movements.

Figure 4.15(a) shows the constellations of the normalized (by the range of *I* channel) *I/Q* signals detected by the two receivers. Compared with the constellation for a single moving object shown in Figure 4.12, such constellations reflect the composite motions due to the scattering of two moving objects.

Figure 4.15(b) shows the constellations of the *I/Q* signals separated by (4.26), (4.27), and (4.28). Figure 4.15(c) shows the recovered click and respiration motions by the extended DACM algorithm. It is seen that the recovered respiration is with an amplitude close to 2 cm and a frequency around 0.35 Hz. The gray line in Figure 4.15(d) shows the recovered click action parasitic on a segment of a very slow RBM, and the black line shows the clicks after the RBM was filtered. The clicks are with an amplitude ranging from −2.54 to 2.31 cm and a frequency around 3.4 Hz, complying with the actual situations.

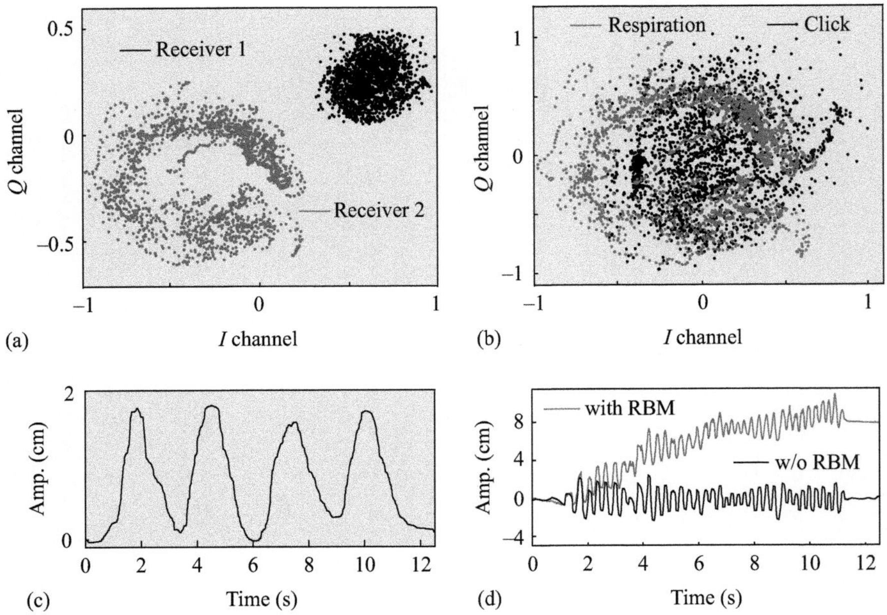

Figure 4.15 Separating "click" hand gesture from the interfering respiration. (a) Normalized I/Q signals detected by the two receivers. (b) Normalized I/Q signals after applying the blind separation algorithm. (c) Recovered respirations by the extended DACM algorithm. (d) Recovered finger clicks with a slow RBM

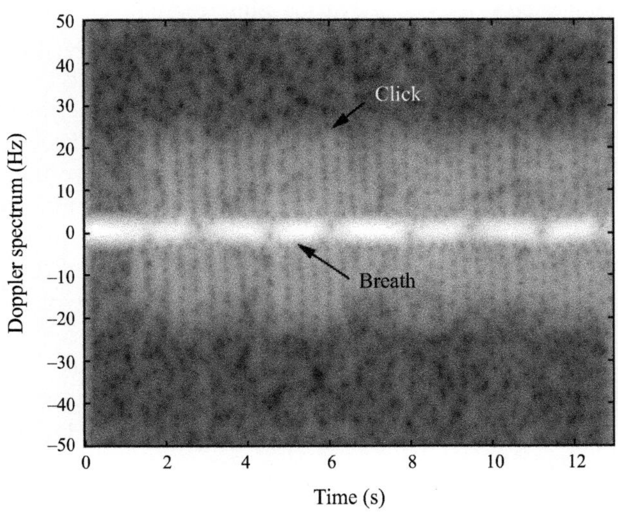

Figure 4.16 The time-frequency analysis of the "click" hand gesture with interfering respirations and RBM

The above experiment verified the effectiveness of the blind motion separation algorithm discussed before. It can be potentially used in HGRs with unwanted interfering movements such as human RBMs and respirations.

Finally, based on the same experimental results shown in Figure 4.15(a), the micro-Doppler spectrogram can be computed by (4.37). The obtained result is shown in Figure 4.16. Note that in the computation, only the signal received by the antenna R2 was used. It is seen that both signatures of the finger clicks and the human breaths can be clearly identified in the spectrogram. In the meantime, the spectrum of the interfering slow RBM exhibits as a noise spectrum located close to the DC frequency.

It is seen that the micro-Doppler analysis can be potentially used in the HGR applications. The unique advantage is that it can work with a SISO-structured CW DRS.

4.5 Summary

Based on the Doppler effect, DRSs have been used in the remote detection of the displacement and velocity of moving objects for decades. Nowadays, based on the progress of RFICs, miniaturized DRSs have been used in indoor, short-range detection of human vital signs and activities. Among many potential applications, an attractive one is the human gesture detection and recognition. Compared with the image processing-based approaches, DRS-based HGRs rely on cost-effective hardware, extremely low data rate, and linearized algorithms. These advantages imply promising potentials to realize practical HGR applications in diverse hardware platforms ranging from MCU-based to computer-based systems.

This chapter focuses on the introduction of the optimal transceiver architecture suitable for the detection of highly dynamic motions, the related algorithms for the linear retrieval of Doppler signals, the tracking and localization of moving objects, the separation of interfering motions and the micro-Doppler spectrum analysis, and the experimental demonstration of how such systems and algorithms can be cooperated to obtain the signature patterns of human gestures. Hopefully, this chapter could help to increase more awareness and interests to the DRS-based HGRs and motivate more researches and developments to improve this technology.

References

[1] Y. Kalkan, "Frequency-only radars and other frequency-based systems with applications," *IEEE microwave magazine*, vol. 17, no. 7, pp. 26–52, 2016.

[2] C. Li, V. M. Lubecke, O. Boric-Lubecke, and J. Lin, "A review on recent advances in Doppler radar sensors for noncontact healthcare monitoring," *IEEE Trans. Microw. Theory Techn.*, vol. 61, no. 5, pp. 2046–2059, 2013.

[3] C. Li, Z. Peng, T. Huang, *et al.*, "A review on recent progress of portable short-range noncontact microwave radar systems," *IEEE Trans. Microw. Theory Techn.*, vol. 65, no. 5, pp. 1692–1706, 2017.

[4] J. Muñoz-Ferreras, Z. Peng, R. Gómez-García, and C. Li, "Review on advanced short-range multimode continuous-wave radar architectures for healthcare applications," *IEEE Journal of Electromagnetics, RF and Microwaves in Medicine and Biology.*, vol. 1, no. 1, pp. 14–25, 2017.

[5] J. Wang, X. Wang, Z. Zhu, J. Huangfu, C. Li, and L. Ran, "1-D microwave imaging of human cardiac motion: An ab-initio investigation," *IEEE Trans. Microw. Theory Techn.*, vol. 61, no. 5, pp. 2101–2107, 2013.

[6] Q. Lv, D. Ye, S. Qiao, *et al.*, "High-dynamic-range motion imaging based on linearized Doppler radar sensor," *IEEE Trans. Microw. Theory Techn.*, vol. 62, no. 9, pp. 1837–1846, 2014.

[7] T. Fan, C. Ma, Z. Gu, *et al.*, "Wireless hand gesture recognition based on continuous-wave Doppler radar sensors," *IEEE Trans. Microw. Theory Techn.*, vol. 64, no. 11, pp. 4012–4020, 2016.

[8] Q. Lv, L. Chen, K. An, *et al.*, "Doppler vital signs detection in the presence of large-scale random body movements," *IEEE Trans. Microw. Theory Techn.*, vol. 66, no. 9, pp. 4261–4270, 2018.

[9] J. Wang, X. Wang, J. Huangfu, C. Li, and L. Ran, "Non-contact distance and amplitude independent vibration measurement based on an extended DACM algorithm," *IEEE Trans. Instrum. Meas.*, vol. 63, no. 1, pp. 145–153, 2014.

[10] B. Spencer, Jr., S. Cho, and S.-H. Sim, "Wireless monitoring of civil infrastructure comes of age," *Structure Mag.*, vol. 13, pp. 12–16, 2011.

[11] B. F. Spencer, Jr., M. E. Ruiz-Sandoval, and N. Kurata, "Smart sensing technology: Opportunities and challenges," *Struct. Control Health Monitor.*, vol. 11, no. 4, pp. 349–368, 2004.

[12] F. Mohd-Yasin, C. E. Korman, and D. J. Nagel, "Measurement of noise characteristics of MEMS accelerometers," *Solid-State Electron.*, vol. 47, no. 2, pp. 357–360, 2003.

[13] H. H. Nassif, M. Gindy, and J. Davis, "Comparison of laser Doppler vibrometer with contact sensors for monitoring bridge deflection and vibration," *NDT E Int.*, vol. 38, no. 3, pp. 213–218, 2005.

[14] M. Pieraccini, M. Fratini, F. Parrini, and C. Atzeni, "Dynamic monitoring of bridges using a high-speed coherent radar," *IEEE Trans. Geosci. Remote Sens.*, vol. 44, no. 11, pp. 3284–3288, 2006.

[15] M. Pieraccini, M. Fratini, F. Parrini, C. Atzeni, and G. Bartoli, "Interferometric radar vs. accelerometer for dynamic monitoring of large structures: An experimental comparison," *NDT E Int.*, vol. 41, no. 4, pp. 258–264, 2008.

[16] M. Fratini, F. Parrini, M. Pieraccini, C. Borri, and C. Atzeni, "Structural oscillation modes identification by applying controlled loads and using microwave interferometry," *NDT E Int.*, vol. 42, no. 8, pp. 748–752, 2009.

[17] A. Zhu, T. Fan, Z. Gu, Q. Lv, C. Li and L. Ran, "Indoor localization based on a single-tone SIMO-structured Doppler radar system," *2018 IEEE Radio and Wireless Symposium (RWS)*, Anaheim, CA, pp. 7–9, 2018.

[18] C. Li, X. Yu, C.-M. Lee, D. Li, L. Ran, and J. Lin, "High-sensitivity software-configurable 5.8-GHz radar sensor receiver chip in 0.13-μm CMOS

for noncontact vital sign detection," *IEEE Trans. Microw. Theory Techn.*, vol. 58, no. 5, pp. 1410–1419, 2010.

[19] J. Li, M. Li, R. J. Evans, E. Skafidas, and B. Moran, "Band-pass sampling techniques for high resolution multi-channel single-chip radar systems," *2012 IEEE Radar Conference*, pp. 499–503, May 2012.

[20] G. K. Felic, R. J. Evans, H. T. Duong, H. V. Le, J. Li, and E. Skafidas, "Single-chip millimeter wave radar," *Microw. J.*, vol. 58, no. 1, pp. 108–116, 2015.

[21] J. G. Proakis, "Digital Communications," *McGraw-Hill Education*, 2007.

[22] Z. Gu, T. Fan, Q. Lv, *et al.*, "Remote blind motion separation using a single-tone SIMO Doppler radar sensor," *IEEE Transactions on Geoscience and Remote Sensing (Early Access)*, pp. 1–11, August 2018.

[23] A. Vergara, N. Petrochilos, O. Boric-Lubecke, A. Host-Madsen, and V. Lubecke, "Blind source separation of human body motion using direct conversion Doppler radar," *IEEE MTT-S Int. Microw. Symp. Dig.*, June 2008, pp. 872–875.

[24] H. Sawada, R. Mukai, S. Araki, and S. Makino, "A robust and precise method for solving the permutation problem of frequency-domain blind source separation," *IEEE Trans. Speech Audio Process.*, vol. 12, no. 5, pp. 530–538, 2004.

[25] V. Chen, "Micro-Doppler effect in radar: phenomenon, model, and simulation study," *IEEE Trans. Aerosp. Electron.*, vol. 42, no. 1, pp. 1–21, 2006.

[26] Y. Kim and H. Ling, "Human activity classification based on micro-Doppler signatures using a support vector machine," *IEEE Transactions on Geoscience & Remote Sensing*, vol. 47, no. 5, pp. 1328–1337, 2009.

[27] V. C. Chen and H. Ling, *Time-frequency Transforms for Radar Imaging and Signal Analysis*, Artech House Inc., Boston, MA, 2001.

[28] G. Wang, C. Gu, T. Inoue, and C. Li, "A hybrid FMCW-interferometry radar for indoor precise positioning and versatile life activity monitoring," *IEEE Transactions on Microwave Theory & Techniques*, vol. 62, no. 11, pp. 2812–2822, 2014.

[29] Q. Wan, Y. Li, C. Li, and R. Pal, "Gesture recognition for smart home applications using portable radar sensors," *36th Annual International Conference of the IEEE Engineering in Medicine and Biology Society*, Chicago, IL, pp. 6414–6417, 2014.

Chapter 5
FMCW radar systems for short-range micro-motion sensing

Zhengyu Peng[1], Changzhi Li[2], Roberto Gómez-García[3], and José-María Muñoz-Ferreras[3]

Short-range micro-motion sensing has a wide range of applications, including noncontact human–machine interface [1–3], smart home, IoT tracking/localization [4,5], etc. Camera-based techniques have been used for motion sensing for decades. Advanced image processing algorithms have provided camera-based sensors with the capability of depth sensing and object tracking [6,7]. However, high computational demands hinder their applications, especially in portable and wearable devices. Moreover, camera-based sensors can be easily blocked by obstacles. On the other hand, radar systems have unique advantages in micro-motion sensing. For instance, microwave or millimeter-wave-based radar sensors can easily penetrate walls and obstacles [8–10]. In addition, the computational resource required by a radar system is much lower than that of a camera-based solution.

Among various radar types, continuous-wave (CW) radar systems feature low transmitted power, simple structure, and high sensitivity. CW radar systems can be categorized into several types based on their waveforms. A Doppler (interferometry) radar transmits a single-tone signal to detect Doppler shifts of moving targets. It features high sensitivity and high accuracy in velocity and motion detection. However, it lacks the capability of measuring absolute ranges of targets [11,12]. In order to detect ranges of targets, a modulated transmitted signal is necessary. The frequency-modulated CW (FMCW) radar is one of the most popular modulated CW radar systems [13–16]. FMCW radar systems have been widely used in advanced driver assistance systems (ADAS) and autonomous vehicles. Researchers are also actively investigating the applications of FMCW radar systems to biomedical, senior care, security, gesture recognition, and so forth. K-band frequency has been used in automotive radar and other areas [17]. Since the regulation authorities have opened up new frequencies for automotive radar, the radar systems in ADAS and autonomous driving have been moved to the 77-GHz band.

[1]Aptiv PLC, USA
[2]Department of Electrical and Computer Engineering, Texas Tech University, USA
[3]Department of Signal Theory and Communications, University of Alcalá, Spain

For other applications, such as biomedical, 24- and 60-GHz Industrial Scientific Medical (ISM) bands are currently used. Researchers are also investigating radar systems in the sub-THz band. Using higher frequencies not only improves the sensitivity and resolution of the radar systems, but also reduces the size.

This chapter gives an overview of the basic requirements and design considerations of an FMCW radar system for short-range micro-motion sensing. At the beginning of the chapter, fundamentals of CW radar systems are introduced and the advantages of FMCW radar are discussed. For an FMCW radar system used in short-range micro-motion sensing applications, the design principles and considerations will be discussed in detail from system to component levels. Advanced beamforming techniques for FMCW radar systems, including phased array, radio frequency (RF) beamforming, and multiple-input and multiple-output (MIMO), are also detailed in this chapter. In addition, basic and advanced radar signal processing methods will also be introduced. Actual radar measurement data are presented to help readers have a better understanding of radar signal processing.

5.1 FMCW radar fundamentals

The basic architecture for a CW radar system is shown in Figure 5.1. The CW radar architecture consists of a transmitter and a receiver. In the transmitter, a signal synthesizer is used to generate CW signals. These CW signals can be a single-tone signal, frequency-shift keying (FSK) modulated signal, stepped frequency CW (SFCW) signal, or FMCW signal. A power divider is used to distribute the signal from the signal generator to both the transmitter antenna and the mixer in the receiver. A power amplifier (PA) can be used to increase the power level of the transmitted signal, which will help to extend the detection range of the radar.

In the receiver, the receiver antenna or antenna array receives the reflected signal from targets. The received signal is processed by a low-noise amplifier (LNA) and several stages of gain blocks. Then, the mixer is used to mix the received signal with the aforementioned replica of the generated signal. The output

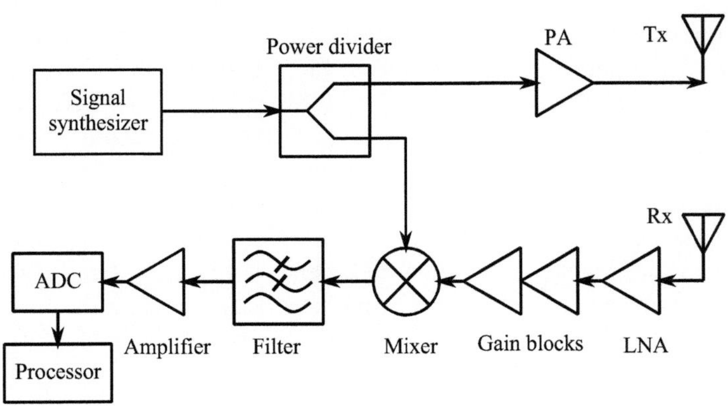

Figure 5.1 Top-level architecture of a CW radar system

of the mixer is the baseband signal, which is subsequently filtered by a low-pass filter and amplified by a baseband amplifier. In order to process the baseband signal, an analog-to-digital converter (ADC) converts the analog baseband into the digital domain. Radar signal processing algorithms can be implemented into the processor, which can be a computer, an embedded digital signal processor, or an field-programmable gate array, to obtain the properties of targets. Thanks to the advance of monolithic microwave integrated circuit (MMIC) and RF complementary metal-oxide semiconductor technologies, it is possible to integrate a radar system into a small printed circuit board (PCB) or even into a chip [11,18].

Based on different types of transmitting waveforms shown in Figure 5.2, CW radar systems can be sorted into two major categories: (1) unmodulated CW radar systems and (2) modulated CW radar systems. An unmodulated CW radar system, which is also called a Doppler or an interferometry radar, transmits and receives a single-tone signal to obtain phase history of targets based on the Doppler effect. This kind of radar systems features high accuracy in displacement and velocity measurement. However, they lack the capability to obtain the absolute ranges of targets. On the other hand, modulated CW radar systems include FMCW radar, SFCW radar, and FSK radar, etc. An FMCW radar transmits linear frequency-modulated CW waveform, an SFCW radar transmits step-modulated CW waveform, and an FSK radar transmits square wave-modulated waveform. The time-domain waveforms and the frequency versus time diagrams of different radar modulations are shown in Figure 5.2. Among these types of modulated CW radar systems, the

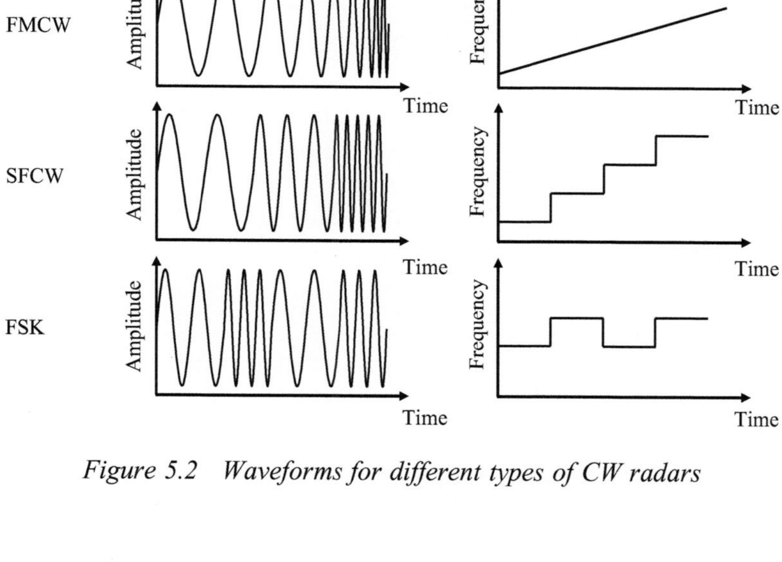

Figure 5.2 Waveforms for different types of CW radars

FMCW radar is the most popular one due to its simple baseband signal processing. An FMCW radar is able to provide the absolute range information of targets. It is also able to obtain the velocities of targets by extracting Doppler information. The simple architecture, easy signal processing, and the capabilities to provide both range and Doppler information of targets make an FMCW radar ideal for short-range micro-motion sensing. The discussion in this chapter will focus on the FMCW radar for short-range micro-motion sensing.

Figure 5.3 illustrates the principle of an FMCW radar in target detection. A linear frequency-modulated CW signal f_t, which is also known as a chirp, is transmitted by the transmitter antenna of the FMCW radar. The signal is reflected back by a target which is located at a range R. The receiver of the radar receives the reflected signal f_r. A mixer in the radar system mixes f_r with a replica of f_t to obtain the baseband signal f_b, which is also known as the beat signal. Figure 5.4 shows the waveforms of an FMCW radar. In the scenario illustrated in Figure 5.3, the round-trip delay t_d between f_t and f_r is

$$t_d = \frac{2R}{c} \tag{5.1}$$

where R is the distance between the radar and the target and c is the speed of light.

The corresponding beat signal f_b, which is the mixed result of f_t and f_r, can be expressed as

$$f_b = B\frac{t_d}{T} \tag{5.2}$$

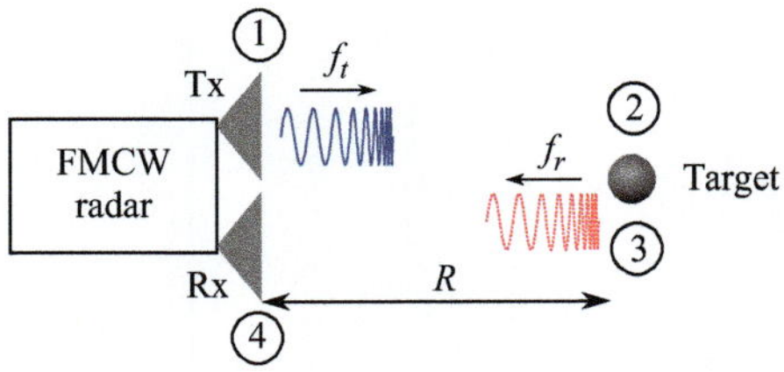

Figure 5.3 Principle of an FMCW radar

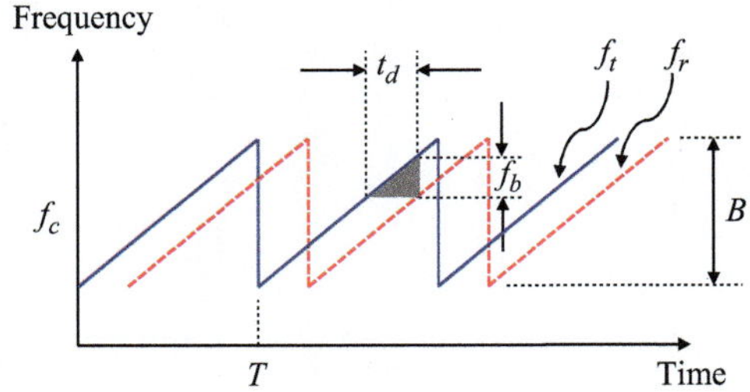

Figure 5.4 Typical FMCW radar waveforms

where B is the bandwidth of the transmitted signal and T is the chirp repetition period (CRP), as shown in Figure 5.4.

After combing (5.1) and (5.2), the range R of the target can thus be obtained as

$$R = \frac{cTf_b}{2B} \tag{5.3}$$

With an FMCW radar, it is simple to obtain absolute range information of targets. Doppler information of moving targets can also be extracted with a coherent FMCW radar to achieve range-Doppler imaging. The details of calculating range and Doppler of a target are introduced in the following sections.

To define the performance of an FMCW radar, three basic parameters are used. The first parameter is the range resolution (ΔR), which defines the radar's capability to distinguish two closely located targets in range, as shown in Figure 5.5. ΔR is mainly determined by the bandwidth of the FMCW radar:

$$\Delta R = \frac{c}{2B} \tag{5.4}$$

As illustrated in Figure 5.6, the detection precision of the radar defines the minimal displacement a radar can differentiate for a single target. The main factor that affects the detection precision is signal-to-noise ratio (SNR).

The detection accuracy is another parameter which describes the error between the detected range and the ground truth. The detection accuracy is related to systematic bias and SNR. Systematic bias includes path delays of the transmission lines in the radar transceiver and delays introduced by transceiver components. These delays usually cause a constant offset in the detection, which can be minimized by careful calibration.

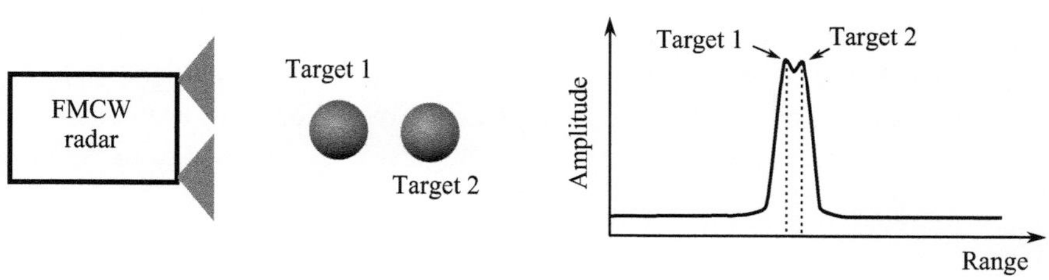

Figure 5.5 Range resolution

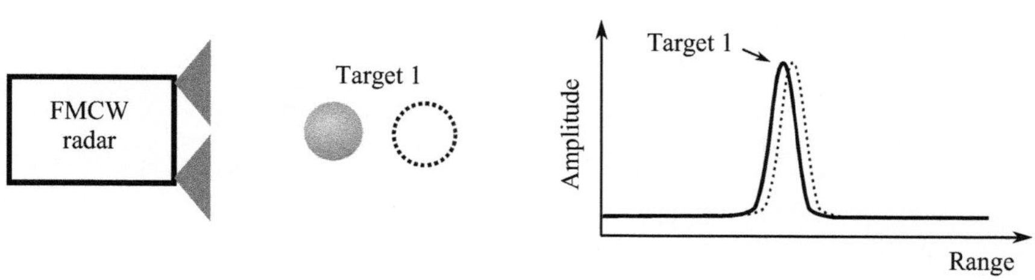

Figure 5.6 Detection precision

5.2 FMCW radar transceiver

The basic architecture of an FMCW radar transceiver is illustrated in Figure 5.7. On the transmitter channel, a chirp generator generates the frequency-modulated chirp signal repeatedly. This chirp signal is amplified by the PA and then transmitted through the Tx antenna. A replica of the chirp signal is fed into the receiver channel as a local oscillator (LO) signal. On the receiver channel, the reflected chirp signal is processed by the LNA and gain blocks. Then, a mixer is used to mix the received chirp signal with the replica of the transmitted chirp signal. This procedure is also known as deramping. The output of the mixer is the beat signal. Usually, a high-pass filter is necessary to remove the low-frequency mixer bias. An ADC is then used to convert the beat signal into the digital domain for further processing.

The following sections will discuss the design considerations of major components in an FMCW radar transceiver.

5.2.1 Chirp generator

The chirp generator is one of the most important components in an FMCW radar transceiver. A chirp generator can be realized with either a free-running voltage-controlled oscillator (VCO) or a phase-locked loop (PLL). Due to limitations in semiconductor processes, it is extremely difficult to achieve a perfect linear frequency tuning curve for a high-frequency VCO. Figure 5.8 shows the frequency tuning curve of a C-band VCO manufactured by analog devices. In order to investigate the consequences of the VCO linearity for an FMCW radar, a third-order polynomial was used to fit this VCO tuning curve:

$$f_{VCO} = 2 \times 10^{-12} V_t^3 - 17.3 V_t^2 + 283.9 V_t + 5595.5 \quad (5.5)$$

where f_{VCO} (in MHz) is the output frequency of the VCO and V_t (in volts) is the tuning voltage. The tuning voltage V_t was swept from 0.15 to 1.4 V, corresponding to an output frequency bandwidth of 320 MHz. By means of (5.5), a simulation has been performed, as shown in Figure 5.9. In Figure 5.9(a), the range between the

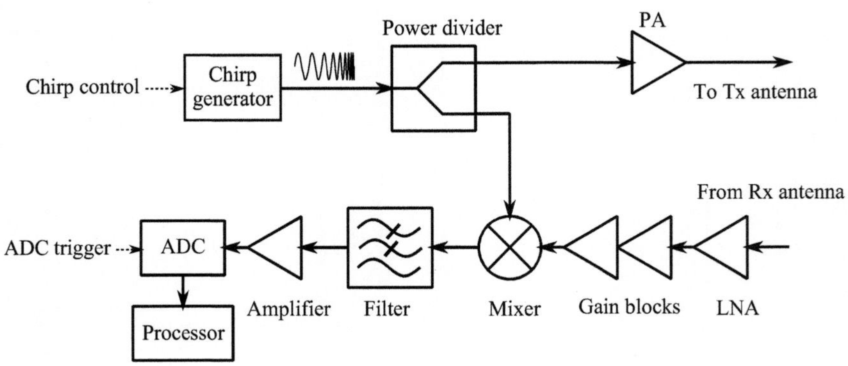

Figure 5.7 Schematic of an FMCW radar transceiver

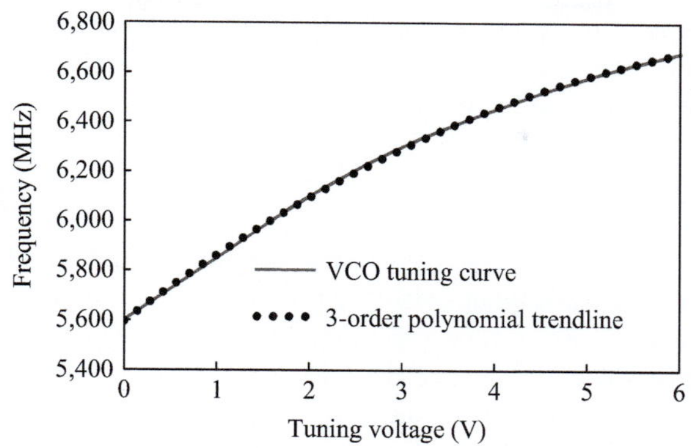

Figure 5.8 The frequency versus tuning voltage of a C-band VCO. From [16]

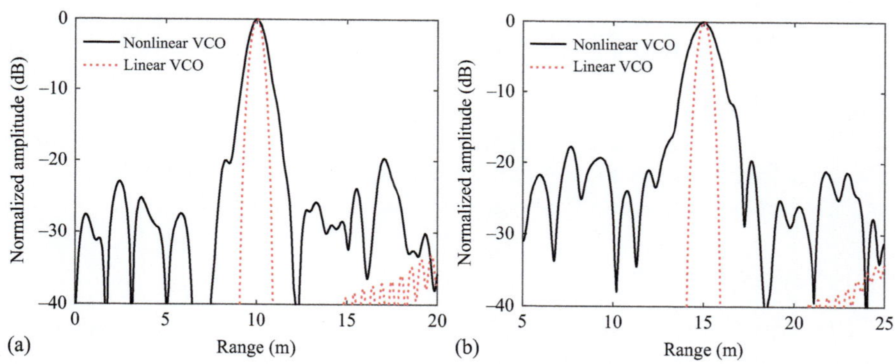

Figure 5.9 Simulated range profiles with a nonlinear VCO and an ideal VCO. (a) A target at 10 m. (b) A target at 15 m. From [16]

radar and the point target was 10 m, and in Figure 5.9(b), the range was 15 m. It can be clearly seen that due to the nonlinearity, the width of the signatures at the target locations are broadened, corresponding to some degradation of the range resolution. Moreover, the side lobes of the signatures are increased. This degradation is acceptable in some short-range applications of the FMCW radar systems. However, in other applications, such as automotive, the degradation of the range resolution and side-lobe level is not acceptable since it increases the false detection rate. A PLL-based chirp generator can achieve a highly linear chirp signal. However, its realization is complex and the cost is much higher.

5.2.2 Coherence

Coherence is another important requirement in an FMCW radar transceiver. The coherence in an FMCW radar transceiver includes two aspects: (1) transmitter and receiver coherence and (2) chirp and beat signal coherence [16]. The coherence between the transceiver and the receiver has been guaranteed by the radar architecture, where the received signal is mixed with a replica of the transmitted signal. With the coherence between the transmitter and the receiver, the time-domain beat signal can be expressed as

$$s_b(t) = s_{TX}(t)s_{RX}^*(t) = \sigma \exp\left(j\left(\frac{4\pi\gamma R(\tau)t}{c} + \frac{4\pi f_c R(\tau)}{c} + \phi\right)\right) \tag{5.6}$$

where $s_b(t)$ is the beat signal, $s_{TX}(t)$ is the transmitted signal, $s_{RX}^*(t)$ is the complex conjugate of the received signal, σ is the amplitude of the beat signal, $\gamma = B \times \text{CRF}$ is the chirp rate, and CRF is chirp repetition frequency. $R(\tau)$ represents the variation of the distance between a target and the FMCW radar, τ is the "slow time," f_c is the center frequency of the frequency ramp, and c is the speed of light. Finally, ϕ is the negligible residual video phase, which can be disregarded.

After performing Fourier transform to the beat signal, the frequency domain representation of the beat signal is

$$S_b(f) = \sigma T \exp\left(j\frac{4\pi f_c R(\tau)}{c}\right) \text{sinc}\left(T\left(f - \frac{2\gamma R(\tau)}{c}\right)\right) \tag{5.7}$$

where $T = 1/\text{CRF}$ is the CRP.

In (5.7), $2\gamma R(\tau)/c$ corresponds to the beat frequency, which is proportional to the range from the radar to the target. Moreover, the slow-time phase history $\phi_d(\tau)$ in (5.7) is simply related to the range evolution of the target $R(\tau)$ by

$$\phi_d(\tau) = \frac{4\pi f_c R(\tau)}{c} \tag{5.8}$$

In order to preserve the slow-time phase history $\phi_d(\tau)$, great efforts should be made to align the ADC start trigger to each chirp. It is usually required to share the clock across the whole system in a PLL-based FMCW radar. For an FMCW radar system with a free-running VCO, the sawtooth voltage sequence, which is used to control the frequency of the VCO, can be used to align the beat signal and maintain the coherence between chirp and beat signal, as shown in Figure 5.10.

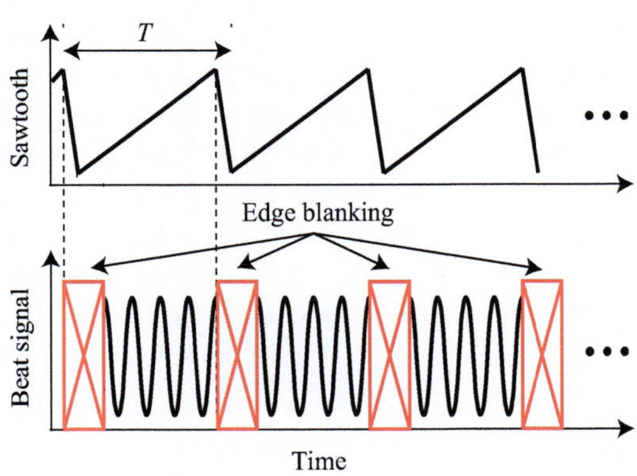

Figure 5.10 Maintenance of coherence between chirp and beat signal for an FMCW radar with a free-running VCO. Adapted from [19]

5.2.3 Link budget

In evaluating the detection capability of an FMCW radar, a link budget analysis can be performed. For an FMCW radar shown in Figure 5.11, the power at different locations can be expressed as follows:

1. Effective radiation power at the transmitter antenna:

$$PG_t \tag{5.9}$$

2. Signal radiation power density at the target location:

$$\frac{PG_t}{4\pi R^2} \tag{5.10}$$

3. Signal power reflected by the target:

$$\frac{PG_t\sigma}{4\pi R^2} \tag{5.11}$$

4. Signal radiation power density at the receiver antenna

$$\frac{PG_t\sigma}{(4\pi)^2 R^4} \tag{5.12}$$

5. Signal power received by the receiver antenna

$$\frac{PG_t\sigma A}{(4\pi)^2 R^4} \tag{5.13}$$

where P is the transmitted power, G_t is the transmitter antenna gain, σ is the radar cross section (RCS) of the target, R is the range, and A is the effective aperture of the receiver antenna.

The noise of the receiver can be calculated as

$$N = kT_0BF_n \tag{5.14}$$

where $k = 1.38 \times 10^{-23}$ W · s/K is Boltzmann's constant, T_0 is the noise temperature, B is the receiver bandwidth, and F_n is the noise figure of the receiver.

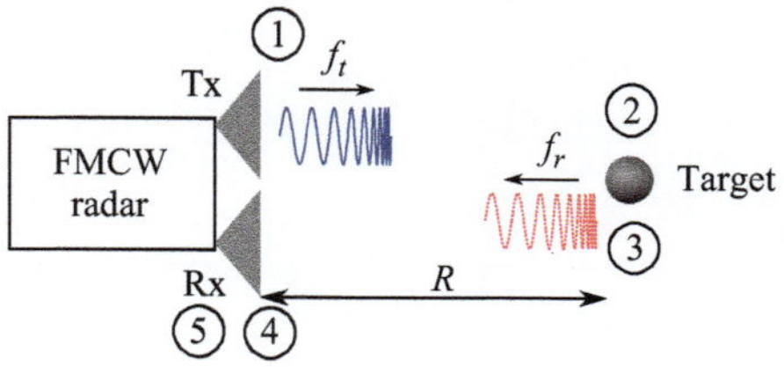

Figure 5.11 Link budget analysis

Thus, the overall SNR of the radar is

$$SNR = \frac{PG_t\sigma A}{(4\pi)^2 R^4 kT_0BF_nL} \tag{5.15}$$

where L is the extra loss or gain introduced in the receiver.

Assume the minimal detectable SNR is $SNR_{\min}$, then the maximum detection range $R_{\det}$ for a target with σ RCS can be obtained as

$$R_{\det} = \sqrt[4]{\frac{PG_t\sigma A}{(4\pi)^2 SNR_{\min} kT_0BF_nL}} \tag{5.16}$$

5.3 Antenna

The performance and properties of the antennas determine some major capabilities of the radar system. For example, the beamwidth of the antenna determines the field of view (FOV) of the radar system, and the antenna gain affects the detection range. Typically, antennas with higher gain increase the detection range of the radar system. However, increasing the gain usually sacrifices the FOV and increases the overall size of the antenna. In addition, increasing the operation frequency of the radar will also help to reduce the size of the antenna. However, a higher frequency has higher path loss and shorter detection range with the same transmitting power. It is essential to balance these characteristics of the antennas depending on the applications. Figure 5.12 shows two examples of antennas used in

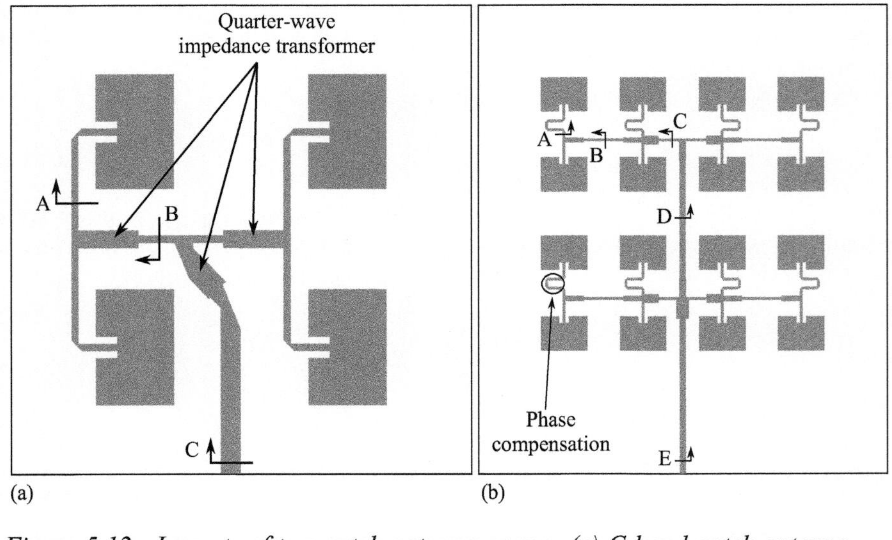

Figure 5.12 Layouts of two patch antenna arrays. (a) C-band patch antenna array. (b) K-band patch antenna array

a portable FMCW radar system [16,20]. Figure 5.12(a) is a 2 × 2 patch antenna array working in the C-band, and Figure 5.12(b) is a 4 × 4 patch antenna array working at K-band. The total size of the C-band array is 65 mm × 65 mm. Its half-power beam widths on E-plane and H-plane are 40.3 and 44.4, respectively. The directivity of the C-band array is 11.8 dBi. On the other hand, the size of the K-band patch array is 45 mm × 50 mm. The half-power beam width on E-plane is 16.5, and on H-plane, the beam width is 18.3. The antenna directivity of this K-band array can reach 19.8 dBi.

With the fixed beam antennas, as shown in Figure 5.12, a mechanical steering system is necessary for a radar to perform a two-dimensional scan. It is known that mechanical steering systems increase the size, weight, and cost while limiting the reliability. In short-range motion sensing, a mechanical steering system is not an option due to its bulky size. The other option to obtain a two-dimensional detection is the beamforming technology, which features lightweight, low profile, and high steering speed. It is optimal for the application of short-range motion sensing.

5.3.1 Beamforming

For an array of antennas, by controlling the phase and amplitude of the narrow-band signal in each antenna channel, the array can form different radiation patterns suitable for various applications (Figure 5.13). The combined pattern of an array can be expressed as the multiplication of the embedded radiation pattern of one element and the array factor. For an N-element linear array, its array factor can be written as

$$AF = \sum_{n=1}^{N} w_n e^{jk\phi_n(\theta)} \tag{5.17}$$

$$\phi_n(\theta) = x_n \sin\theta \tag{5.18}$$

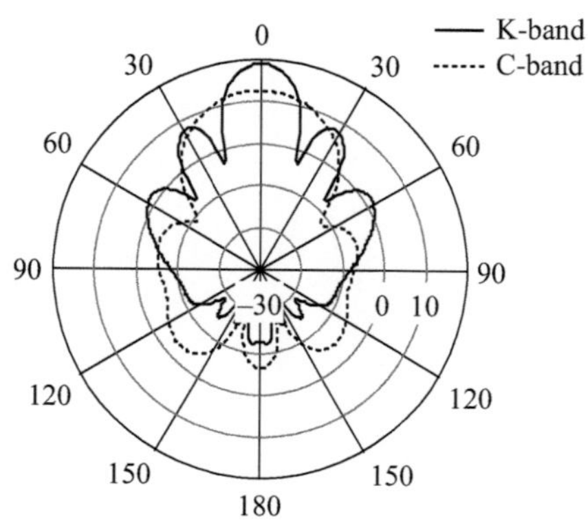

Figure 5.13 Radiation pattern of the C-band and K-band patch antenna arrays at E-plane

where N is the number of elements, k is the wave number, which equals to $2\pi/\lambda$, λ is the wavelength, x_n is the location of the nth element, and θ is the broadside angle. w_n is the complex weight of the nth element, which corresponds to the phase and amplitude of the excitation.

If the antenna elements are uniformly spaced with a separation of $\lambda/2$, AF can be simplified to

$$AF = \sum_{n=1}^{N} w_n e^{j(n-1)\pi \sin\theta} \tag{5.19}$$

For an eight-element linear array with uniformed excitation, where

$$\mathrm{w} = [w_1, w_2, ..., w_8] = [1, 1, ..., 1] \tag{5.20}$$

the corresponding radiation pattern has a main lobe pointing to 0° (broadside) with 12.8 dB side-lobe level, as shown in Figure 5.14.

The direction of the main lobe can be tuned to a different angle by adding different phase components into w, which means different antenna channels have different phase shifts. For example, let

$$\mathrm{w} = \left[1, \mathrm{e}^{-j\pi \sin(-30^\circ)}, \mathrm{e}^{-j2\pi \sin(-30^\circ)}, ..., \mathrm{e}^{-j7\pi \sin(-30^\circ)}\right] \tag{5.21}$$

The main lobe of the radiation pattern will be directed to −30°, and its side-lobe level is 12.8 dB, as shown in Figure 5.15.

Moreover, it is also possible to change both the phase and amplitude for each antenna channel to have more flexibility in controlling the array radiation pattern. A simple example is applying a Chebyshev window to the weight array in (5.21), as shown

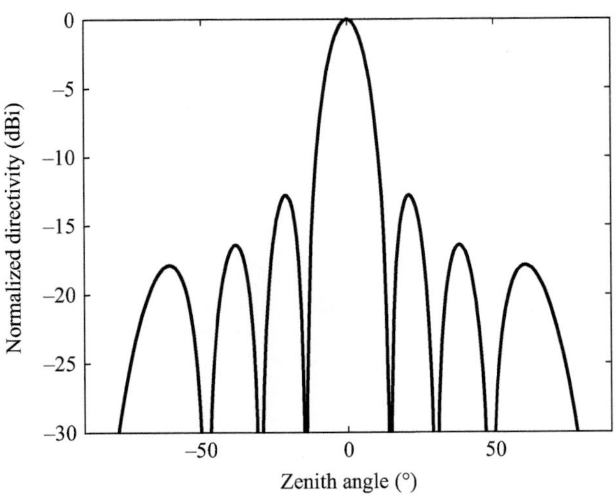

Figure 5.14 $\lambda/2$ spaced eight-element array factor with uniform excitation (broadside direction)

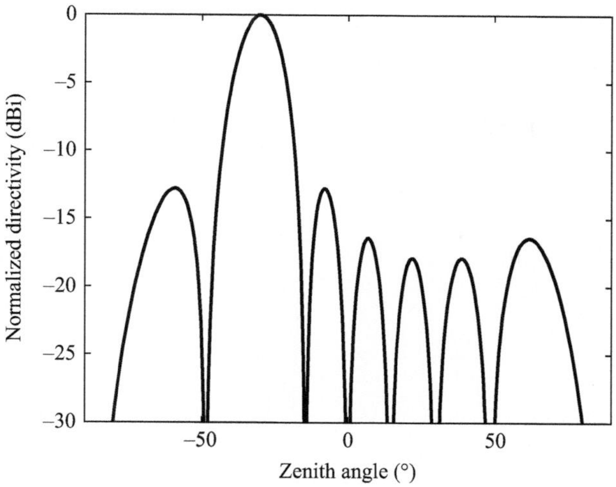

Figure 5.15 *λ/2 spaced eight-element array factor with uniform excitation (−30° direction)*

Table 5.1 *Weighting for a Chebyshev window and a −30° pointing*

w_1	w_2	w_3	w_4
0.38	$0.58e^{-j\pi\sin(-30^\circ)}$	$0.84e^{-j2\pi\sin(-30^\circ)}$	$e^{-j3\pi\sin(-30^\circ)}$
w_5	w_6	w_7	w_8
$e^{-j4\pi\sin(-30^\circ)}$	$0.84e^{-j5\pi\sin(-30^\circ)}$	$0.58e^{-j6\pi\sin(-30^\circ)}$	$0.38e^{-j7\pi\sin(-30^\circ)}$

in Table 5.1. The corresponding figure is illustrated in Figure 5.16. It can be seen that a radiation pattern pointing to −30° with 30 dB side-lobe level can be obtained.

Due to the limitation of the antenna size and mutual coupling, the spacing between the antenna elements has to be larger than $\lambda/2$ sometimes. Increasing the spacing of the array can help to narrow the beamwidth of the main lobe; however, grating lobes may also be introduced. An example of grating lobe is illustrated in Figure 5.17. In this example, the array is still an eight-element linear array; however, the spacing between elements is $3\lambda/4$. In the radiation pattern, it can be seen that besides the desired main lobe at −30°, a grating lobe with the same amplitude as the main lobe appears at around 60°. In practice, if there is a target at the direction of the main lobe or the grating lobe, it is not possible for the radar to distinguish the actual direction of the target. In this situation, the FOV of the radar has to be limited within the angles around the broadside direction without the grating lobes to avoid target aliasing.

Array beamforming can be realized either in the RF domain (RF beamforming) or the digital domain (digital beamforming). For RF beamforming, a phase shifter is required for each antenna channel to control the phase of the RF signal.

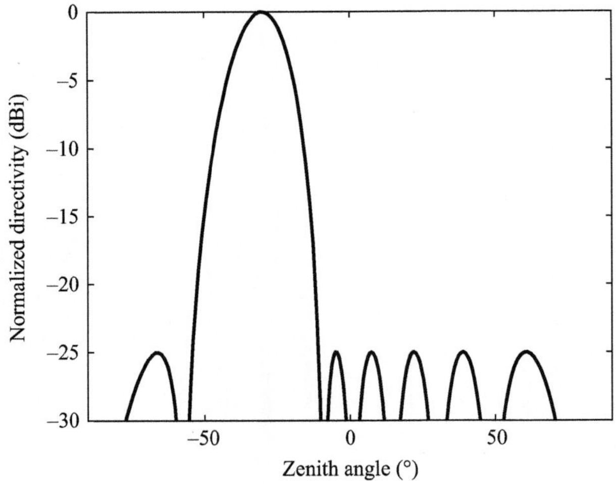

Figure 5.16 *λ/2 spaced eight-element array factor with uniform excitation (Chebyshev window and −30° pointing)*

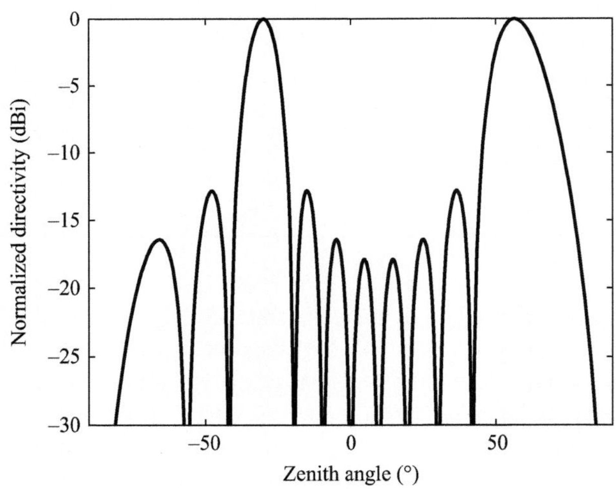

Figure 5.17 *Grating lobe*

A simplified architecture of an RF beamforming receiver is illustrated in Figure 5.18. The RF beamforming receiver consists of an antenna array. Each antenna element in the array is connected with respective LNA and RF gain modules, as well as some necessary filters. Then, a phase shifter is connected to each channel to control the phase of the RF signal. After the phase shifter, all the channels are combined together and downconverted to baseband for further processing. The advantages of an RF beamforming architecture is the relatively low computational

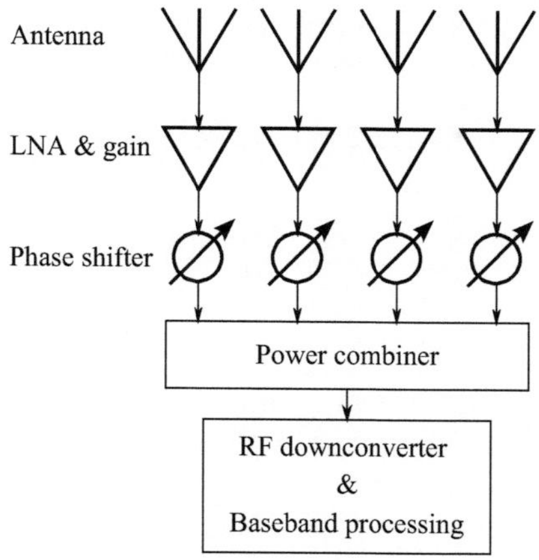

Figure 5.18 RF beamforming

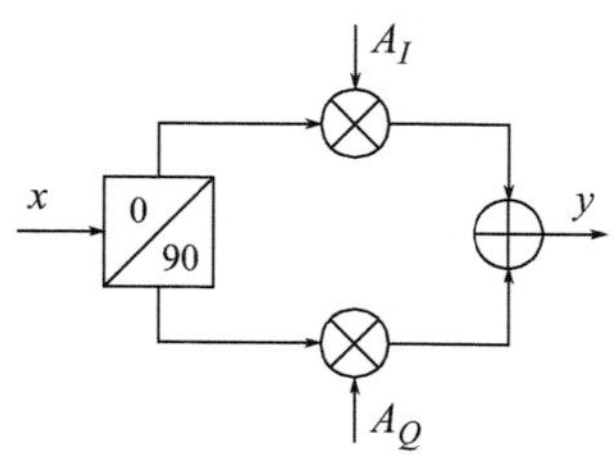

Figure 5.19 Conceptual representation of a vector modulator

demand for baseband processing and low power consumption. However, it suffers from beamforming accuracy due to phase errors from phase shifters. In addition, in high-frequency implementations, especially at frequencies above K-band, the costs of phase shifters are very high. New RF beamforming techniques are emerging for higher frequencies. Using delay lines, such as Butler matrix, is one of the simplest approaches at high-frequency RF beamforming [21,22]. However, these methods only have few steering angles due to the limitation of size and signal routing. Other solutions include LO phase-shifting [23,24], vector modulator [19,25], etc. The LO phase-shifting technique requires a complex multiphase VCO and multiphase select switches. The steering angles are limited by the number of different phase paths and switches. It is extremely difficult and very expensive to integrate many phase paths for a multiphase VCO. The vector modulator technique, which is based on the concept of vector sum, has been widely used in integrated circuits (IC) for high-frequency phase shifters. The basic concept of a vector modulator is shown in Figure 5.19. An RF input signal $x = \sin(2\pi f_c t)$ is first divided into an in-phase (I) signal and a quadrature (Q) signal by a quadrature power divider. f_c is the carrier frequency, and t is time. The amplitudes of I and Q signals are controlled separately through variable gain amplifiers or attenuators. Then, these I and Q signals are combined together. The output signal y can be written as

$$y = A_I \sin(2\pi f_c t) + A_Q \cos(2\pi f_c t) = \mathrm{Re}\{A e^{j\phi} e^{-j2\pi f_c t}\} \tag{5.22}$$

$$A = \sqrt{A_I^2 + A_Q^2} \tag{5.23}$$

$$\phi = \arctan\frac{A_Q}{A_I} \tag{5.24}$$

By comparing the output signal y with the input signal x, it can be clearly seen that the output signal has a phase shift of ϕ and an amplitude change of A. Based on this concept, integrated phased-array chips working at frequencies higher than 24 GHz have been developed by researchers. A PCB realization of a 24-GHz RF beamforming radar system has also been developed based on vector modulator with RF attenuators. 2-D target mapping result in short-range localization application has been reported.

With the advance of high-speed ADCs and digital to analog converters (DACs), digital beamforming techniques have been widely used in modern communication systems, as well as high-end automotive radar systems. Digital beamforming technique applies the phase shift and amplitude control in the digital domain, which overcomes the phase error issue in an RF beamforming system. Figure 5.20 shows an example of a four-channel digital beamforming receiver. In this receiver, the RF signal received by each antenna channel is processed by a dedicated RF front end, which includes LNAs, gain blocks, filters, and a frequency downconverter. The downconverted signal is then sampled by an ADC into the digital signal. The phase and amplitude of each channel are controlled in the digital domain by using basic digital operators. Digital beamforming architecture has high beamforming accuracy

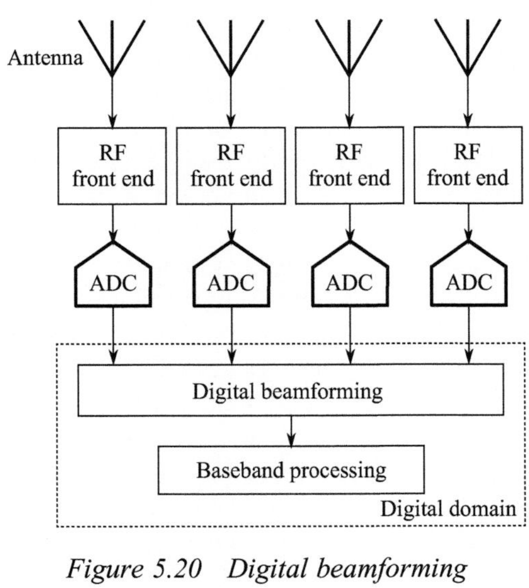

Figure 5.20 Digital beamforming

and flexibility. However, due to the massive use of ADCs or DACs, the power consumption is usually high. Researchers have been working on solutions to reduce the required number of ADCs or DACs in a digital beamforming system. One of the methods consists in implementing the concept of code modulation multiplexing to share ADCs or DACs among different channels to reduce the hardware intensity.

5.3.2 Two-way pattern and MIMO

For a radar system, it is necessary to consider both the transmitter and the receiver radiation patterns to evaluate the detection coverage. The effective radiation pattern of a radar system is the combination of the transmitter pattern and the receiver. The combined pattern is called two-way pattern, which is the multiplication of the transmitter pattern and the radiation pattern. By properly utilizing the two-way pattern, it can help to improve the performance of the radar system, such as the angular resolution. Figure 5.21 shows a sample array configuration for a radar system with two transmitter antennas and four receiver antennas. The separation between the two transmitter antennas is 2λ, and the receiver antennas are $\lambda/2$ spaced. The radiation patterns of the transmitter array and the receiver array when the main lobes are steered to $-20°$ are illustrated in Figure 5.22. Since the

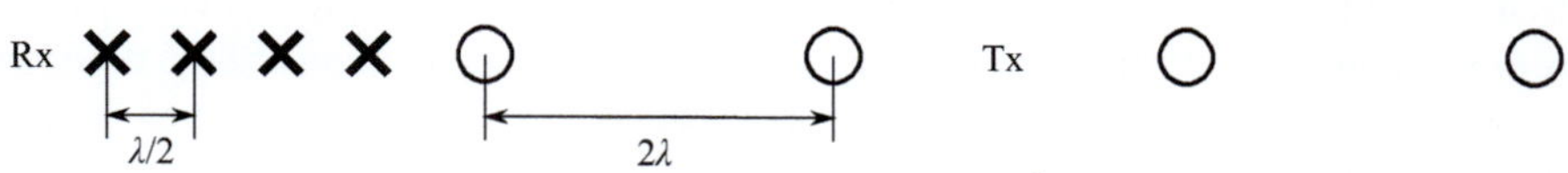

Figure 5.21 Spatial configuration of a linear receiver array without grating lobes and a linear transmitter array with grating lobes

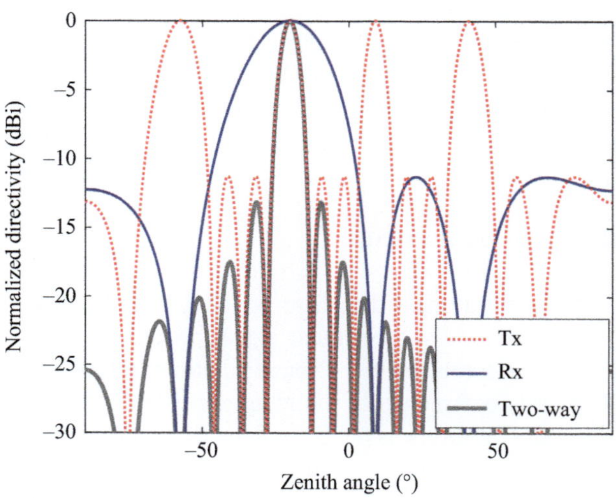

Figure 5.22 Radiation patterns for the linear arrays in Figure 5.21 and two-way radar radiation pattern

transmitter array is a sparse array, three grating lobes can be observed at −58°, 9°, and 42°. The half-power beamwidth of the main lobe in the transmitter pattern is about 8°. On the other hand, the pattern of the receiver has one wide main lobe without any grating lobes. It can be also observed that the directions of the three grating lobes on the transmitter patterns are located exactly at the directions of three nulls on the receiver patterns. Thus, the signals transmitted through the grating lobes are rejected by the receiver array. The effective pattern, namely the two-way pattern, of the transceiver array can be obtained by multiplying the transmitter pattern with the receiver pattern, as shown in Figure 5.22. This two-way pattern has a main lobe with 8° half-power beamwidth, and it does not have grating lobes.

Typically, in order to obtain the capability of forming the two-way pattern, a radar system should have the beamforming on both the transmitter and the receiver. Conventional methods include implementing RF beamforming on both the transmitter and the receiver, or using digital beamforming on both the transmitter and the receiver. However, these conventional methods dramatically increase the hardware intensity and cost.

MIMO is a novel technique to utilize the benefit of the combined patterns. On the transmitter of a MIMO radar, orthogonal waveforms are generated and transmitted through different transmitter antenna channels. In an FMCW radar system, the orthogonal waveforms can be realized by using phase modulation or time division multiplexing (TDM). By using phase modulation, all the antenna channels can transmit simultaneously. At the same time, on the receiver, same phase codes are used to demodulate the received signal and separate the signal from each transmitter antenna channel. The beamforming for the transmitter and the receiver can be realized at the same time on the receiver. On the other hand, TDM method requires each transmitter antenna transmit at its own time slot. The receiver channels separate signals from different transmitter antenna channels based on different time. Since TDM method does not require modulation or demodulation, it has much simpler hardware architecture than the phase modulation method. However, transmitting signals at different time slots increases the CRP of an FMCW radar, which decreases the unambiguous Doppler frequency in range-Doppler maps.

5.4 Radar signal processing

In this section, the measurement data from an FMCW radar prototype is used to demonstrate the basic procedures of radar signal processing [16]. The radar prototype used in the measurement is a C-band multimode CW radar, as shown in Figure 5.23. The basic parameters of the radar are listed in Table 5.2. The center frequency of this radar prototype is 6 GHz and its bandwidth is 320 MHz. The chirp repetition of the radar prototype is 3.5 ms. The baseband signal of the radar prototype is sampled by the audio card of a computer. The sampling rate is 192 ksps. With these configurations, the range resolution of the radar is 0.67 m with a maximum range of 16.4 m.

Figure 5.23 FMCW C-band radar prototype. From [16]

Table 5.2 Parameters of the C-band radar prototype

Center frequency	6 GHz	Transmitted power	8 dBm
Antenna directivity	11.8 dBi	Beamwidth (E-plane)	40.3°
Beamwidth (H-plane)	44.4°	Bandwidth	320 MHz
Chirp repetition period (T)	3.5 ms	Sampling rate (f_s)	192 ksps
Range resolution	0.67 m	Maximum range	16.4 m

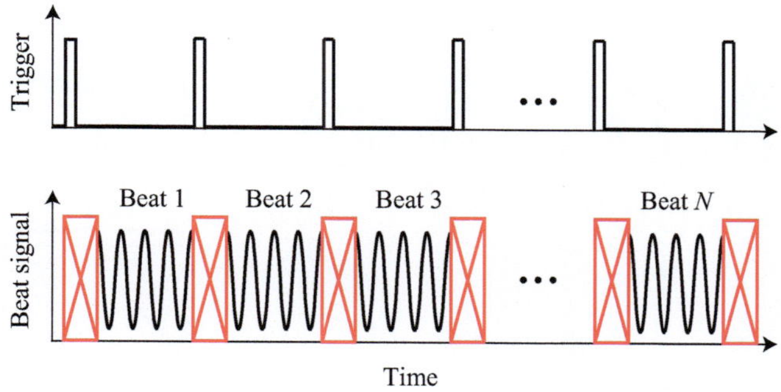

Figure 5.24 Formatting of the beat signal according to the trigger

5.4.1 Range profile

The C-band radar prototype samples two-channel data, as shown in Figure 5.24. The first channel is the chirp trigger pulse, which aligns the chirp sequences on the transmitter. The other channel is the beat signal sequence. To process the beat signal, the beat signal sequence is sliced into sections, i.e., Beat 1, Beat 2, ..., and Beat N, according to the trigger sequence. Edges of each beat signal sections need to be discarded to avoid edging effects of chirps. An example of raw baseband data from the C-band is illustrated in Figure 5.25.

The procedure to obtain a range profile from a section of the beat signal, Beat n, is illustrated in Figure 5.26. First, a window is applied to this time-domain beat signal. A Chebyshev window and a Taylor window are the most widely used

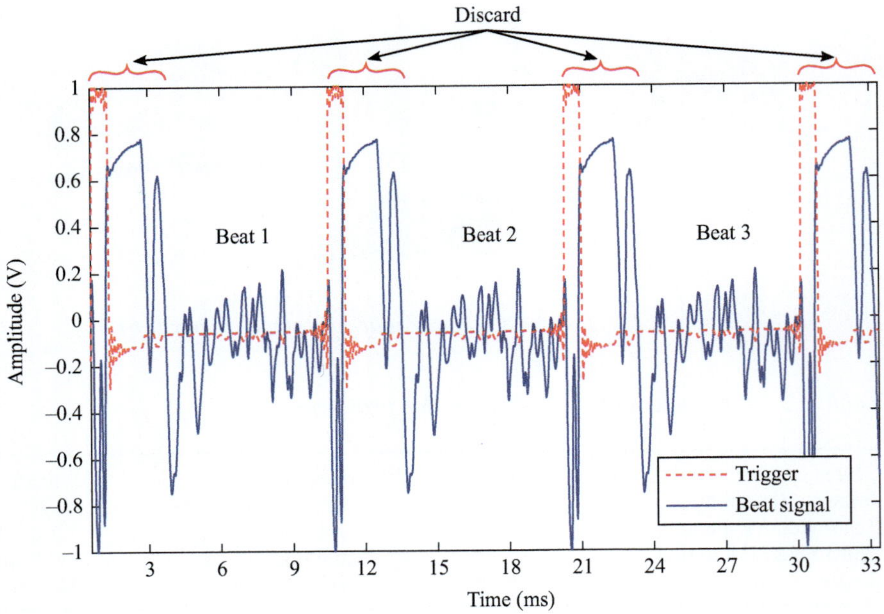

Figure 5.25 Discarding of samples in each FMCW period

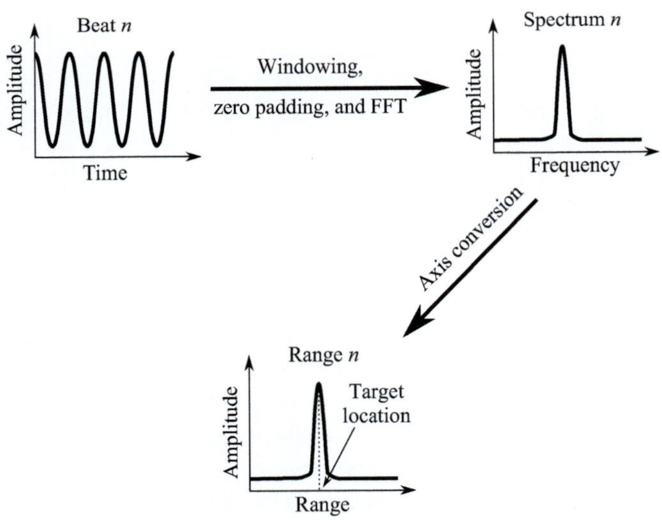

Figure 5.26 Obtaining of range profile

windows for radar signal processing. A Chebyshev window provides the narrowest possible main lobe for a specified side-lobe level. A Taylor window allows you to make trade-offs between the main lobe width and the side-lobe level. A Taylor window also avoids edge discontinuity, which is an issue for a Chebyshev window.

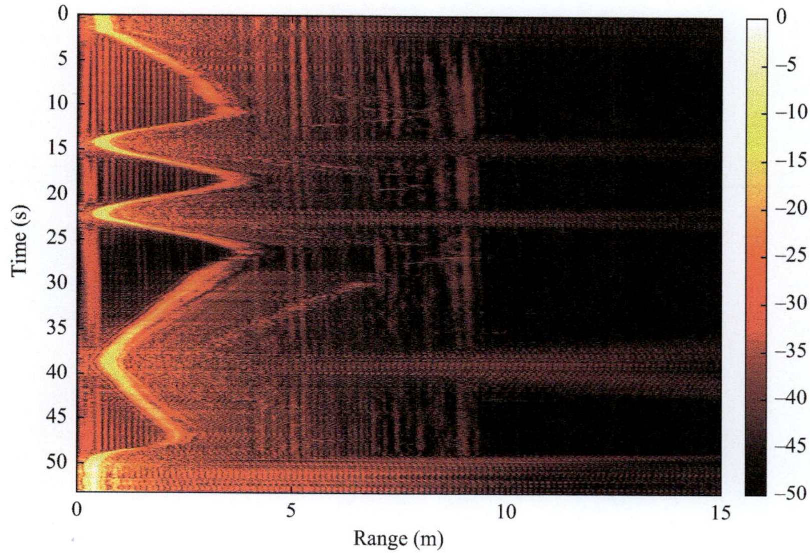

Figure 5.27 Range-slow-time map for a pedestrian. Adapted from [26]

After applying the windows and necessary zero padding, fast Fourier transform (FFT) is then used to convert the time-domain signal into the frequency domain. Since the frequencies of the beat signal and the ranges of the targets are one-to-one corresponded, the range profile can be obtained through a simple axis conversion.

Figure 5.27 shows an experiment with a human target walking back and forth in front of the radar prototype. The horizontal axis corresponds to the range of the human target, and the vertical axis is the measurement time. The bright curve is the range history of the human target.

5.4.2 Human-aware detection

It is not difficult for an FMCW radar to differentiate a moving target from stationary targets. If the desired target is also not moving, it may be challenging to discriminate the target. However, for stationary human target detection with an FMCW radar system, some special properties can be observed from the radar signal. These special properties in the radar signal can be used to discriminate the human target from other stationary objects. When a human subject is standing or sitting still, he or she cannot be completely stationary, because of physiological motions such as respiration. Since the human body is not a perfect reflector, a small motion of the human body caused by the respiration will introduce a significant variation in the amplitude of the reflected radar signal. This kind of effect is referred to as the "vital-Doppler" effect.

The procedure of "vital Doppler"-based "human-aware" range detection is illustrated in Figure 5.28. Due to the vital-Doppler effect, the measured range profiles at different beat signal sections have variations for a human target. On the

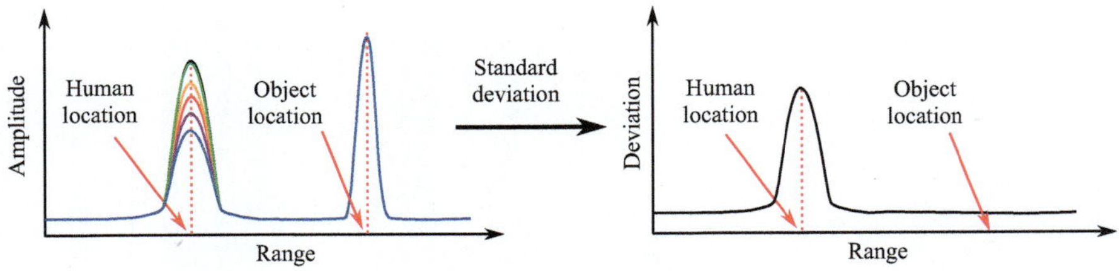

Figure 5.28 The vital-Doppler effect

Figure 5.29 Human-aware experiment setup. From [16]

other hand, the measured results from a stationary object are relatively stable. If the standard deviation of the different measurements is calculated, a peak at the location of the human subject can be observed in the standard deviation versus range diagram, as shown in Figure 5.28. Thus, the human subject can be discriminated from other stationary objects.

An experiment was demonstrated to show the "human-aware" detection in Figures 5.29 and 5.30. As shown in Figure 5.29, in this experiment, the C-band

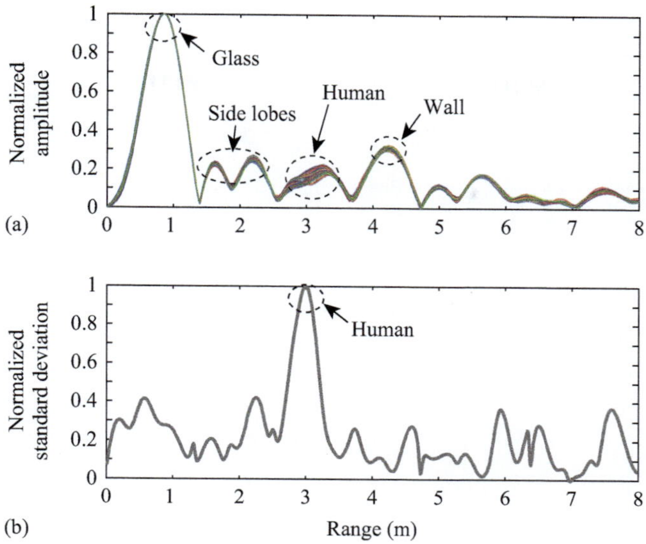

Figure 5.30 Range profiles and standard deviation for the vital-Doppler effect experiment. From [16]

radar was located 0.8 m in front of a glass wall. The dimension of the glass wall was about 3 m × 10 m. A human target stood still between the glass wall and the concrete wall. Figure 5.30(a) plots multiple range profiles into a single figure. Large variations at the location of the human target can be observed, which was caused by the vital-Doppler effect. By calculation of the standard deviation of these measurements, the human subject can be discriminated, and other stationary objects are eliminated, as shown in Figure 5.30(b).

5.4.3 Range-Doppler imaging

In a crowded environment, strong clutter from stationary makes it very difficult for an FMCW radar to discriminate wanted moving targets. Range-Doppler imaging in an FMCW radar can be used to extract Doppler information from moving targets. With range-Doppler imaging, two-dimensional diagrams, including range and Doppler, can be derived. These two-dimensional diagrams help to isolate moving targets from stationary clutter.

Figure 5.31 details the major steps to obtain the range-Doppler diagrams. The first step of range-Doppler imaging is obtaining and arranging the raw data matrix, which consists of arrays of beat signals. The time interval along the horizontal axis in the raw data matrix is $1/f_s$, where f_s is the sampling rate of the ADC. $1/f_s$ is also called fast time. On the vertical axis, the time interval is T, which is the CRP. T is also named as slow time. The range profile can be easily obtained by performing FFT along the fast time after windowing and zero padding. A series of range-Doppler diagrams can be obtained by windowing the range profile and performing

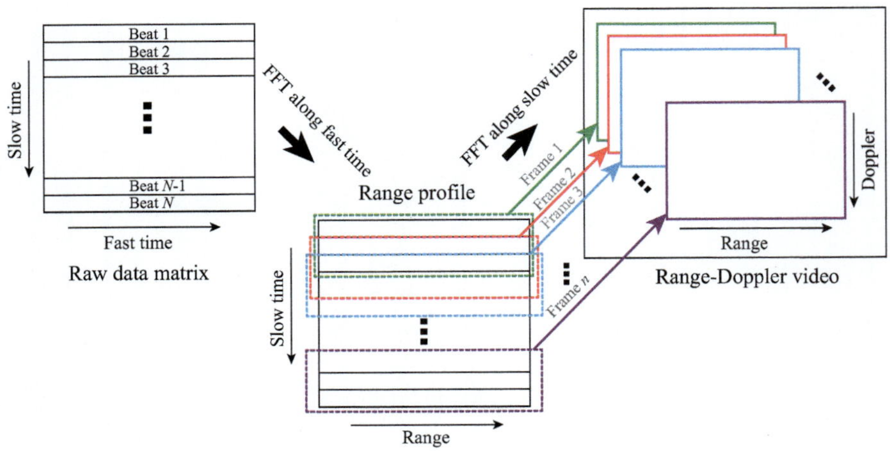

Figure 5.31 Obtaining the video of range-Doppler maps. From [20]

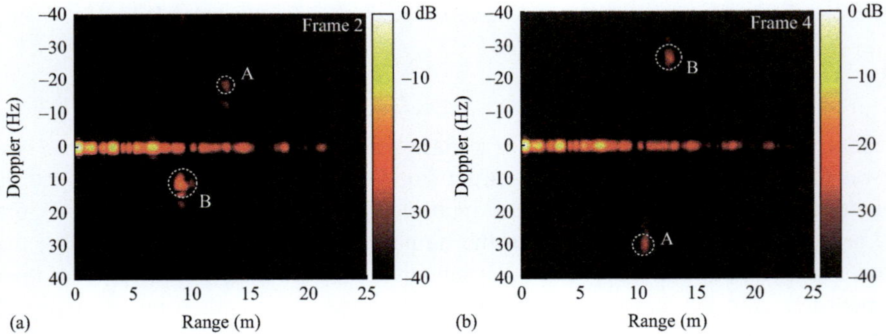

Figure 5.32 Examples of range-Doppler maps for two pedestrians. From [16]

FFT along the slow time. These range-Doppler diagrams can be combined into a video for further processing.

Figure 5.32 shows an experiment of range-Doppler imaging with two human subjects (Subjects A and B) walking in front of the C-band radar prototype. Figure 5.32(a) shows the measurement when Subject A was walking toward the radar and Subject B was walking away from the radar. On the other hand, Figure 5.32(b) illustrates the time when Subject B was walking toward the radar and Subject A was leaving the radar. It can be clearly seen that the target has a negative Doppler frequency when it is moving toward the radar. Moreover, the target has a positive Doppler frequency when it is moving away from the radar. All the stationary targets in this experiment, such as the clutter from walls and the ground, has zero Doppler. In addition, micro-Doppler from the human subjects can also be observed surrounding the signatures of the moving subjects. Researchers have been investigating the potential applications of range-Doppler imaging in various areas, such as fall detection, gesture recognition, and active shooter detection.

References

[1] Lien J, Gillian N, Karagozler ME, *et al.* Soli: ubiquitous gesture sensing with millimeter wave radar. ACM Trans Graph. 2016;35(10). Available from: http://dl.acm.org/citation.cfm?id=2925953 http://doi.acm.org/10.1145/2897824.2925953.

[2] Fan T, Ma C, Gu Z, *et al.* Wireless hand gesture recognition based on continuous-wave Doppler Radar sensors. IEEE Transactions on Microwave Theory and Techniques. 2016;64(11):4012–4020. Available from: http://ieeexplore.ieee.org/document/7592916/.

[3] Peng Z, Li C, Muñoz-Ferreras JM, *et al.* An FMCW radar sensor for human gesture recognition in the presence of multiple targets. In: IEEE MTT-S International Microwave Bio Conference (IMBIOC); 2017. pp. 1–3. Available from: http://ieeexplore.ieee.org/document/7965798/.

[4] Li C, Lubecke VM, Boric-Lubecke O, *et al.* A review on recent advances in Doppler radar sensors for noncontact healthcare monitoring. IEEE Transactions on Microwave Theory and Techniques. 2013;61(5):2046–2060. Available from: http://ieeexplore.ieee.org/document/6504804/.

[5] Peng Z, Muñoz-Ferreras JM, Gómez-García R, *et al.* FMCW radar fall detection based on ISAR processing utilizing the properties of RCS, range, and Doppler. In: IEEE MTT-S International Microwave Symposium (IMS). IEEE; 2016. pp. 5–7. Available from: http://ieeexplore.ieee.org/document/7540121/.

[6] Zhang S, Spie M, and Huang PS. High-resolution, real-time three-dimensional shape measurement. Optical Engineering. 2006;45(12):123601. Available from: https://www.spiedigitallibrary.org/journals/Optical-Engineering/volume-45/issue-12/123601/High-resolution-real-time-three-dimensional-shape-measurement/10.1117/1.2402128.short http://opticalengineering.spiedigitallibrary.org/article.aspx?doi=10.1117/1.240.

[7] Wang Y, Liu K, Hao Q, *et al.* Period coded phase shifting strategy for real-time 3-D structured light illumination. IEEE Transactions on Image Processing. 2011;20(11):3001–3013. Available from: http://ieeexplore.ieee.org/abstract/document/5770217/.

[8] Wang FK, Horng TS, Peng KC, *et al.* Seeing through walls with a self-injection-locked radar to detect hidden people. In: IEEE MTT-S International Microwave Symposium (IMS); 2012. pp. 1–3.

[9] Wang FK, Horng TS, Peng KC, *et al.* Detection of concealed individuals based on their vital signs by using a see-through-wall imaging system with a self-injection-locked radar. IEEE Transactions on Microwave Theory and Techniques. 2013;61(1):696–704.

[10] Liu X, Leung H, and Lampropoulos GA. Effect of wall parameters on ultra-wideband synthetic aperture through-the-wall radar imaging. IEEE Transactions on Aerospace and Electronic Systems. 2012;48(4):3435–3449.

[11] Li C, Yu X, Lee CM, *et al.* High-sensitivity software-configurable 5.8-GHz radar sensor receiver chip in 0.13-μm CMOS for noncontact vital sign detection. IEEE Transactions on Microwave Theory and Techniques. 2010;58(5): 1410–1419. Available from: http://ieeexplore.ieee.org/document/5438822/.

[12] Gu C, Wang G, Li Y, *et al.* A hybrid radar-camera sensing system with phase compensation for random body movement cancellation in Doppler vital sign detection. IEEE Transactions on Microwave Theory and Techniques. 2013;61(12):4678–4688. Available from: http://ieeexplore.ieee.org/document/6662489/.

[13] Mitomo T, Ono N, Hoshino H, *et al.* A 77 GHz 90 nm CMOS transceiver for FMCW radar applications. IEEE Journal of Solid-State Circuits. 2010;45(4):928–937. Available from: http://ieeexplore.ieee.org/lpdocs/epic03/wrapper.htm?arnumber=5437479.

[14] Pohl N, Jaeschke T, and Aufinger K. An ultra-wideband 80 GHz FMCW radar system using a SiGe bipolar transceiver chip stabilized by a fractional-N PLL synthesizer. IEEE Transactions on Microwave Theory and Techniques. 2012;60(3):757–765. Available from: http://ieeexplore.ieee.org/lpdocs/epic03/wrapper.htm?arnumber=6140934.

[15] Scheiblhofer S, Schuster S, and Stelzer A. High-speed FMCW radar frequency synthesizer with DDS based linearization. IEEE Microwave and Wireless Components Letters. 2007;17(5):397–399. Available from: http://ieeexplore.ieee.org/document/4167932/ http://ieeexplore.ieee.org/xpl/freeabs_all.jsp?arnumber=4167932.

[16] Peng Z, Muñoz-Ferreras JM, Tang Y, *et al.* A portable FMCW interferometry radar with programmable low-IF architecture for localization, ISAR imaging, and vital sign tracking. IEEE Transactions on Microwave Theory and Techniques. 2017;65(4):1334–1344.

[17] Peng Z, Lixin Ran L, Li C, *et al.* A 24-GHz low-cost continuous beam steering phased array for indoor smart radar. In: IEEE 58th International Midwest Symposium on Circuits and Systems (MWSCAS). IEEE; 2015. pp. 1–4. Available from: http://ieeexplore.ieee.org/document/7282120/ http://ieeexplore.ieee.org/lpdocs/epic03/wrapper.htm?arnumber=7282120.

[18] Droitcour AD, Boric-Lubecke O, Lubecke VM, *et al.* Range correlation and I/Q performance benefits in single-chip silicon Doppler radars for noncontact cardiopulmonary monitoring. IEEE Transactions on Microwave Theory and Techniques. 2004;52(3):838–848. Available from: http://ieeexplore.ieee.org/document/1273725/.

[19] Peng Z, Ran L, and Li C. A κ-band portable FMCW radar with beamforming array for short-range localization and vital-Doppler targets discrimination. IEEE Transactions on Microwave Theory and Techniques. 2017;65(9):3443–3452. Available from: http://ieeexplore.ieee.org/document/7858751/.

[20] Peng Z, Muñoz-Ferreras JM, Gómez-García R, *et al.* 24-GHz biomedical radar on flexible substrate for ISAR imaging. In: IEEE MTT-S International Wireless Symposium (IWS). IEEE; 2016. pp. 1–4. Available from: http://ieeexplore.ieee.org/document/7585400/.

[21] Lee MS, and Kim YH. Design and performance of a 24-GHz switch-antenna array FMCW radar system for automotive applications. IEEE Transactions on Vehicular Technology. 2010;59(5):2290–2297. Available from: http://ieeexplore.ieee.org/lpdocs/epic03/wrapper.htm?arnumber=5430954.

[22] Lee W, Kim J, and Yoon YJ. Compact two-layer rotman lens-fed microstrip antenna array at 24 GHz. IEEE Transactions on Antennas and Propagation. 2011;59(2):460–466. Available from: http://ieeexplore.ieee.org/lpdocs/epic03/wrapper.htm?arnumber=5654577.

[23] Hashemi H, Guan X, Komijani A, *et al.* A 24-GHz SiGe phased-array receiver – LO phase-shifting approach. IEEE Transactions on Microwave Theory and Techniques. 2005;53(2):614–625. Available from: http://ieeexplore.ieee.org/lpdocs/epic03/wrapper.htm?arnumber=1393205.

[24] Guan X, Hashemi H, and Hajimiri A. A fully integrated 24-GHz eight-element phased-array receiver in silicon. IEEE Journal of Solid-State Circuits. 2004;39(12):2311–2320. Available from: http://ieeexplore.ieee.org/lpdocs/epic03/wrapper.htm?arnumber=1362839.

[25] Yu T, and Rebeiz GM. A 22–24 GHz 4-element CMOS phased array with on-chip coupling characterization. IEEE Journal of Solid-State Circuits. 2008;43(9):2134–2143. Available from: http://ieeexplore.ieee.org/lpdocs/epic03/wrapper.htm?arnumber=4626012.

[26] Peng Z, and Li C. A portable 24-GHz FMCW radar based on six-port for short-range human tracking. In: IEEE MTT-S International Microwave Workshop Series on RF and Wireless Technologies for Biomedical and Healthcare Applications (IMWS-BIO). IEEE; 2015. pp. 81–82. Available from: http://ieeexplore.ieee.org/document/7303787/ http://ieeexplore.ieee.org/lpdocs/epic03/wrapper.htm?arnumber=7303787.

Chapter 6

Noncontact noninvasive monitoring of small laboratory animal's vital sign activities using a 60-GHz radar

Tien-Yu Huang[1], Linda Hayward[2], and Jenshan Lin[1]

6.1 Background

6.1.1 Development of animal experiment

For decades, laboratory experiments are conducted with animals to advance the understanding of anatomy, physiology, pathology, and pharmacology. According to a survey of the National Library of Medicine, the number of academic published papers in biomedicine involving mice and rats has dramatically increased since 1965. Despite animal experiments receive criticisms, the practice of using animals in biomedical research still underlines that animal experimentation can be tremendously helpful for furthering medical and biological science. Laboratory animals are crucial in evaluating the safety of consumer products. As a matter of fact, drugs, vaccines, food additives, and cosmetics have to get the Food and Drug Administration (FDA) approval before they can be marketed. Pharmaceutical companies are required to obtain animal study results with long-term toxicity studies before starting human tests. Animals are also used for testing new surgical procedures and new treatments before they can be tested on humans. Researchers and scientists rarely use animals of unknown organisms in laboratory tests. Therefore, most of animal experiments are conducted with mice and rats due to their similarity to human biology, genetic consistency, short lifespan, and frequent reproduction [1,2].

The laboratory experiments performed on rats include production and quality control, toxicology and safety tests, and fundamental biomedical research. It is not only because rats are mammals, but also because they share many physiological and genetic similarities with humans and the effects of substances should be the same on humans as on rats. In addition, their short lifespan and intensive social maturity phase can help scientists quickly obtain the test results in a number of their

[1]Department of Electrical and Computer Engineering, University of Florida, USA
[2]Department of Physiological Sciences, University of Florida, USA

generations, therefore providing a full evaluation on the new medicine or chemical compound and further reducing the genetic abnormal possibility and potential damage to humans.

Monitoring physiological signs in animal experiment is a fundamental requirement for the development of new therapeutic strategies, such as monitoring disease progression in real time and testing new therapies. The physiological responses include respiration rate (RR), heart rate (HR), and displacement change. A major concern for monitoring physiological response lies in the fact that lab animals require surgical implantation and postoperative recovery, which yields unintended influence on physiological parameters, and further interfering with experimental results. Besides, physiological signs associated with pain presented in fluctuations in RR, HR, blood pressure (BP), and body temperature. Monitoring signs of pain in a laboratory setting is also critical for both animal welfare and drug efficacy studies.

The conventional methods use invasive physiological and anatomical techniques to access the fundamental physiological responses and reveal the problem of the body function, such as body-restrained contact sensor, electrocardiography (ECG), and plethysmography [3,4]. Physically restraining the animals and their motions helps achieve better quality of physiological data. Implant telemetry and indwelling arterial catheter [5,6] are preinvasive methods and allow long-term monitoring in conscious freely moving animals. However, those methods require surgery, have high initial cost and short transmitting range, cause stress to animals, and need time for animals to recover. Developing a noninvasive method for detecting an animal's physiological response would be a vast improvement over current behavioral or invasive methods for monitoring cardiorespiratory function.

6.1.2 Radar for cardiorespiratory monitoring of laboratory animal

The development of Doppler radar technique links biomedical research with noncontact sensing and brings up various applications [7–10]. For those in favor of using animals in a moral way such as reducing animal use, ensuring the best treatment of the animals used in a study, or painless method, the Doppler radar can be used with the goal of developing a noninvasive and noncontact monitoring system for measuring laboratory rat's cardiorespiratory movement. Measuring a small animal's vital signs with radar [11–14] is more challenging compared to humans due to smaller body displacements. Since higher frequency and shorter wavelength is more sensitive to small displacement, to detect heartbeat and respiration of small animals, the millimeter-wave Doppler radar that can detect small vibrations would be an ideal choice to use.

Body movements caused by respiration and heartbeat activities can be quantified with displacement and frequency. Some techniques for measuring displacement of chest-wall motion including impulse cardiogram, capacitance transducer, and magnetic-field sensor have been reported [15–17]. However, they are contact measurements and might have potential effect on measurement accuracy. The laser speckle interferometry does not require body contact [18], but the surface of subject

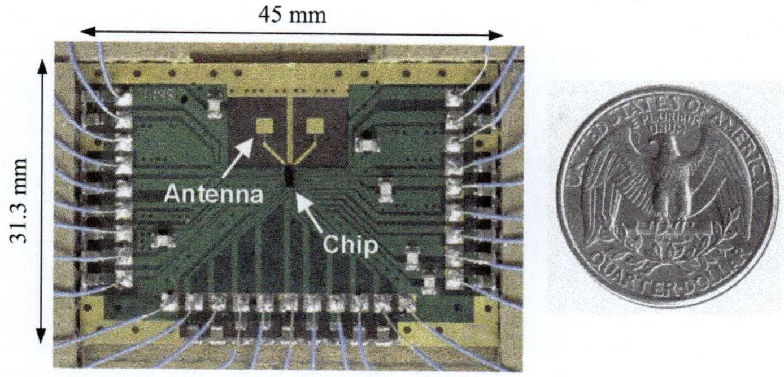

Figure 6.1 The 60-GHz SiP radar and antenna integrated on RT/duroid 5870 laminate PCB. [2013] IEEE. Reprinted, with permission, from Ref. 19

has to be smooth and no clothes are allowed. The existing potential impacts like surgical implantation, risk of infection, unintended drug effect prior to the major drug test, and postoperative recovery are critical adverse effects in the animal experiment. These unintended impacts can be avoided by using radar. The radar technique provides noninvasive and noncontact monitoring with no need for cleaning and disposables, and no risk of infection. It reduces animal use by providing long-term monitoring of the onset and the progression of disease within the same animal, and it's easy to obtain physiological data that helps study biological effects of drug candidates and their therapeutic effectiveness.

In this chapter, we will present test results of monitoring cardiac and respiratory activity of laboratory rats using a 60-GHz system-in-package (SiP) integrated micro-radar [19] to perform noncontact and noninvasive measurement. Figure 6.1 shows the photograph of the 60-GHz SiP radar including transceiver chip fabricated in 90-nm complementary metal–oxide–semiconductor (CMOS) technology, two patch antennas, and DC bias wiring on the printed circuit board (PCB). A guideline of implementing the displacement acquisition method for real animal subjects will be provided, and the issues of harmonics and intermodulation interferences in millimeter-wave radar detected signal will be discussed. Consideration of radar positions and body orientations in animal care or lab environment will be investigated along with conscious, anesthetic, and drug-delivered agents. This system allows the lab animals' cardiac and respiratory activity to be monitored continuously without the need of surgical implants and will eliminate the fatality rate associated with the surgical and recovery process.

6.2 Radar detection position and body orientation

6.2.1 Radar cross section

In radar system, the transmitted power through an antenna is incident upon a target which is assumed to be in the main beam direction of the antenna. If the incident wave hits the target whose purpose is not to extract energy from the incident wave but whose presence creates a scattered field, the quantity is known as the radar cross section (RCS) of a target. Since the target will scatter the incident power in various directions, the RCS is defined as the equivalent area that intercepts that amount of incident power and scatters isotropically in all directions to produce the same scattered power density at the radar receiver [20]. In equation form,

$$RCS(\sigma) = 4\pi r^2 \frac{S_r}{S_t}$$

where S_r is the scattered power density at the radar receiver, S_t is the incident power density at the target, and r is the distance between the radar and the target. The RCS is a far-field parameter used to measure the detectability of the target by radar or characterize the scattering properties of a radar target. It depends on the geometry, orientation, and constituent of the target, the angles of incidence and reflection, as well as the frequency and polarization of the incident wave [21].

It is true that the RCS of a target does not have a simple relationship to the physical area. However, the larger the target size, the larger the cross section is likely to be [22]. In addition, if the target shapes more with round corners, without flat or concave surfaces (ex. sphere), and uses material treatment in flare spots, the RCS will be lower. For complex-shaped targets, the RCS will depend on the direction as viewed by the radar. In general, the RCS in vital sign detection is desired to be higher, which means the magnitude of the wave backscattered off the moving target is higher.

The RCS of complex targets is complicated functions of the observing aspect, the radar polarization, and the operating frequency. A measurement of RCS of a man from different aspects using two types of polarizations was reported, and it shows the RCS difference between two polarizations decreased as frequency increased [23]. For human vital sign sensing, the RCS of human subject also draws many attentions and lots of effort have been contributed to the studies of body modeling, orientation, and positioning. Since human body type, size, and weight varied from subject to subject, an RCS evaluation of body of revolution was presented and the modeling of a human torso was also demonstrated [24].

An X-band life-detection system for detecting a human subject lying on the ground and sitting behind a wall was published in 1980s [25]. Recently, a technique for measuring the cardiopulmonary effective RCS of a human subject was reported [26]. It shows the measurement of the RCS of a human in different orientations with 2.4 and 5.8 GHz operating frequencies, and measured the displacement magnitude from different positions. The results show that the displacement magnitude was smaller in prone position compared to supine position, while the RCS was higher in prone than in supine position. An evaluation of vital sign sensing with different human subjects was also presented [27]. It compares ten human subjects with different ages, heights, and weights in four lying positions and two sitting positions. The result shows that different body types (physical) and breathing behaviors also complicated the respiratory detection whether they were lying, sitting, breathing normally, or holding their breath.

While studies of test subjects show advantages on noncontact clinical sensing and possible orientation identification, there are also researchers focusing on different test objectives. Experiments detecting vital sign signal from four sides of a human body showed that the best performance for HR detection was measuring from the back side of a human body [28]. A recently reported interferometric driver health monitoring system embedded the radar sensor in the seat to monitor the driver's vital sign [29]. A sleep sensing system that detects sleep-related physical and physiological variables also shows its capability to classify sleep and awake stages and evaluates sleep quality. Both applications require orientation consideration to achieve optimum position for vital sign detection. On the other hand, researchers also compared radar- and camera-based vital sign sensors in different body orientations, and the results show the radar has better HR accuracy when measuring from the front and the back of a body [30].

Despite that the detection is based on the reflected radar signal which could be easily affected by the surroundings or test subjects and the analysis is complicated, the advantages of remote sensing as compared to existing techniques still trigger researchers to explore radar applications in different fields.

6.2.2 Measurements of laboratory rat

Rats are the most commonly used organism in animal testing. It is not only because they and humans are both mammals, but also because their anatomy are similar to highly developed vertebrates such as humans. Externally, rat's body can be divided into several regions: head, neck, chest, belly, area where legs attach, and tail [31]. Rat's structural organization is common with all other vertebrates. Their skeletal system supports body and protect for their vital organs. Whether rats crawl, stand, or sleep (curled up), their back (where the vertebrae column and ribs are) are always the largest surface of its body that can be seen.

On the other hand, the internal anatomy of rat is separated into head, neck and thoracic, and abdominal regions [32] as shown in Figure 6.2. The heart is centrally located in the thoracic cavity where the lungs reside. During inhaling, the muscles between the ribs contract and raise the ribs, which increases the size of the chest. During exhaling, the rib muscles relax and decrease the size of the chest. The rat's heart is small but functions as human's, and its basic function is circulating blood to the lungs and body cells. For an adult laboratory rat, the average RR and HR are 85 breaths/min (BR/min) and 330–480 beats per min (BPM), respectively.

In animal testing, an arterial catheter is placed in femoral artery of the rat. The arterial pressure signal is used to measure the BP. By identifying peaks in the arterial pressure corresponding to the time of peak systolic arterial pressure, the heartbeat can be calculated. In radar measurements, the respiratory and cardiac activities are captured through radar-detected movements.

The aim of radar experiments is to evaluate the effectiveness of using radar to measure from different positions in animal testing. In order to study the influence of radar positioning and body orientations, two groups of animals were included in experiments: conscious rats and anesthetic rats. By separating rats into two test

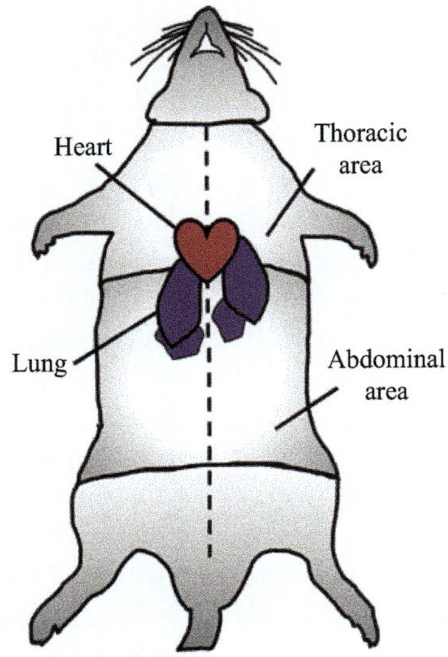

Figure 6.2 Anatomy of rat—thoracic organs. The heart is located in the center of the thoracic portion, and the large size lungs can be also observed

conditions, the consequences of reflected waves due to body orientations in conscious and unconscious agents can be revealed.

In the conscious experiments, two models of rats were used: normotensive Wistar Kyoto (WKY) rats and spontaneously hypertensive rat (SHR). SHR is one of the most common animal models used to study genetic hypertension in humans, and WKY rats have been used as control animals in studies of SHR [33]. Since WKY and SHR have very different HR and RR, it is easier for us to distinguish their signals. In the laboratory setting, cages are set up as in the animal care environmental settings. Three experiments were performed with radar:

1. one WKY recording from cage side, one SHR in the next cage;
2. one SHR recording from cage side, two WKYs in the next cage;
3. one WKY recording from cage top, one SHR in the next cage.

For the first measurement of radar on the cage side, one WKY was housed in the cage with radar monitoring, and one SHR in the next cage was sleeping. The radar was placed in the front of the cage which was closely attached to the sidewall of the cage. The experimental setup is shown as Figure 6.3. Whether the rats in each cage were moving or not, the radar was kept in the same position during the experiment, and signals were recorded for 5 min. The received signals were sampled by a data acquisition (DAQ) module and sent to a laptop for signal processing using MATLAB®.

The radar-recorded time-domain raw data and corresponding spectrum are shown in Figure 6.4. During the experiment, both WKY and SHR were very stable and did not have noticeable body movements. It can be observed from

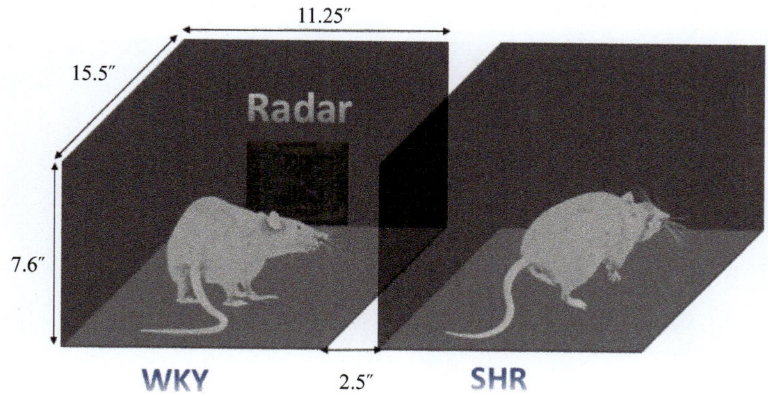

Figure 6.3 First conscious rat measurement: one WKY with radar recording from the cage side, and one SHR housed in the next cage

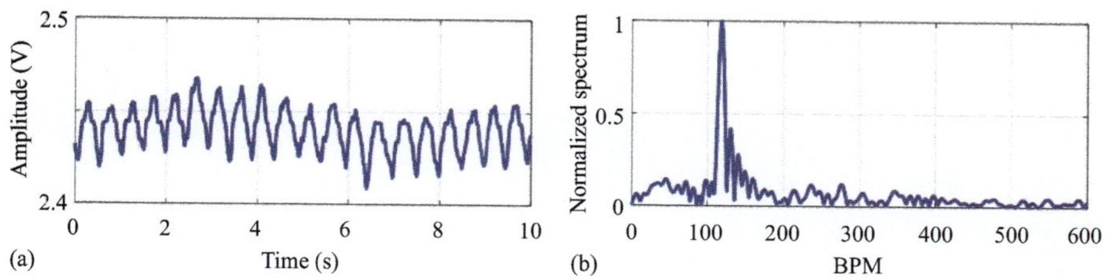

Figure 6.4 Radar-recorded signal of WKY from the cage side. (a) 10-s time-domain waveform. (b) Normalized spectrum

Figure 6.4(a) that a 10-s raw data shows a very slow movement possibly due to natural body drifting which would cause a peak close to DC. Therefore, a Butterworth band-pass filter was used to remove the DC and the spectrum was normalized to unity as shown in Figure 6.4(b). The radar-detected RR is 118 BR/min, HR is 277 BPM, and the respiration displacement is calculated to be 0.17 mm in this 10-s time window.

In the second measurement of radar on the cage side, one SHR was housed in the cage with radar monitoring, and two WKYs housed in the next cage were playing. The experimental setup was shown in Figure 6.5. During the experiment, even though the SHR was sleeping, the body movements from two WKYs in the next cage still caused interferences in the radar-detected signal. The measurement results are shown in Figure 6.6. It can be observed from Figure 6.6(a) that the waveform in this 10-s time window was not only the respiratory movements from one SHR, but also modulated by some random signals from two WKYs. The same Butterworth band-pass filter was used, and the spectrum is shown in Figure 6.6(b). The radar-detected RR is 70 BR/min, HR is 245 BPM, and the calculated respiration displacement is around 0.62 mm. Compared to WKY, SHR has lower RR and HR, and larger respiration displacement associated with slower respiratory movements.

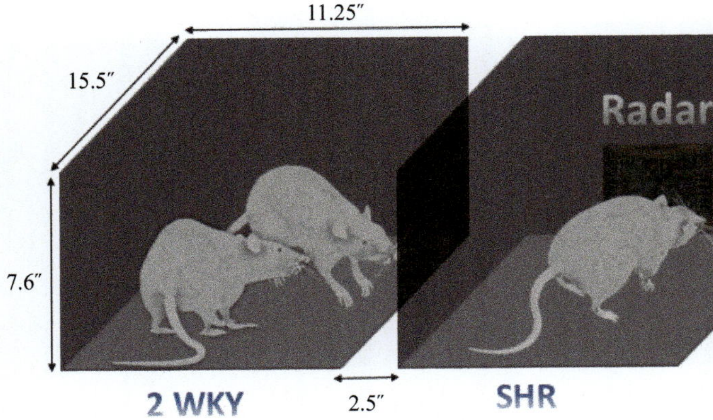

Figure 6.5 Second conscious rat measurement: one SHR with radar recording from the cage side, and two WKYs housed in the next cage

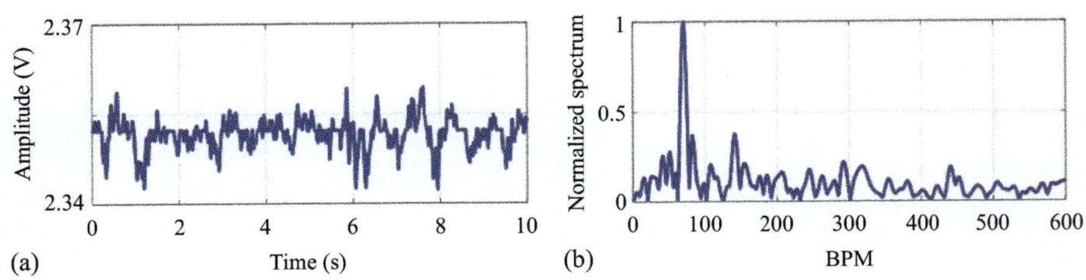

Figure 6.6 Radar-recorded signal of SHR from the cage side. (a) 10-s time domain waveform. (b) Normalized spectrum

In the third measurement, the same WKY was used to examine the effect of radar positions. In the laboratory setting, one WKY was housed in the cage with radar monitoring and one SHR was housed in the other cage. As illustrated in Figure 6.7, the radar measured from the cage top. During the experiments, both WKY and SHR in their cages were sleeping. Therefore, the radar detected signals were less affected by the movements from the other cage.

The measurement results are shown in Figure 6.8. It can be observed from Figure 6.8(a) that the signal strength measured from the top is smaller than the side recording in Figure 6.4(a). It is likely due to the distance between the rat and radar. If the side-mounted radar is closer to the rat compared to the top-mounted radar, the detected signal from the side will be larger. On the other hand, if the rat resides at the center of the cage, which is closer to the cage center top and away from the cage side, the detected signal from the top will be larger. Figure 6.8(b) shows the spectrum. The radar-detected RR is 118 BR/min, HR is 287 BPM, and the calculated respiration displacement is around 0.52 mm. Although the respiration displacement when measured from the top (radar pointing to the rat's back) is larger, comparing Figures 6.4 and 6.8, the HR peak is also clearer when measured from the

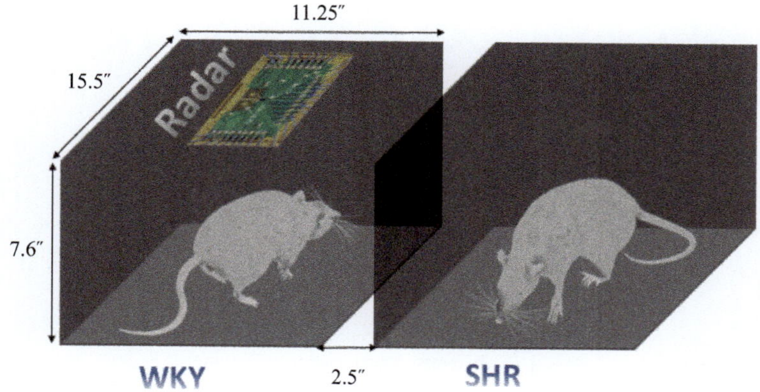

Figure 6.7 Third conscious rat measurement: one WKY with radar recording from the cage top, and one SHR housed in the next cage

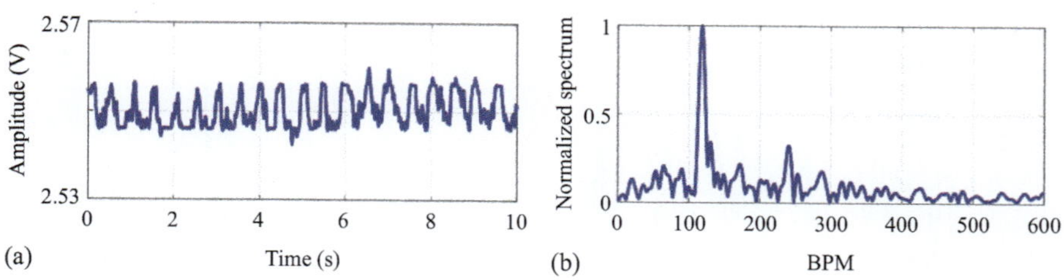

Figure 6.8 Radar-recorded signal from the cage top for one WKY. (a) 10-s time domain waveform. (b) Normalized spectrum

top. The better performance of measuring HR with radar pointing to the rat's back versus other sides is similar to the previously reported results of radar measurements from four sides of human body [28].

Figure 6.9 shows the measurements of conscious rats under three test conditions. Comparison of the time-domain waveforms shows the largest signal strength is observed by placing radar on the side of the cage. Nevertheless, the largest detectable respiration displacement and clearest HR peak can be observed by placing radar on the top in three experiments.

In the anesthesia experiments, an adult rat under anesthesia was housed in a cage and measured by radar. In the experiments, anesthesia of a healthy adult rat was induced with urethane (1.3 mg/kg), and a radar was placed in the front of the cage with antenna pointing to the rat for measuring vital signs. The baseband signal was then sent through a DAQ module to a laptop recorded with LabVIEW user interface and using MATLAB for signal processing analysis. Four measurements were performed in sequence using radar:

1. rat in supine position (face-up), radar on the top pointing to belly;
2. rat in supine position, radar on the side pointing to body side;

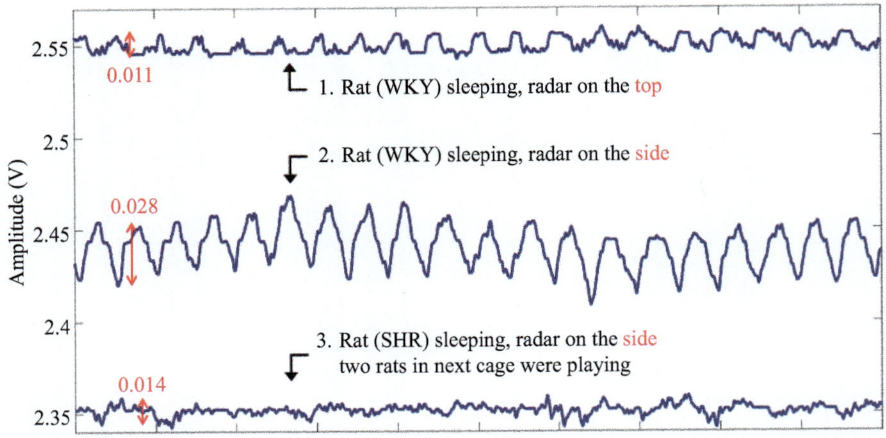

Figure 6.9 Radar-recorded time-domain waveforms of conscious agents

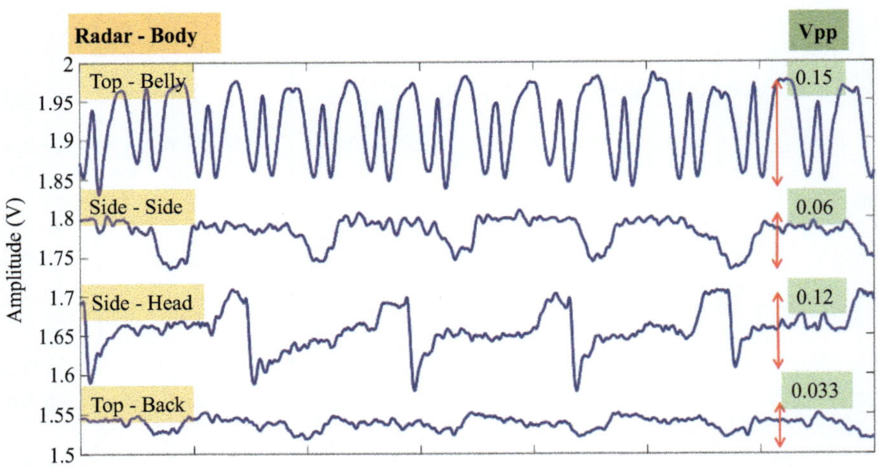

Figure 6.10 Radar-recorded time-domain waveforms of anesthetic agents from four sides of body: belly, side, head, and back (time-domain)

3. rat in supine position, radar on the side pointing to head;
4. rat in prone position (face-down), radar on the top pointing to back.

During the measurements, the anesthetic rat was lying down on the back. The measurement results are shown in Figure 6.10. Comparing the time-domain signals recorded from different body orientations, the strongest signal strength was measured from the top and pointed to its belly which is 0.15 Vpp (peak-to-peak voltage), while the smallest signal strength was measured from the top but pointed to its back. Figure 6.11 illustrates the test scenarios and the radar-measured results. It confirmed an important feature that the radar-measured displacement will not be

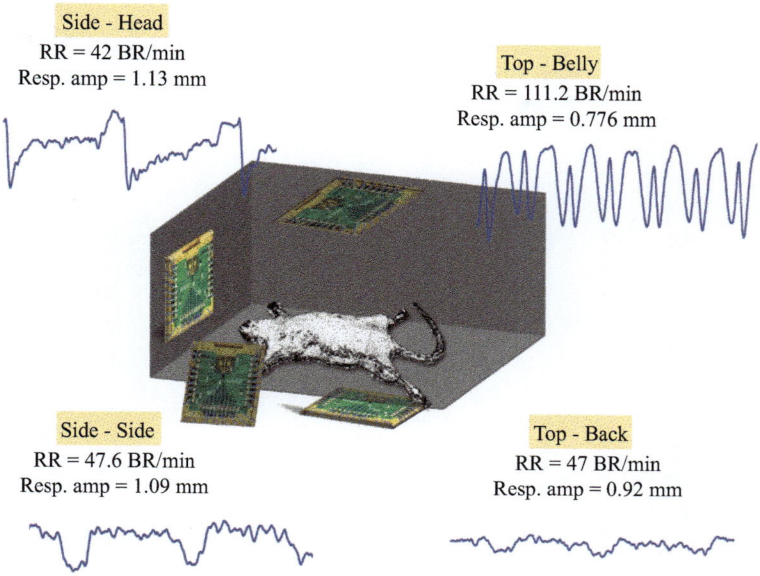

Figure 6.11 Measurements of anesthetic rat from four sides of body: belly, side, head, and back

affected by the signal strength. In other words, the detected displacement is associated with the vibration movement but not the signal strength [34]. Therefore, no calibration is needed when the radar-to-subject detection distance changes during the experiment.

In Figure 6.12, the spectrums of the 7-s time-domain signals of the four radar-to-body combinations are shown. The detected RR peak and radar-measured respiration displacement are also indicated in the figures. As shown in Figure 6.12(a) and (b), the top-to-belly (radar-to-body) measured RR is 111.2 BR/min with calculated respiration displacement equals to 0.776 mm, and the side-to-side measured RR is 47.6 BR/min with calculated respiration displacement equals to 1.09 mm. The results show that the amplitude of respiratory movement is larger with slower RR. It is because the shorter breath results in faster RR and smaller chest-wall movement. Comparing Figure 6.12(b), (c), and (d), with the similar RRs, the calculated respiration displacements are also similar, which means the extracted movement is not affected by body orientations when the rat is under anesthesia. However, the respiration displacement changed significantly with body orientations when the rat was in consciousness. On the other hand, the top-to-belly position shows a strong heartbeat signal which is due to the rat was just falling into unawareness and the RR was still high. After a few minutes, the RR decreased as the rat was deepened into anesthesia, which can be observed from Figure 6.12(b)–(d).

Despite that the single-patch antenna used in the 60-GHz radar is not a high-directivity antenna, the radar can still pick up the vital sign signals from the target in the animal care environmental setting. Table 6.1 shows the summarized

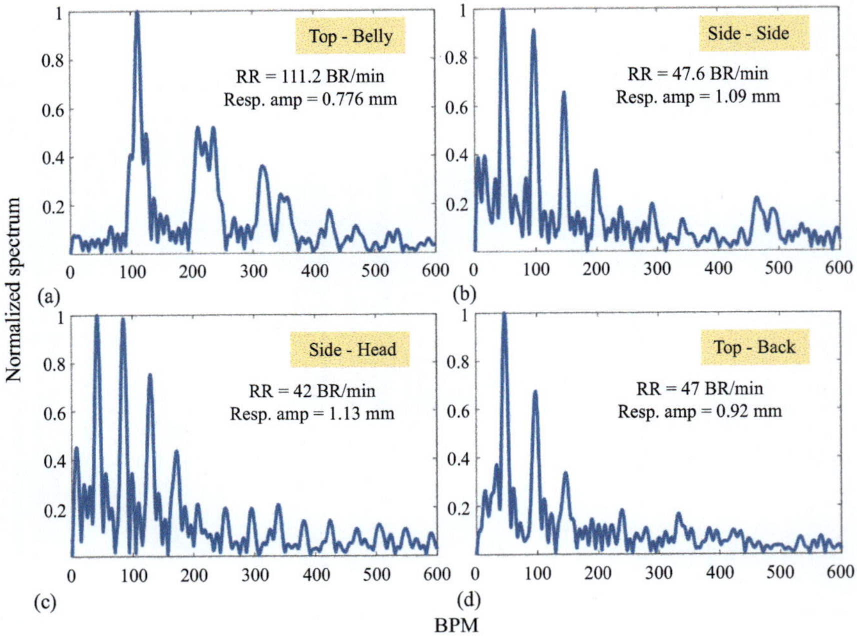

Figure 6.12 Spectrum of an anesthetic rat with radar measured from (a) belly, (b) side, (c) head, and (d) back

Table 6.1 Summary of measurements with different radar positions and body orientations

Radar position on the cage	Antenna pointing direction to the rat	Testing condition	RR (BR/min)	Respiratory displacement (mm)
Side	—	Consciousness	118	0.17
Side	—	Consciousness	70	0.62
Top	—	Consciousness	118	0.52
Top	Belly	Anesthesia	111.2	0.776
Side	Body side	Anesthesia	47.6	1.09
Side	Head	Anesthesia	42	1.13
Top	Back	Anesthesia	47	0.92

measurements. It is found that in conscious rats (free-running) experiment, the signal strength was small when radar was placed on the top, but the measured respiration displacement is larger as compared to placing radar on the cage side. On the other hand, for measurements of anesthetic rat lying on the ground, the largest signal strength was observed when radar was placed on the cage top and pointing to its belly, and the smallest signal strength was observed when radar was placed on the top but pointing to its back. It also proved that radar-measured displacement will not be affected by the signal strength, so no calibration is needed. Compared with the results of conscious rats, the extracted movement is not much affected by body orientations when the rat is under anesthesia. Finally, for both laboratory rats and humans, the best performance for radar heartbeat detection is measured from the back of the body.

6.3 Signal processing for cardiorespiratory movement

6.3.1 Methodology of displacement acquisition

A cardiorespiratory motion consists of respiration and heartbeat signals, and can be modeled as a two-tone periodic movement $x(t)$, expressed as

$$x(t) = m_r \sin(2\pi f_r t) + m_h \sin(2\pi f_h t)$$

where f_r is the respiration frequency, f_h is the heartbeat frequency, and m_r and m_h are the displacements due to respiration and heartbeat, respectively. By mixing received signal with part of transmitted signal as reference, the phase-modulated baseband signal is approximately

$$B(t) \approx \cos\left[\frac{4\pi x(t)}{\lambda} + \Delta\phi\right]$$

If the displacement m is comparable to the wavelength λ, the small-angle approximation cannot be applied. Since the quadrature I/Q radar is used, by applying the complex signal demodulation (CSD) technique [35], the detected signal was generated by combining I and Q outputs. However, the nonlinear transfer function produces harmonics. As a result, the baseband signal consists a number of harmonics of the fundamental frequency with amplitudes expressed with a series of the Bessel functions.

$$\begin{aligned} B(t) &= I(t) + jQ(t) \\ &= \sum_{p=-\infty}^{\infty} \sum_{q=(x-f_h p)/f_r}^{\infty} J_p(a_h) J_q(a_r) \cdot e^{j2\pi xt} \cdot e^{j\phi} \end{aligned}$$

where $J(a)$ is the first kind Bessel function, $a_r = 4\pi m_r/\lambda$ and $a_h = 4\pi m_h/\lambda$, p and q are integers that satisfy $x = f_h p + f_r q$, and x represents the harmonic frequency.

Amplitude of harmonics in baseband spectrum is determined by residual phase ϕ, wavelength λ, and displacement of vibration m. The term $e^{j\phi}$ can be eliminated, since quadrature architecture is used, and the effect of residual phase on amplitude can be neglected. Figure 6.13 illustrates the relationship of the displacement and harmonics. For a fixed displacement, different orders of Bessel function correspond to different values, and result in different amplitudes shown on the detected spectrum. Theoretically, for a purely sinusoidal motion, the displacement can be determined from the ratio of harmonics; for a complex periodic motion, multiple harmonic pairs can be used to create more equations to solve the unknown amplitudes of different frequency components.

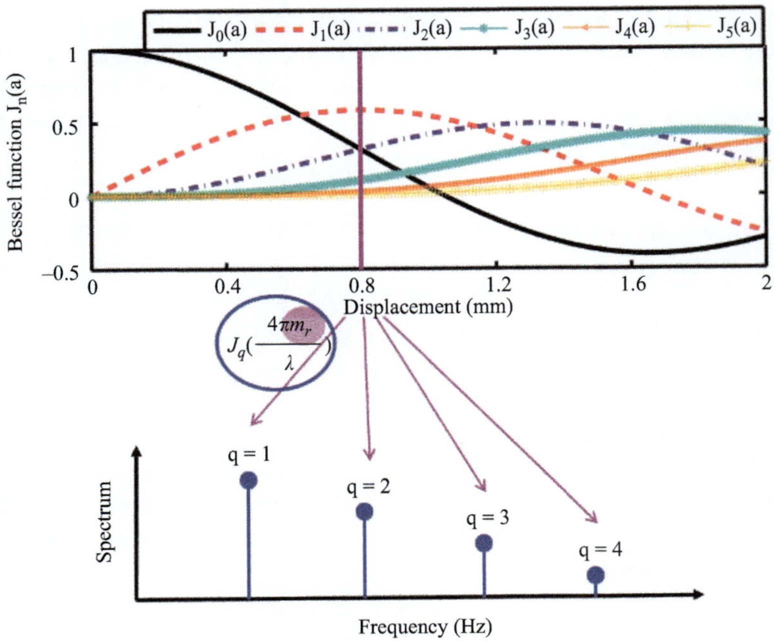

Figure 6.13 Illustration of the vibration displacement, Bessel functions, and the generated harmonics

For the cardiorespiratory motion, the harmonics contain frequency components from both RR and HR, and the amplitude H_x can be expressed as

$$H_x = \left| \sum_{p=-\infty}^{\infty} \sum_{q=(x-f_h p)/f_r}^{\infty} J_p(a_h) J_q(a_r) \right|$$

Harmonics shown in the baseband spectrum can be grouped into two categories [36]: the respiration and its harmonics, and the frequency mixing products of RR and HR, as labeled in Figure 6.14. The mixing products involving higher harmonics of HR will fall outside of the baseband of interest and are therefore not considered here. In order to obtain a more accurate harmonic ratio that contains the strongest frequency components from RR, the first selected pair for respiration displacement extraction is the H_{RR} and H_{2RR}, and the ratio is shown as

$$\frac{H_{RR}}{H_{2RR}} = \frac{|J_0(a_h)J_1(a_r)|}{|J_0(a_h)J_2(a_r)|} = \frac{|J_1(a_r)|}{|J_2(a_r)|}$$

The two harmonics are chosen because they contain the information involving different $J_q(a_r)$ with the same $J_p(a_h)$. Therefore, the same $J_0(a_h)$ can be canceled out, leaving the $J_q(a_r)$ only for calculating m_r. The selection of harmonics with

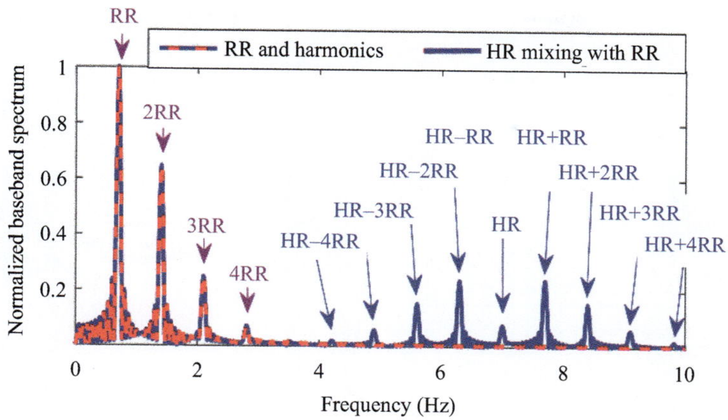

Figure 6.14 Baseband spectrum with identified respiration demodulation-generated (DG) harmonics and mixing products between HR and RR. [2017] IEEE. Reprinted, with permission, from Ref. 36

$x > 3RR$ may be too small to be useful, and the three strongest harmonics are sufficient for displacement extraction and can be expressed as

$$H_{RR}:H_{2RR}:H_{3RR} = |J_1(a_r)|:|J_2(a_r)|:|J_3(a_r)|$$

Heartbeat displacement is extracted by utilizing the stronger mixing products between HR and respiration harmonics. Three mixing products, H_{HR-RR}, H_{HR-2RR}, and H_{HR-3RR}, are chosen and paired up with the respiration DG harmonics for calculating m_h, as shown in the following:

$$\frac{H_{RR}}{H_{HR-RR}} = \frac{H_{2RR}}{H_{HR-2RR}} = \frac{H_{3RR}}{H_{HR-3RR}} = \frac{|J_0(a_h)|}{|J_1(a_h)|}$$

Another possible source of harmonics is the nonlinearity of radio-frequency and analog circuits in receiver. These harmonics, however, can be controlled and eliminated by keeping signal level small enough to stay within the linear region. In order to take advantage of the nonlinear Doppler phase demodulation effect, we should know that the detection accuracy depends on the accuracy of harmonic ratio, which means the ratio of m/λ will be important [28]. The larger the ratio, the stronger the nonlinear effect, resulting in more harmonics generated on the spectrum. At a fixed wavelength, if the displacement is too small compared to the wavelength, the detection accuracy would be degraded because it might not be able to generate sufficiently strong harmonics for obtaining the ratio.

Experiment that uses a 60-GHz CMOS radar for measuring a laboratory rat is depicted in Figure 6.15. An arterial catheter (FC-M-37L, Braintree Scientific Inc., Braintree, MA, USA) was placed in femoral artery of an anesthetic adult rat and then the rat was placed in a whole-body plethysmograph (Chamber volume 4 L; Buxco Electronic Inc., Wilmington, NC, USA). The arterial catheter was threaded

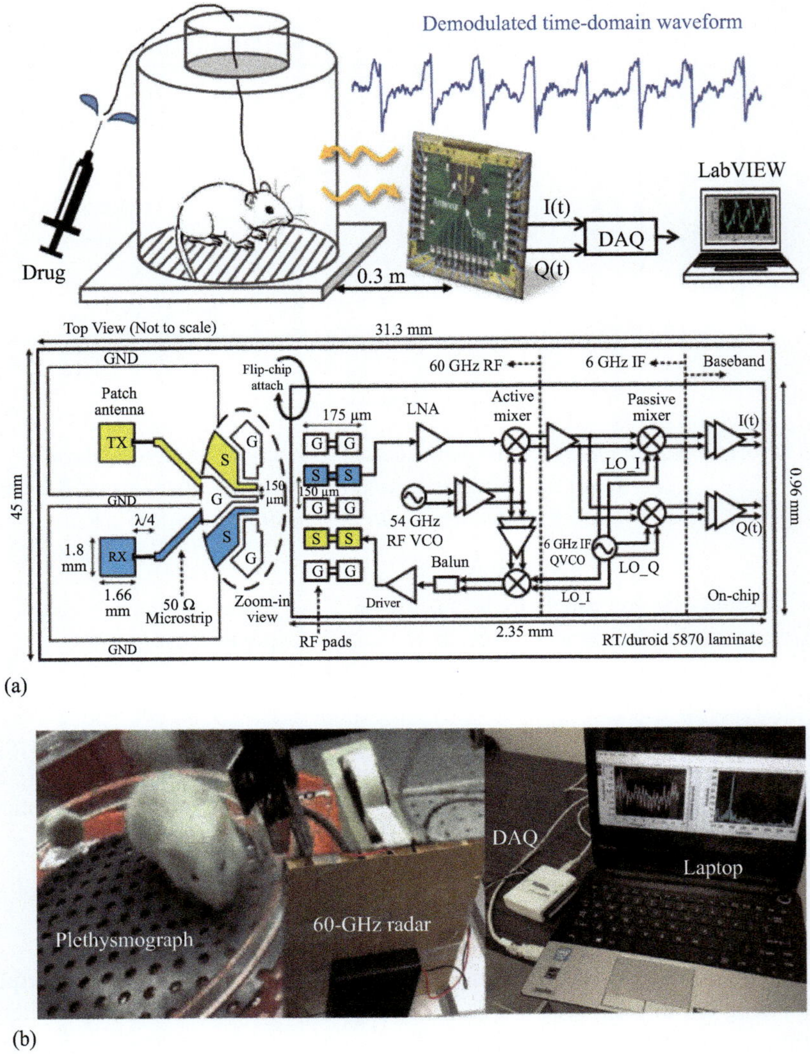

Figure 6.15 Experimental setup. (a) Measuring a laboratory rat in a plethysmography chamber using a 60-GHz radar connected with DAQ module and a laptop. (b) Photograph of the measurement setup with radar facing the rat. [2017] IEEE. Reprinted, with permission, from Ref. 36

through the top of the chamber and connected to a calibrated pressure transducer (Stoelting, Woodale, IL, USA). Oxygen (100%) was delivered to the chamber at a rate of 2 L/min. The plethysmograph was fitted with a pressure transducer (TRD5700Buxco Electronic Inc.). The arterial pressure signal and analog output from the plethysmograph pressure transducer were recorded online (sampling frequency 200 Hz) via a DAQ interface [Power1404, Cambridge Electronic Design (CED), Cambridge, UK] connected to a PC equipped with a complimentary data analysis software system (Spike2, CED, Cambridge, UK). Off-line Spike2 subroutines were used to identify peaks in the arterial pressure and plethysmograph pressure traces, which corresponded to the time of peak systolic arterial pressure and peak inspiratory pressure, respectively. HR and RR were calculated as 60/(peak-to-peak interval). Respiratory movement, BP, and HR are simultaneously recorded with Spike2 (200 Hz) for reference. This commercially available instrumentation cannot measure the cardiopulmonary displacement, but can only measure HR and RR.

The radar was placed at 0.3 m from the chamber to detect the rat's movements, and the received quadrature signals were sampled by a National Instruments DAQ module and then sent to a laptop for signal processing. Anesthesia was induced with urethane (1.3 mg/kg). Two drug tests with dropped HR (phenylephrine) and raised HR (atropine) were performed to measure an anesthetic rat using radar to extract both frequencies and displacements of its respiration and heartbeat using the method previously described.

Since the rat body needs some settling time to react after drug injection and the raised and dropped RRs varied with time significantly, the time window was shortened from default 15 to 5 s to see the instant changes. Figures 6.16 and 6.17 show the respiratory movement variations with 200-s total observation time, and the drug-injected points were labeled. The reflexively slowed HR and RR were induced with phenylephrine, which raised the BP and reflexively decreased HR. It can be observed from Figure 6.16(b) that RR dropped from 100 to 75 BR/min after injection and stayed relatively unchanged for 50 s. During the 50-s period, the displacement fluctuated and increased from minimum 0.9 to maximum 1.092 mm (20% increase). After that, the RR slowly recovered to its baseline with constant fluctuations in displacement.

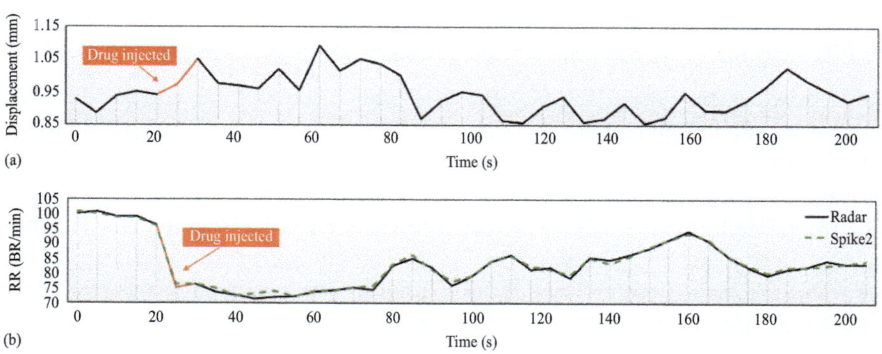

Figure 6.16 Phenylephrine-induced (dropped HR). (a) Respiratory displacement variation. (b) RR variation. [2017] IEEE. Reprinted, with permission, from Ref. 36

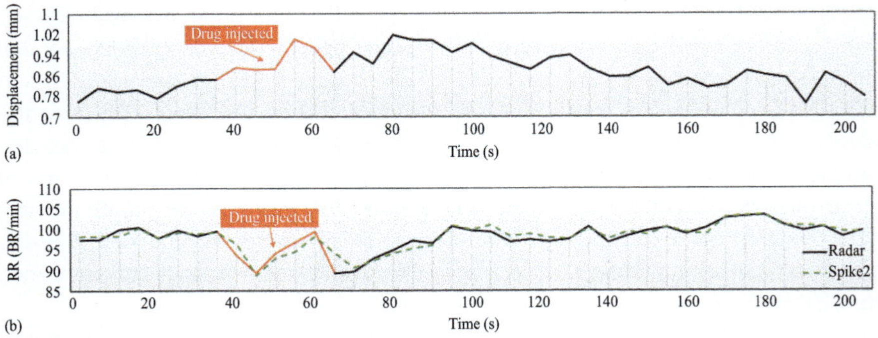

Figure 6.17 Atropine-induced (raised HR). (a) Respiratory displacement variation. (b) RR variation. [2017] IEEE. Reprinted, with permission, from Ref. 36

In Figure 6.17, the raised HR was induced with atropine, which blocked parasympathetic drive to the heart, resulting in raised HR and increased RR. The drug took around 20 s to activate as color labeled. RR slowly sped up when the drug became effective, and respiratory displacement slowly decreased by 26% following the raised RR. It was found that the HR amplitude became weaker if drug was induced in the experiment. The drug effect makes respiratory movement much larger, which suppresses the HR amplitude in the detected spectrum. In addition, the large respiratory displacement will also make J_0 in the HR-mixing products close to the zero-crossing point, which further degrades the HR amplitude in spectrum.

Two drug tests are performed to verify the monitoring function with controlled raised and dropped HR. By observing the time periods before, during, and after drug injection, results of rat's physiological response to drug reveals the drug efficacy and body reaction over time. The measured cardiorespiratory variations can provide useful information without using implant devices, and the implementation could be an improvement over current invasive methods.

6.3.2 Adaptive harmonics spectrum cancelation for HR

With growing interests of Doppler radar sensing, a number of detection methods have been contributed from many researchers in this field. The unfolding of advanced motion detection focuses on enhancing the desired signal reflected from the target while eliminating noise and interference caused by unwanted reflections or artificial motions. Techniques that have been developed to improve the detection accuracy and system performance including random body movement cancelation with multiple radars detect from different angles [37], a spectral estimation using RELAX algorithm [38], and a time-frequency analysis [39,40].

Monitoring the vital signs of small animals by the millimeter-wave radar causes strong nonlinear effect in the detected spectrum due to the Doppler phase demodulation, which degrades the accuracy for HR detection. The main obstacle for noncontact, noninvasive vital sign detection has been accurate and consistent measurement of HR without interference from respiration movements. It is true that the RR and HR can be identified as long as they are clearly shown in the spectrum. However, in some cases, the heartbeat is weak or the respiratory movement is too strong to overwhelm the HR detection, or the HR peak is overlapped by the respiration harmonics in the detected spectrum. All these possibilities make the HR measurement challenging.

The working principle of comb filter with adaptive notch gain that estimates the periodic noise signal by using the noise correlation to adjust the notch gain was reported by [41]. Using this concept, an adaptive spectrum cancelation method for removing respiration harmonics is proposed so that the unwanted harmonic interferences can be selectively canceled out from the detected spectrum as long as the respiration information is known. The method is based on Fourier transform of the detected signal that utilizes the harmonics generated in the spectrum due to the demodulation method to extract the displacement of respiratory movement by the ratio of harmonic amplitudes [34,42]. Using the extracted RR and calculated displacement, the cancelation factor can be determined and constructed as shown in Figure 6.18, which consists of multiple notch frequencies at respiration and its harmonics [43]. After applying the filter in the original spectrum, the respiration harmonics can be removed without removing the heartbeat signal.

The detected baseband signal in I and Q channels are combined by the CSD method expressed as

$$
\begin{aligned}
B(t) = I + jQ &= \cos\left[\frac{4\pi x_h(t)}{\lambda} + \frac{4\pi x_r(t)}{\lambda} + \phi\right] \\
&= J_0(a_h)\sum_{q=-\infty}^{\infty} J_q(a_r)\cdot e^{j2\pi(f_r q)t}\cdot e^{j\phi} \\
&\quad + J_1(a_h)\sum_{q=-\infty}^{\infty} J_q(a_r)\cdot e^{j2\pi(f_r q+f_h)t}\cdot e^{j\phi} + \cdots
\end{aligned}
$$

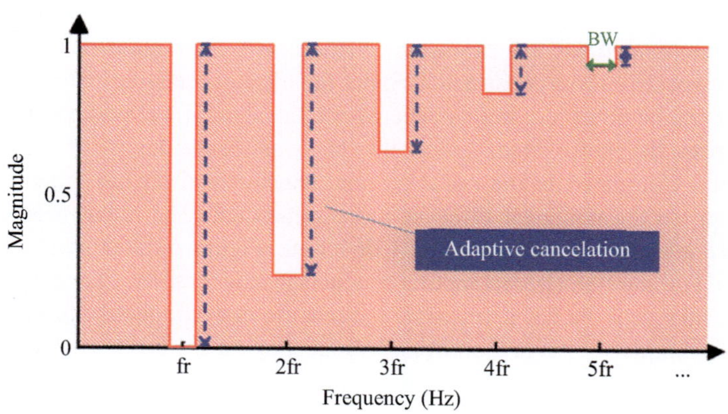

Figure 6.18 Adaptive spectrum cancelation method for removing respiration harmonics. [2016] IEEE. Reprinted, with permission, from Ref. 43

After taking the fast Fourier transform of the baseband signal, f_r can be determined. Since the respiratory movement is not an ideal single tone or symmetric movement, the first and second harmonics are chosen for m_r extraction which can be expressed as

$$m_r = \frac{J_0(a_h)J_1(a_r)}{J_0(a_h)J_2(a_r)} = \frac{J_1(a_r)}{J_2(a_r)}$$

Using the respiration spectrum to design a comb notch filtering function, the notch attenuation at each notch frequency is adaptive to the respiration displacement. After applying the spectrum-based cancelation method in the detected signal, the output of the filter can be expressed as

$$Y(t) = J_1(a_h) \sum_{q=-\infty}^{\infty} J_q(a_r) \cdot e^{j2\pi(f_r q + f_h)t} \cdot e^{j\phi} + J_2(a_h) \sum_{q=-\infty}^{\infty} J_q(a_r) \cdot e^{j2\pi(f_r q + 2f_h)t} \cdot e^{j\phi} + \cdots$$

The RR harmonics are removed, and only the HR mixing products with RR remain. The method will not remove the heartbeat even the HR is overlapping with the RR harmonics. The bandwidth of the filter stop-band depends on the observation window. If the observation interval of the signal is limited, the spectrum resolution will be low and suffer from spectral leakages. As a result, the bandwidth can be adjusted following the resolution to filter out the unwanted harmonics.

Measurements of conscious rat, anesthetized rat, and drug-induced hypertensive (high BP) rat with and without the spectrum cancelation method are compared to verify the proposed technique. Spectrums before and after filtering under three test scenarios, i.e. consciousness, anesthesia, and drug-induced dropped HR (hypertensive) rat under anesthesia, are shown in Figures 6.19(a), 6.20(a), and 6.21(a), respectively. As can be seen from Figure 6.19(a), for conscious rat experiment, the radar-measured RR is 1.58 Hz and the calculated m_r is 0.742 mm. Using the extracted RR and calculated displacement from radar-recorded data, the canceling spectrum was constructed. Spectrum after cancelation with HR identified at 6.1 Hz can be observed. In this experiment, the signal level was not good enough to suppress the noise floor since the rat was conscious with random body movement. Compared to the conscious rat, anesthetized rat has larger chest-wall movement, so the nonlinear effect is more significant. The radar-measured RR is 1.632 Hz and the calculated m_r is 1.061 mm. Figure 6.20(a) shows the radar-measured HR at 6.088 Hz. HR mixing products in low frequency are enlarged by the spectral leakage of the RR harmonics, so there are more frequency mixing products in the low-frequency range than in high-frequency range. For hypertensive rat, the radar-measured RR is 1.213 Hz and the calculated m_r is 0.991 mm. In Figure 6.21(a), the HR peak cannot be identified from the original spectrum. After cancelation, the HR appeared at 3.638 Hz, which is exactly covered by the third harmonic of RR, and the frequency mixing products in high-frequency spectrum can be seen.

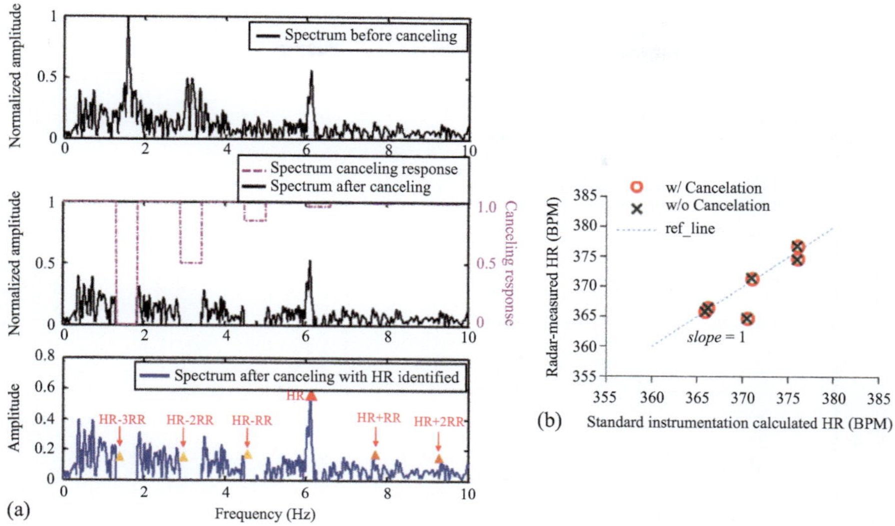

Figure 6.19 Conscious rat with random body movement. (a) Measurement spectrum. (b) Correlation diagram. [2016] IEEE. Reprinted, with permission, from Ref. 43

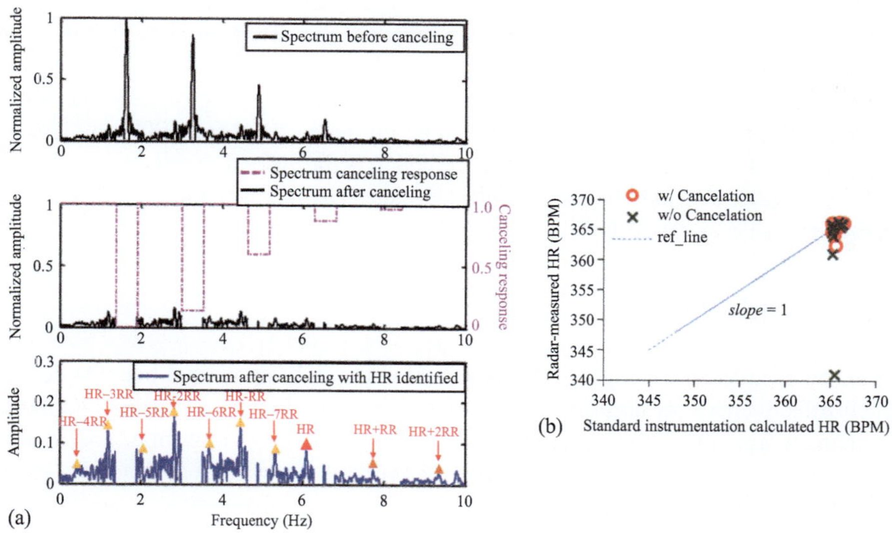

Figure 6.20 Rat under anesthesia. (a) Measurement spectrum. (b) Correlation diagram. [2016] IEEE. Reprinted, with permission, from Ref. 43

The correlation diagrams of the instrumentation-calculated HR and radar-calculated HR are shown in Figures 6.19(b), 6.20(b), and 6.21(b) for conscious rat, anesthetized rat, and drug-induced hypertensive rat, respectively. The red circle and green cross marks are radar measurements with and without cancelation,

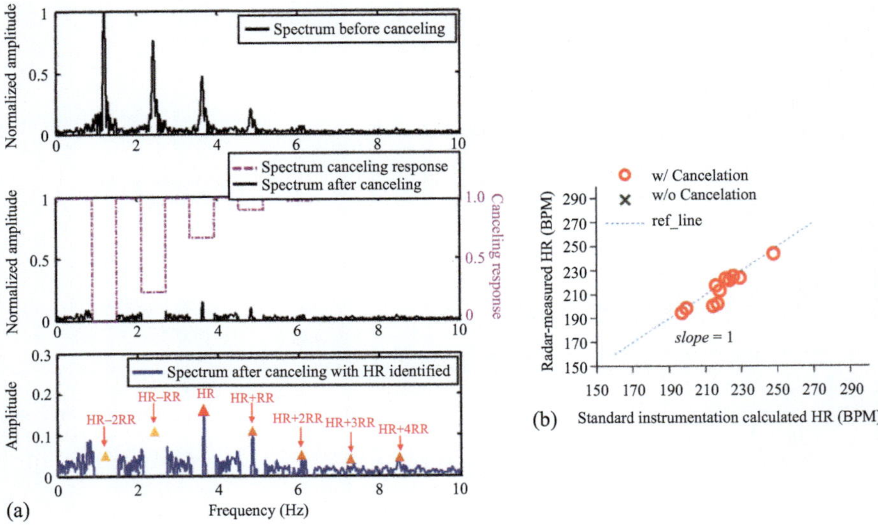

Figure 6.21 Hypertensive rat. (a) Measurement spectrum. (b) Correlation diagram. [2016] IEEE. Reprinted, with permission, from Ref. 43

respectively. Note that the green cross marks fall out of chart in Figure 6.21(b). In conscious rat experiment, the average error of HR detection is 0.39% with or without cancelation. Nevertheless, the error of HR detection in anesthetized rat experiment can be reduced from 0.82% to 0.19% after applying cancelation. Most importantly, the error of HR detection can be further reduced from 90% to 1.8% in drug-induced hypertensive rat experiment which is the case of overlapping HR and RR harmonics. The method successfully corrected the radar-measured HR that had large error due to strong interference from respiration harmonics.

6.4 Conclusion

The use of radar technologies for biomedical research has potential benefits due to their effectiveness for noninvasively monitoring physiological responses and more humane treatments of the living animals. It provides a useful tool for a more rapid, efficacious, and cost-effective extraction of animal's physiological data with great potential for clinical animal research, such as studying behavioral science (feeding rate or licking rate), monitoring the onset and progression of disease, and assessment of physiological responses to new treatments or drugs in a stress-free manner. For instance, monitoring sleeping respiratory rate in pets (dogs and cats) is very helpful in accessing for the incidence or reoccurrence of left-sided congestive heart failure (CHF). It is important for animals with common cardiac diseases to have early diagnosis of CHF by monitoring their sleeping respiratory rate. A self-injection-locked radar [44] has been used to monitor livestock animal's vital signs [45] to help farmers gauge an animal's overall health and identify potential issues in the early stage. Most recently, in addition to monitoring RR and HR, radar measuring the beat-to-beat BP [46] offers a great advantage of noninvasive respiratory and cardiac monitoring to scientists in future biomedical research.

References

[1] K. E. Quesenberry and J. W. Carpenter, Ferrets, Rabbits and Rodents: Clinical Medicine and Surgery. St Louis, MO: Saunders, Elsevier, 2004.

[2] National Research Council, Microbial Status and Genetic, Evaluation of Mice and Rats: Proceedings of the 1999 US/Japan Conference. Washington, DC, USA: Nat. Academies Press, 2000.

[3] M. E. Shinder and J. S. Taube, "Active and passive movement are encoded equally by head direction cells in the anterodorsal thalamus," J. Neurophysiology, vol. 106, no. 2, pp. 788–800, 2011.

[4] A. A. Steiner, V. F. Turek, M. C. Almeida, *et al.*, "Nonthermal activation of transient receptor potential vanilloid-1 channels in abdominal viscera tonically inhibits autonomic cold-defense effectors," J. Neurosci., vol. 27, pp. 7459–7468, 2007.

[5] M. P. Koeners, C. P. C. Ow, D. M. Russell, *et al.*, "Telemetry-based oxygen sensor for continuous monitoring of kidney oxygenation in conscious rats," Amer. J. Physiology, vol. 304, no. 12, pp. 1471–1480, 2013.

[6] K. P. McKeown and A. A. Shoukas, "Chronic isolation of carotid sinus baroreceptor region in conscious normotensive and hypertensive rats," Amer. J. Physiology, vol. 275, no. 1, pp. 322–329, 1998.

[7] J. C. Lin, "Noninvasive microwave measurement of respiration," Proc. IEEE, vol. 63, no. 10, p. 1530, 1975.

[8] B. K. Park, O. Boric-Lubecke, and V. M. Lubecke, "Arctangent demodulation with DC offset compensation in quadrature Doppler radar receiver systems," IEEE Trans. Microw. Theory Techn., vol. 55, no. 5, pp. 1073–1079, 2007.

[9] F. Wang, C. Li, C. Hsiao, *et al.*, "A novel vital-sign sensor based on a self-injection-locked oscillator," IEEE Trans. Microw. Theory Techn., vol. 58, no. 12, pp. 4112–4120, 2010.

[10] J. Kranjec, S. Begu, J. Drnovek, and G. Gerak, "Novel methods for non-contact heart rate measurement: a feasibility study," IEEE Trans. Instrum. Meas., vol. 63, no. 4, pp. 838–847, 2014.

[11] C. J. Gordon and J. S. Ali, "Measurement of ventilatory frequency in unrestrained rodents using microwave radiation," Respiration Physiology, vol. 56, pp. 73–79, 1984.

[12] D. Kropveld and R. A. F. M. Chamuleau, "Doppler radar device as a useful tool to quantify the liveliness of the experimental animal," Med. Biol. Eng. Comput., vol. 31, pp. 340–342, 1993.

[13] N. Hafner, J. C. Drazen, and V. M. Lubecke, "Fish heart rate monitoring by body-contact Doppler radar," IEEE Sensors J., vol. 13, no. 1, pp. 408–414, 2013.

[14] T. Zeng, C. Mottband, D. Molliconeb, and L. D. Sanforda, "Automated determination of wakefulness and sleep in rats based on non-invasively acquired measures of movement and respiratory activity," J. of Neuroscience Methods, vol. 24, pp. 276–287, 2012.
[15] A. A. Deliyannis, P. M. S. Gillam, J. P. D. Mounsey, and R. E. Steiner, "The cardiac impulse and the motion of the heart," Brit. Heart J., vol. 26, no. 3, pp. 396–411, 1964.
[16] G. Ramachandran, S. Swarnamani, and M. Singh, "Reconstruction of out-of-plane cardiac displacement patterns as observed on the chest wall during various phases of ECG by capacitance transducer," IEEE Trans. Biomed. Eng., vol. 38, no. 4, pp. 383–385, 1991.
[17] K. Mohri, T. Jinnouchi, and K. Kawano, "Accurate mechanocardiogram sensors using amorphous star-shaped core multivibrator combined with a magnet," IEEE Trans. Magn., vol. Mag-23, no. 5, pp. 2212–2214, 1987.
[18] G. Ramachandran and M. Singh, "Three-dimensional reconstruction of cardiac displacement patterns on the chest wall during the P, QRS and T-segments of the ECG by laser speckle interferometry," Med. Biol. Eng. Comput., vol. 27, no. 5, pp. 525–530, 1989.
[19] T. Kao, Y. Yan, T. Shen, A. Chen, and J. Lin, "Design and analysis of a 60-GHz CMOS Doppler micro-radar system-in-package for vital-sign and vibration detection," IEEE Trans. Microw. Theory Techn., vol. 61, no. 4, pp. 1649–1659, 2013.
[20] D. K. Cheng, Field and Wave Electromagnetics, 2nd ed., Boston, MA, Addison-Wesley Publishing Company, Inc., 1989.
[21] C. Balanis, Antenna Theory Analysis and Design. Hoboken, NJ: John Wiley & Sons, Inc., 2005, Ch. 2, pp. 110–112.
[22] M. L. Skolnik, Introduction to Radar Systems, 3rd ed., New York, NY: McGraw-Hill Inc., 2001.
[23] F. V. Schultz, R. C. Burgener, and S. King, "Measurement of the radar cross section of a man," IEEE Proceedings, vol. 46, pp. 476–481, 1958.
[24] T. Wu, "Radar cross section of arbitrarily shaped bodies of revolution," IEEE Proceedings, vol. 77, no. 5, pp. 735–740, 1989.
[25] K.-M. Chen, D. Misra, H. Wang, H.-R. Chuang, and E. Postow, "An X-band microwave life-detection system," IEEE Trans. Biomed. Eng., vol. 33, no. 7, pp. 697–701, 1986.
[26] J. E. Kiriazi, O. Boric-Lubecke, and V. M. Lubecke, "Dual-frequency technique for assessment of cardiopulmonary effective RCS and displacement," IEEE Sensors Journal, vol. 12, no. 3, pp. 574–582, 2012.
[27] J. Kuutti, M. Paukkunen, M. Aalto, P. Eskelinen, and R. E. Sepponen, "Evaluation of a Doppler radar sensor system for vital signs detection and activity monitoring in a radio-frequency shielded room," Measurement, vol. 68, pp. 135–142, 2015.
[28] C. Li, Y. Xiao, and J. Lin, "Experiment and spectral analysis of a low-power Ka-band heartbeat detector measuring from four sides of human body," IEEE Trans. Microw. Theory Techn., vol. 54, pp. 4464–4470, 2006.

[29] G. Vinci, T. Lenhard, C. Will, and A. Koelpin, "Microwave interferometer radar-based vital sign detection for driver monitoring systems," IEEE MTT-S Intern. Conf. Microw. Intelligent Mobility (ICMIM), pp. 1–4, April 2015.
[30] L. Ren, L. Kong, F. Foroughian, H. Wang, P. Theilmann, and A. E. Fathy, "Comparison study of noncontact vital signs detection using a Doppler stepped-frequency continuous-wave radar and camera-based imaging photoplethysmography," IEEE Trans. Microw. Theory Techn., pp. 1–11, 2017.
[31] X. Bai, L. Yu, Q. Liu, *et al.*, "A high-resolution anatomical rat atlas," Journal of Anatomy, vol. 209, no. 5, pp. 707–708, 2006.
[32] J. Hau and S. J. Schapiro, Handbook of Laboratory Animal Science, 2nd ed.: Essential Principles and Practices, vol. 1, Boca Raton, FL, CRC Press, 2002.
[33] P. B. Jr. H'Doubler, M. Petersen, W. Shek, H. Auchincloss, W. M. Abbott, and R. W. Orkin, "Spontaneously hypertensive and Wistar Kyoto rats are genetically disparate," Lab Animal Sci., vol. 41, no. 5, pp. 471–473, 1991.
[34] Y. Yan, L. Cattafesta, C. Li, and J. Lin, "Analysis of detection methods of RF vibrometer for complex motion measurement," IEEE Trans. Microw. Theory Techn., vol. 59, no. 12, pp. 3556–3566, 2011.
[35] C. Li and J. Lin, "Complex signal demodulation and random body movement cancellation techniques for non-contact vital sign detection," IEEE MTT-S Int. Microw. Symp. Dig., pp. 567–570, 2008.
[36] T.-Y. Huang, L. F. Hayward, and J. Lin, "Noninvasive measurement and analysis of laboratory rat's cardiorespiratory movement," IEEE Trans. Microw. Theory Techn., vol. 65, no. 2, pp. 574–581, 2017.
[37] C. Li and J. Lin, "Random body movement cancellation in Doppler radar vital signs detection," IEEE Trans. Microw. Theory Techn., vol. 56, no.12, pp. 3143–3152, 2008.
[38] C. Li, J. Ling, J. Li, and J. Lin, "Accurate Doppler radar noncontact vital sign detection using the RELAX algorithm," IEEE Trans. Instru. Meas., vol. 59, no. 3, pp. 687–695, 2010.
[39] V. C. Chen and H. Ling, "Joint time-frequency analysis for radar signal and image processing," IEEE Signal Process. Mag., vol. 16, no. 2, pp. 81–93, 1999.
[40] V. C. Chen, "Analysis of radar micro-Doppler with time-frequency transform," IEEE Workshop Statist Signal Array Process., pp. 463–466, Aug. 2000.
[41] Y. Sugiura, A. Kawamura, and Y. Iiguni, "A comb filter with adaptive notch gain for periodic noise reduction," Signal Inf. Process. Assoc. Annu. Summit Conf., pp. 1–5, Dec. 2012.
[42] T.-Y. Huang, J. Lin, and L. Hayward, "Non-invasive measurement of laboratory rat's cardiorespiratory movement using a 60-GHz radar and nonlinear Doppler phase modulation," IEEE Int. Microw. RF Wireless Techn. Biomed. Healthcare Appl., pp. 83–84, Sept. 2015.
[43] T. Huang, L. Hayward, and J. Lin, "Adaptive harmonics comb notch digital filter for measuring heart rate of laboratory rat using a 60-GHz radar," IEEE MTT-S Int. Microw. Symp., pp. 1–4, May 2016.

[44] F.-K. Wang, T.-S. Horng, K.-C. Peng, J.-K. Jau, J.-Y. Li, and C.-C. Chen, "Single-antenna Doppler radars using self and mutual injection locking for vital sign detection with random body movement cancellation," IEEE Trans. Microw. Theory Techn., vol. 59, no. 12, pp. 3577–3587, 2011.

[45] C. Li, Z. Peng, T.-Y. Huang, *et al.*, "A review on recent progress of portable short-range noncontact microwave radar systems," IEEE Trans. Microw. Theory Techn., vol. 65, no. 5, pp. 1692–1706, 2017.

[46] H. Zhao, X. Gu, H. Hong, Y. Li, X. Zhu, and C. Li, "Non-contact beat-to-beat blood pressure measurement using continuous wave Doppler radar," IEEE MTT-S Int. Microw. Symp., pp. 1413–1415, June 2018.

Chapter 7

Dynamic monopulse radar sensor for indoor positioning and surgical instrument positioning

Chia-Chan Chang[1], Jen-Chieh Wu[2], Yen-Chih Chang[3], Peng-Yu Chen[4], and Sheng-Fuh Chang[1]

7.1 Introduction

The demand for target position sensing has exponentially increased in the past decade. From outdoors to indoors, people continue to exhibit a desire of picturing the surroundings, thereby leading to many emerging applications and services. The context-aware application is a good example [1], where the system can implement services for users based on their locations, allowing them to facilitate the access to nearby resources. Other examples include tracking assets in a large warehouse, guiding people in a public area, monitoring patients in hospitals, or home health-care for the elderly. The ability to precisely locate or track the targets of individuals is essentially required for those systems.

There are two major technologies used in positioning: fingerprinting [2,3] and triangulation. Although the fingerprinting technology has been proved more accurate than triangulation, it takes a considerable amount of time to create a fingerprinting map, which needs to be reconstructed every time the environment features change. The triangulation is relatively simple, using distance measurement (lateration) [4,5] or angle measurement (angulation) [6–9]. The lateration method acquires distance information from deriving the received signal strength (RSS), the time of arrival, round-trip time of flight, or time difference of arrival. However, the measured deviation of signal propagation due to time of flight is difficult to calculate, especially in a small space. Furthermore, since more than two positioning nodes are usually required to acquire the desired information, synchronization, and calibration between different nodes sometimes can be very challenging. Therefore, by means of receiving signal strength indicator (RSSI) to estimate the relative distance is one of the most well-known approaches [10,11]. However, electromagnetic wave reflection and scattering caused by obstructions or target

[1]Department of Electrical Engineering, National Chung-Cheng University, Taiwan
[2]Foxconn Industrial Internet Corporation, China
[3]Realtek Semiconductor Corporation, Taiwan
[4]Wistron Corporation, Taiwan

movements could generate multipath fading, resulting in inaccurate and imprecise distance estimation.

On the other hand, the angulation estimation is usually realized by computing the angle of arrival (AoA) (also referred to as direction of arrival), wherein phased antenna arrays with beam steering capability are commonly used as sensing technology. Since the angular resolution relies heavily on the beam resolution of antenna arrays, it usually requires a large array and precise beam steering.

Various kinds of technologies, such as infrared (IR) [12], light [13], vision analysis [14], ultrasound, or sound [15], have been utilized for positioning systems. Those solutions have their benefits and limitations. Their complexity, computing methods, deployment cost, and accuracy are quite different as well. Within those technologies, the radio-frequency (RF)-based system attracts considerable attention owing to the rapidly developed communication services. Many RF-based indoor positioning systems and locating methodology have been introduced, such as Bluetooth [16], wireless fidelity (Wi-Fi) [17,18], RF identification [19], and worldwide interoperability for microwave access (WiMAX). The radar technologies are also widely involved for position sensing. For instance, the ultra-wideband radar and frequency-modulated continuous wave-based radar systems are commonly used for distance detection [20,21]. Those systems follow easy adaptation to existing accessible wireless devices, emerging to be the next-generation mobile application.

Another emerging trend for radar sensing has been aimed toward biomedical applications, such as vital sign monitoring [22–24], motion detection [25–27], tissue temperature sensing [28], or cancer detection [29]. In addition to the human target, the medical instrument is another area that has the demand of RF radar techniques. Medical positioning/guiding [30] is an assistance tool to help the surgeons precisely position the instruments to the desired surgical site, especially for those surgeries requiring delicate procedures with high precision, such as minimally invasive surgery (MIS). Nowadays, MIS has become very popular as a treatment choice since it is performed through several tiny incisions instead of one large opening. Accordingly, the patients will experience less pain and blood loss, which can lower the risk of infection and speed up the recovery.

Since the MIS incisions are extremely small with very limited field of view (FOV), the surgeons can only monitor the internal procedure relying on the endoscope images. However, during surgery the endoscope lens can get fouled by blood or rinsing fluid and the chromatic aberration from the camera may also increase the difficulty in judgment. Accordingly, the surgeons require extensive training and practice to improve their skill in manipulating the instruments. With the assistance of guiding devices, the surgeon is able to precisely bring the instruments to the desired surgical site, and perform the delicate procedure with increasing successful rate.

Numerous technologies have been applied in medical guiding/positioning devices [31–33], including ultrasound, laser, and optical. Each technique has its own advantages as well as limitations. For instance, the ultrasound imaging shows potential when direct vision is not possible, but limited signal-to-noise ratio can lower its resolution. On the other hand, the optical imaging provides significant accuracy associated with good light conditions and high camera quality. Some surgical navigation systems require the installation of reflective markers as references. However, the path of optical rays cannot be obstructed by other objects, and the calibration procedure is usually time-consuming. Computed tomography (CT) or X-ray scanning is also commonly used during the operation, but the major concern is the harmful radiation exposure for both patients and surgeons. Electromagnetic approaches, such as using permanent magnets, which can relieve the blockage issue, have been also applied to instrument tracking [34] or wireless capsule endoscopy (CE) localization [35], but close sensing is required in most cases.

In this chapter, we propose a novel dynamic monopulse radar (DMR) sensor with new localizing algorithm for both indoor positioning and surgical instrument positioning. The proposed DMR system consists of a steerable comparator, RSSI modules, and a microcontrol unit (MCU). Based on the engagement between radar techniques and phased array techniques, this proposed system can provide two-dimensional (2-D) angulation positioning using only a single positioning node.

7.2 Indoor positioning system

7.2.1 Selecting-and-averaging algorithm

The classic monopulse radar was originally developed for target tracking during WWII [36]. Since then, it has been widely employed for target angle measurement and is in common use today. Conventionally, a monopulse radar is composed of two main methodologies: phase comparison and amplitude comparison. The phase-comparison method requires precise phase coherence between the antennas, leading to great difficulty in circuitry implementation. Instead, the amplitude-comparison method receives the power imbalance from the comparator circuitry, which makes the angular localization easier. Figure 7.1(a) displays the diagram of a traditional 1-D monopulse radar system. By comparing the powers received from the sum port (Σ) and difference port (Δ) of the comparator (such as a rat-race coupler or branch-line coupler), the angulation information of the unknown target can be estimated by mapping the received Σ-Δ power ratio into the prestored Σ–Δ curve (also called the error detection curve). Since the phase distributions at the Σ-port and Δ-port are usually fixed, a mechanical motor is needed for continuous beam steering and tracking.

However, in a realistic indoor environment, the multipath reflections could seriously alter the power distribution in the space, leading to the deterioration of the angular accuracy. In order to solve this issue, a DMR sensor with selecting-and-averaging algorithm is proposed in this work [37]. Different from the traditional monopulse comparator, the proposed steerable comparator has an additional pair of electrically tunable phase shifters, as shown in Figure 7.1(b). The radiation beam can be switched into different directions by applying a progressive phase shift on those phase shifters. Each switched beam can generate its own corresponding Σ-Δ curve. With more data being collected, we can estimate the angular position more

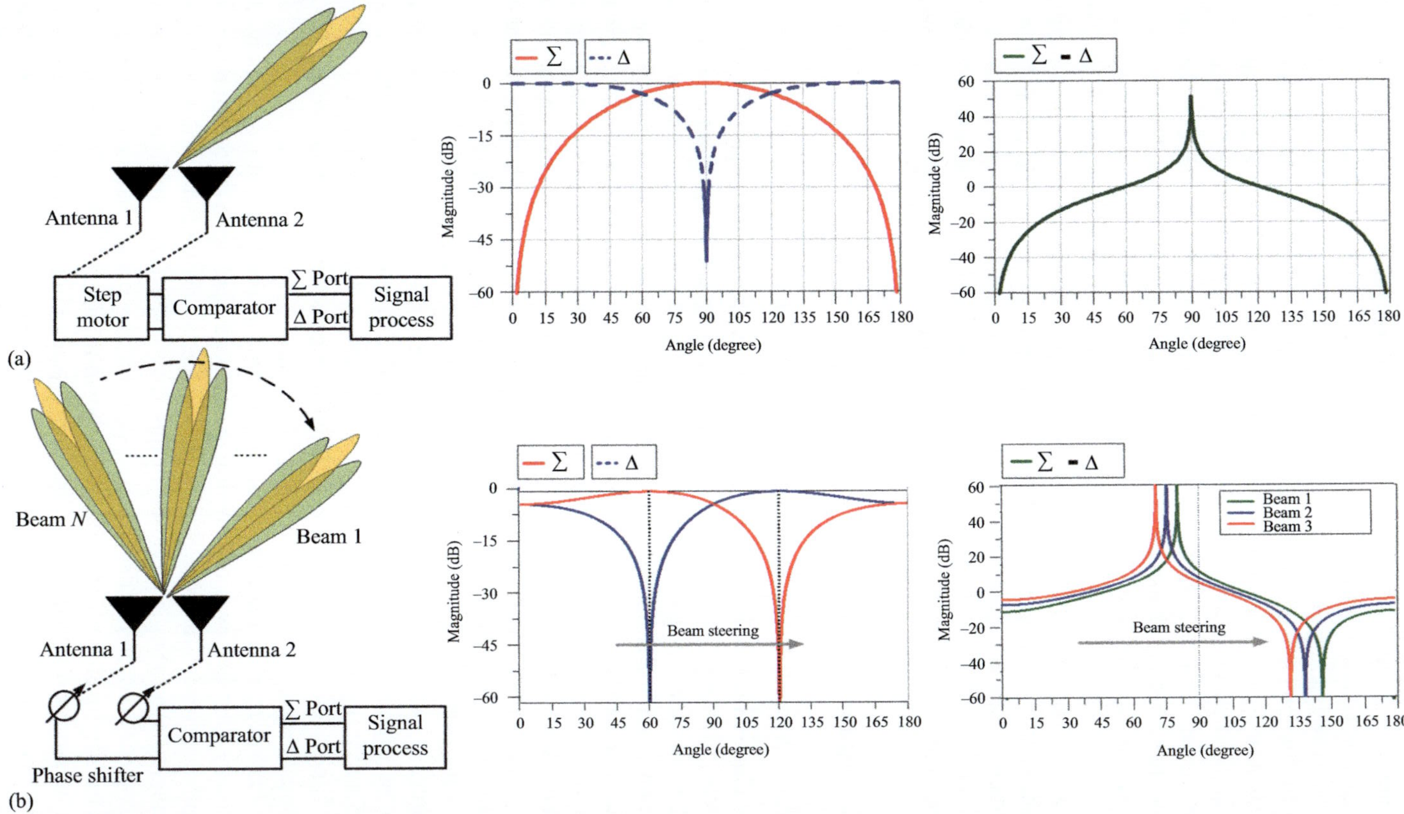

Figure 7.1 The operating concept of (a) traditional monopulse radar system and (b) proposed DMR system

accurately by using a so-called selecting-and-averaging algorithm. The positioning concept is described in the following procedure.

Assuming that the target is located at angular direction of θ_T and the system receives the target signal with beam being switched from Beam-1 to Beam-N in sequence, for each beam state, we can map the RSS power ratio reading on its corresponding Σ-Δ curve, resulting in two estimated angles. As an example shown in Figure 7.2, two angles θ_1 and θ_2 can be obtained based on the RSS readings from Beam-1's Σ-Δ curve. When the beam is switched to Beam-2, the new RSS readings will result in another two estimated angles θ_3 and θ_4 on Beam-2's Σ-Δ curve. The algorithm will select and average the two most correlated data while eliminating the other two less-correlated data. In this illustration, the angles θ_2 and θ_3 will be selected and averaged to estimate the target angle θ_T is $(\theta_2+\theta_3)/2$. Accordingly, N switched beams can result in N Σ-Δ curves with $2N$ estimated angles. The target angular position can thus be determined based on the average of the N most correlated data. With more data being averaged, the angular positioning accuracy can be further improved.

Note that symmetrically distributed Σ-Δ curves are highly desired to ensure consistent performance in either the leftward or rightward area. For that reason, the progressive phase shift for each switched beam is suggested based on (7.1):

$$\delta_n = \left(\frac{90^\circ}{N}\right) \times (2n - 1) - 90^\circ \ , \ n = 1, 2, ..., N \tag{7.1}$$

where N is the number of switched beams, δ ($n = 1, \ldots, N$) represents the phase difference between two phase shifters required for each switched beam state. Based on (7.1), the suggested phase shifts for each beam state ($N = 2$ to $N = 8$) are also summarized in Figure 7.2.

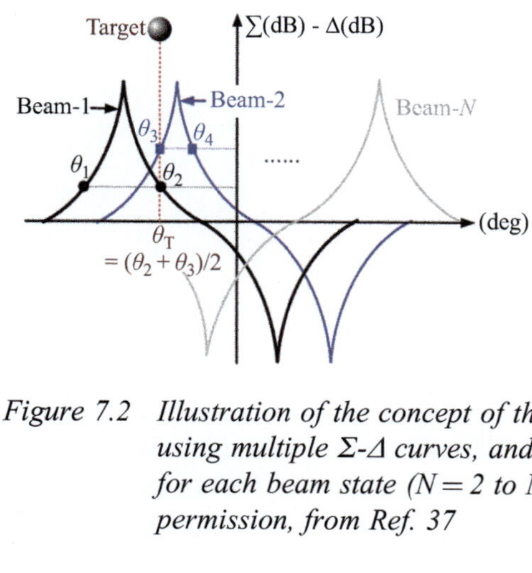

Assignment for progressive phase shifts								
N	$\delta 1$	$\delta 2$	$\delta 3$	$\delta 4$	$\delta 5$	$\delta 6$	$\delta 7$	$\delta 8$
2	45.0°	135.0°						
3	30.0°	90.0°	150.0°					
4	22.5°	67.5°	112.5°	157.5°				
5	18.0°	54.0°	90.0°	126.0°	162.0°			
6	15.0°	45.0°	75.0°	105.0°	135.0°	165.0°		
7	12.9°	38.6°	64.3°	90.0°	115.7°	141.4°	167.1°	
8	11.3°	33.8°	56.3°	78.8°	101.3°	123.8°	146.3°	168.8°

Figure 7.2 Illustration of the concept of the selecting-and-averaging algorithm using multiple Σ-Δ curves, and the suggested progressive phase shift for each beam state (N = 2 to N = 8). [2010] IEEE. Reprinted, with permission, from Ref. 37

7.2.2 2-D positioning concept

There are two approaches to achieve 2-D positioning using the proposed DMR sensor. As seen in Figure 7.3, two 1-D DMRs can be deployed at the corners on the same wall with known distance. The FOV of each DMR is overlapped to form a desired positioning area. The DMRs simultaneously detect the events through the switched beams, and the target's position can be determined by the intersection of two angle direction lines from both DMR sensors. Comparing to the conventional phased array antenna (PAA)-based AoA positioning, the angular resolution in this case is no longer limited by the beam resolution since we used Σ-Δ curves to estimate the angles. Our previous work [37] has demonstrated that a mean error distance of 0.33 cm for 75% cumulative probability can be achieved in a 5 m × 5 m space using this deployment.

Figure 7.3(a) proposed another deployment for 2-D positioning, where only a single DMR sensor is needed instead of two. The system is required to acquire not only the azimuth angle ϕ_A, but also the elevation angle θ_E information of the target. Based on the same concept of 1-D system, a 2-D comparator is designed to connect a 2 × 2 antenna array. As depicted in Figure 7.4, the elevation angle can be estimated from the Σ-Δ power ratio resulting from antenna (1,2)–antenna (3,4), and the azimuth angle can be obtained from the Σ-Δ power ratio resulting from antenna (2,3)–antenna (1,4). The beams are steered horizontally to estimate the target's azimuth angle φ_A and are steered vertically to estimate the target's elevation angle θ_E. The phase assignment for both horizontal switched beams and vertical switched beams can also be calculated by (7.1) Note that the number of distributed Δ-Σ_A curves is not necessary to be equal to the number of distributed Σ-Δ_E curves if the required angular accuracies in those two planes are different.

However, the realistic Σ-Δ curves won't be as symmetric as the theoretical predictions. The circuit imperfections as well as multiple reflections can cause the

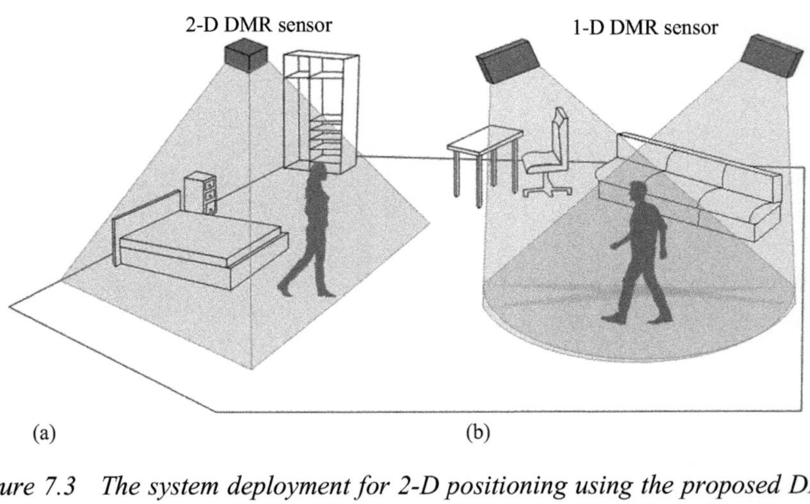

Figure 7.3 The system deployment for 2-D positioning using the proposed DMR sensors. It can be either (a) single positioning node or (b) two positioning nodes

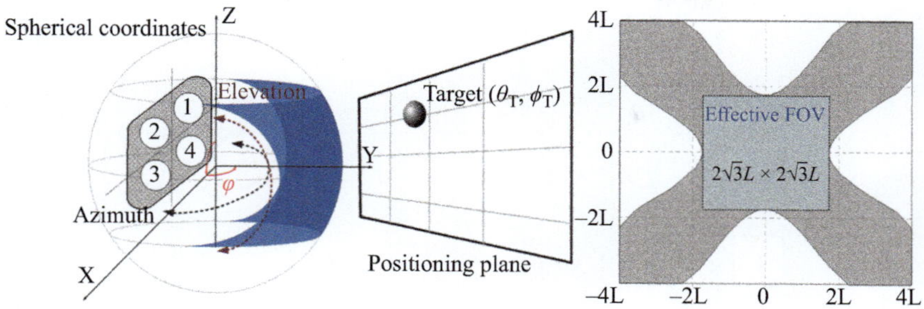

Figure 7.4 Angle estimation of a 2-D array system, and its conversion from spherical coordinates to the planar positioning plane

unequal, asymmetric Σ-Δ curves, which will be shown later in Figure 7.8. Since the multipath interference is unpredictable and scenario-dependent, the primary Σ-Δ curves should be premeasured in an anechoic chamber for calibration purposes as well as the follow-up data processing.

Note that the angles θ_E are determined from the Σ-Δ curves on *y-z* plane ($\phi = 90°$), while φ_A is determined from the Σ-Δ curves on *x-y* plane ($\theta = 90°$). In most cases, the target may not be located on both planes. Therefore, we need to convert the measured (θ_E, φ_A) to the corresponding (θ_T, φ_T). To determine the target's true position, the target position (r, θ_T, φ_T) can be projected on the *x*- and *z*-axes in rectangular coordinates, given as

$$x = r \sin \theta_T \cos \varphi_T \tag{7.2}$$

$$z = r \cos \theta_T \tag{7.3}$$

The estimated target's AoA (θ_E, φ_A) is also projected on the *x*- and *z*-axes, and given as

$$x = r \cos \varphi_A \tag{7.4}$$

$$z = r \cos \theta_E \tag{7.5}$$

Comparing (7.2) with (7.4) and (7.3) with (7.5), we can conclude

$$\theta_E = \theta_T \tag{7.6}$$

$$\varphi_A = \cos^{-1}\left(\frac{\cos \varphi_T}{\sin \theta_E}\right) \tag{7.7}$$

Since the array patterns can be steered vertically as well as horizontally, the effective positioning area is limited by the steering range of both θ_E ($\theta_{E1} \leq \theta_E \leq \theta_{E2}$) and φ_A ($\varphi_{A1} \leq \varphi_A \leq \varphi_{A2}$). Figure 7.4(b) shows the planar positioning region after being conformally mapped from the spherical coordinates into a planar positioning plane. Note that the area size is also proportional to the distance L between the radar receiver and positioning plane. Since the planar positioning region is X-shaped, for

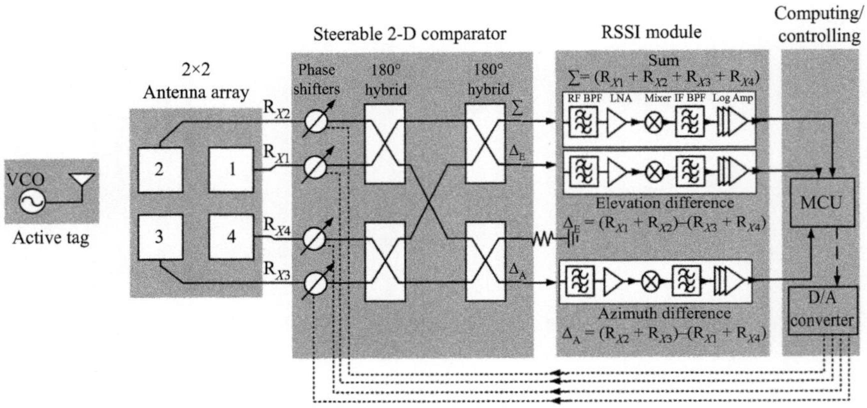

Figure 7.5 Circuit block diagram of proposed RSS-based indoor localization system. [2010] IEEE. Reprinted, with permission, from Ref. 37

convenient demonstration, we define a square zone as the effective FOV with size of $2\sqrt{3}L \times 2\sqrt{3}L$.

7.2.3 System hardware design

As shown in Figure 7.5, the proposed 2-D DMR sensor consists of a 2×2 planar antenna array, a 2-D steerable comparator, three RSSI modules, an MCU module, and an active tag. In this work, 2.4 GHz is chosen as the operating frequency. The detailed design for each sub-circuit is described in the following.

7.2.3.1 2-D steerable comparator

The 2-D Σ-Δ comparator designed in this system consists of four 180° hybrids and four 180° tunable phase shifters. The 2-D steerable comparator is composed of four tunable 180° phase shifters and four 180° hybrids, where the 180° hybrids are constructed by the 90° branch-line coupler and a 90°-Schiffman phase shifter. Similar to the conventional monopulse system, a 2×2 antenna array along with the 2-D comparator can generate one sigma signal (Σ) and two delta signals (Δ_E and Δ_A). The target's angular data, both elevation and azimuth direction, are derived from Σ-Δ_E and Σ-Δ_A curves. Assuming that the received signals from the antennas are R_{X1} to R_{X4}, thus the Σ, Δ_E, and Δ_A signals can be derived:

$$\Sigma = R_{X1} + R_{X2} + R_{X3} + R_{X4} \tag{7.8}$$

$$\Delta_E = (R_{X1} + R_{X2}) - (R_{X3} + R_{X4}) \tag{7.9}$$

$$\Delta_A = (R_{X2} + R_{X3}) - (R_{X1} + R_{X4}) \tag{7.10}$$

In order to steer the beams, at least 150° phase tuning is required to cover our desired FOV. The tunable phase shifter is thus designed based on asymmetric reflection-type configuration (50 Ω/12.5 Ω port impedance) to achieve wider phase tuning. The measured phase tuning range from the implemented circuit is ~180° with insertion loss variation less than 0.86 dB.

7.2.3.2 Circular-polarized antenna array

Comparing to a linear polarized antenna, a circular-polarized antenna is more effective in resisting multiple-reflection signals owing to its strictness to the rotating sense, making it more suitable in an environment with rich interference. Therefore, a dual-fed circular patch antenna is chosen as the antenna element to generate left-handed circular polarization (LHCP) for this 2×2 planar array. The patch is lifted over a ground plane with air thickness of 4.5 mm so that wider impedance and axial-ratio (AR) bandwidth can be achieved. The measured return loss is lower than 10 dB from 2.03 to 2.58 GHz. For a single antenna element, maximum gain occurs in the broadside direction with a 3-dB beamwidth of ~100°. The AR is less than 3 dB within $\pm 35°$ spatial region, while the isolation between left-handed and right-handed is more than 15 dB.

7.2.3.3 RSSI module and MCU

The RSSI module contains a 2.3-GHz band-pass filter, a low-noise amplifier (Hittite HMC287), a down-converting mixer (Maxim MAX2681), a two-stage 10.7-MHz intermediate frequency (IF) filter (Murata SFELF10M7GA), and a power detector (Analog Device AD8310). In order to measure the strength of an incoming signal, the 2.4-GHz RF power is first down-mixed to a 10.7-MHz IF signal. The resultant IF signal is then converted to a DC voltage through the power detector. The measured dynamic range is approximately 70 dB, covering from -35 to -105 dB. It also shows a great linearity with 24.8 mV/dB resolution.

The programmable MCU is employed not only to control the tunable phase shifters for beam switching but also to calculate the target's positions based on the RSSI readings. In addition, the MCU can also control local oscillator (LO) frequency shifting so that multiple targets with different assigned frequencies can be detected and recognized simultaneously.

7.2.3.4 Positioning tag

The positioning tag consists of a dual-fed circular patch antenna, a miniaturized branch coupler, a frequency synthesizer (Analog Device AD4360), and an MCU (Microchip PIC10F200). The circular patch antenna has the same design as the DMR array element. The identification of each tag is based on its synthesizer frequency, which is also controlled by the MCU. To allow multiple-target detection, individual tags can be assigned to different synthesizing frequencies with 1.3 MHz interval within 2.03–2.58 GHz. Accordingly, the current system is capable of detecting 100 targets simultaneously. The entire tag circuit is implemented on FR4 substrate with overall size of 6 cm $\times$ 6 cm.

Figure 7.6(f) shows the implemented circuit photos for each sub-circuit and the assembly of the entire 2-D DMR sensor with the system size of 30 cm $\times$ 30 cm $\times$ 20 cm.

7.2.4 Multipath interference analysis

In an actual indoor environment, the multipath reflections could significantly degrade the localization accuracy. Although increasing the number N of Σ-Δ curves

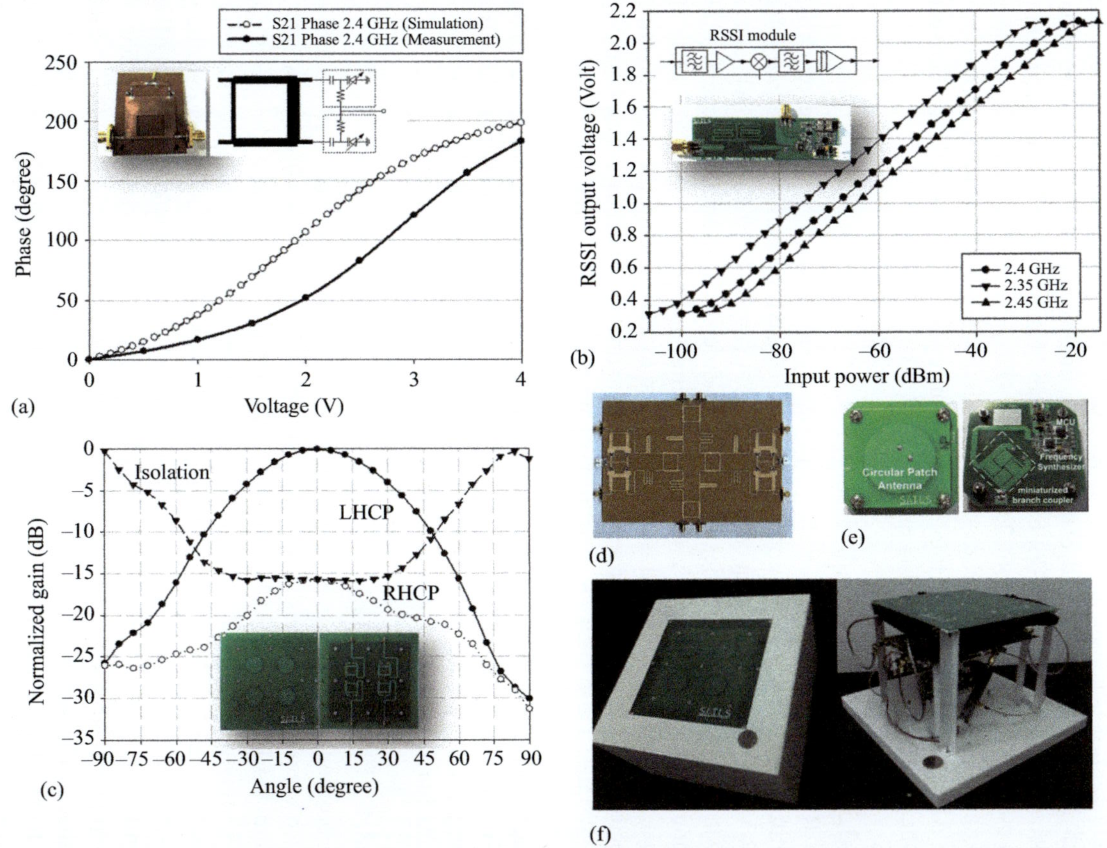

Figure 7.6 Implemented circuit photos: (a) tunable phase shifter, (b) RSSI module, (c) circular-polarized antenna array, (d) 2-D steerable comparator, (e) active tag, and (f) system assembly. [2010] IEEE. Reprinted, with permission, from Ref. 37

can help to reduce the multipath effect through the selecting-and-averaging process, the influence of multipath reflections should be further investigated. Assume that there are received signals at each antenna (R_{x1} and R_{x2}), including the direct signal from the target and mixed multipath signal. The resultant Δ and Σ from the comparator can thus be represented by

$$\Sigma = R_{X1}e^{j\delta} + R_{X2}e^{-j90^\circ} \tag{7.11}$$

and

$$\Delta = R_{X1}e^{j(\delta-90^\circ)} + R_{X2} \tag{7.12}$$

A *K* factor, which is defined as the power ratio (in dB) of target signal to multipath signals, is given by

$$K \equiv 10\log\left(\frac{A_T^2}{E\left[|A_M|^2\right]}\right)(\text{dB}) \tag{7.13}$$

In the above, A_T and A_M represent the amplitude of the target signal and the amplitude of the Rayleigh-distributed multipath signal [38], respectively. Based on this assumption, the amplitude A_M can be represented by $\alpha\sqrt{N_r^2 + N_i^2}$, where N_r and N_i are zero-mean Gaussian random variables with unit variance and $\alpha = \left(A_T/\sqrt{2}\right) \times 10^{\frac{-K}{20}}$.

The simulation results with different number of *N* are shown in Figure 7.7(a), where *K* is set to 10 dB and the antenna spacing is 0.5λ. Within the FOV from 45° to 135°, it shows that the conventional monopulse radar suffers from the dramatic error variation, especially in the line-of-sight region. Instead, the proposed dynamic Σ-Δ radar can generate smoother error variations, and the variations become smaller as *N* is increased.

Figure 7.7(b) shows the cumulative probability when the environment *K* value is 4 dB. It can be seen that the proposed system improves the error angles by more than 2° from $N=2$ to $N=8$ at the 75th percentile. Once again, it reveals that choosing larger *N* can improve the angle estimation accuracy. This result demonstrates the accuracy enhancing capability using the proposed selecting-and-averaging positioning algorithm, especially in a rich multipath environment.

The performance of the proposed system with different environment *K* values is also evaluated. Figure 7.7(c) shows the positioning error angle by different *K* values within the effective area, where *N* (and *M*) are chosen as 8 herein. As predicted, the system performs the best angle estimation accuracy within the central region of the FOV, whereas the error is increased gradually when it approaches the boundaries. Moreover, the angle error becomes larger when the multipath signals are relatively strong ($K=4$). Based on these results, the cumulative probability is also evaluated, where the 75th percentile of the cumulative probability has the maximum error angles of 27°, 13°, and 7° under $K=4$, 10, and 16 B conditions, respectively.

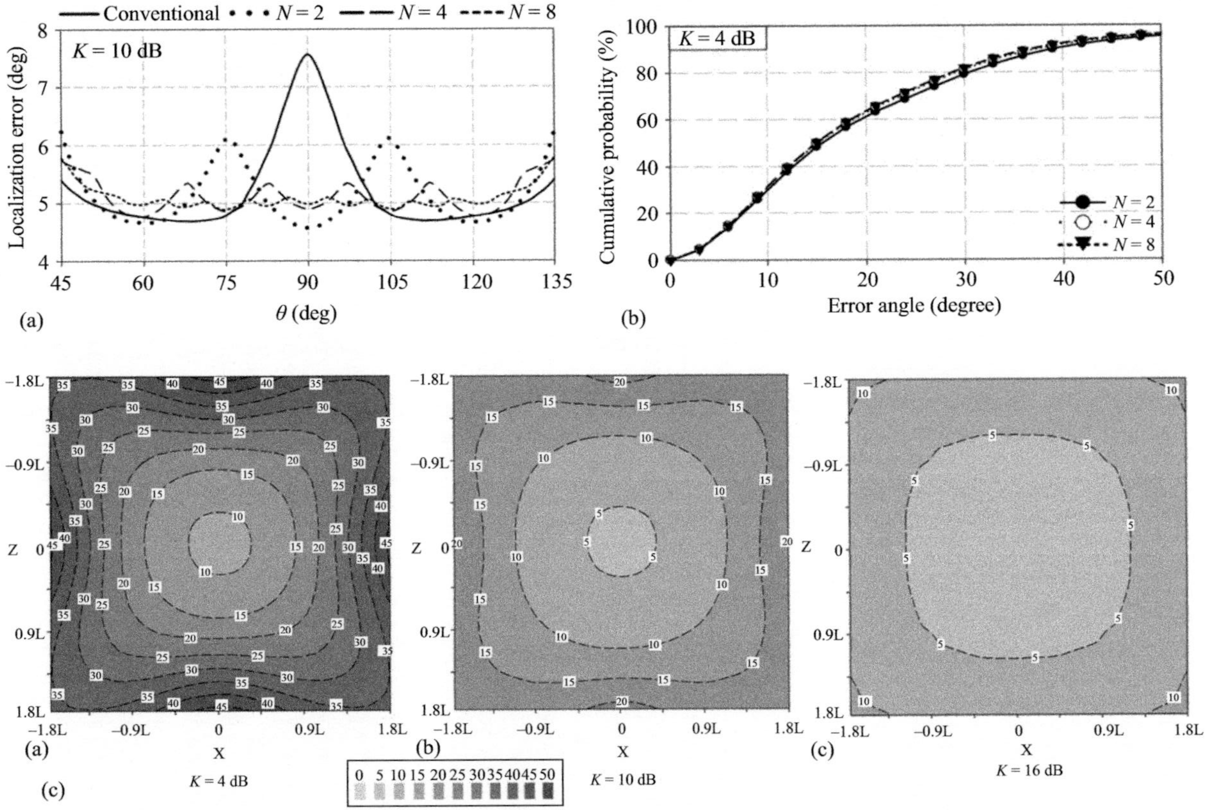

Figure 7.7 *Performance evaluation in multipath indoor environments. (a) The localization error as a function of the angle of rotation. (b) Cumulative probability for different N values. (c) The analysis of the location error angle (K = 4, 10, and 16 dB). [2010] IEEE. Reprinted, with permission, from Ref. 37*

7.2.5 Indoor positioning demonstration

The assembled tunable 2×2 phased-antenna array is placed in an anechoic chamber for measurement. The switched beam patterns are precisely controlled to generate reference $R_\Sigma - R_\Delta$ power ratio curves, where those curves will be prestored and later used for power mapping along with target angle estimation. Figure 7.8 shows the measured $R_\Sigma - R_\Delta$ power ratio curves resulted from the case of $N = 2$, 4, and 8 switching beams. Once again, it can be observed that those curves are symmetrically distributed within the effective positioning FOV.

The proposed 2-D DMR positioning system has been tested in the laboratory. The radar receiver is installed on the ceiling, looking down at the space beneath the

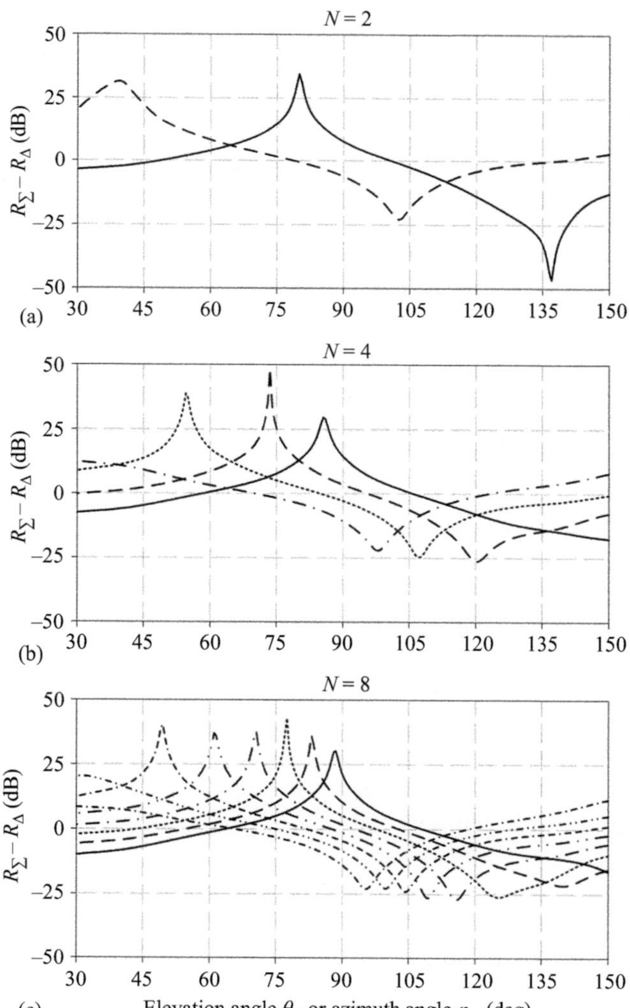

Figure 7.8 Measured $R_\Sigma - R_\Delta$ power ratio curves: (a) $N = 2$, (b) $N = 4$, and (c) $N = 8$

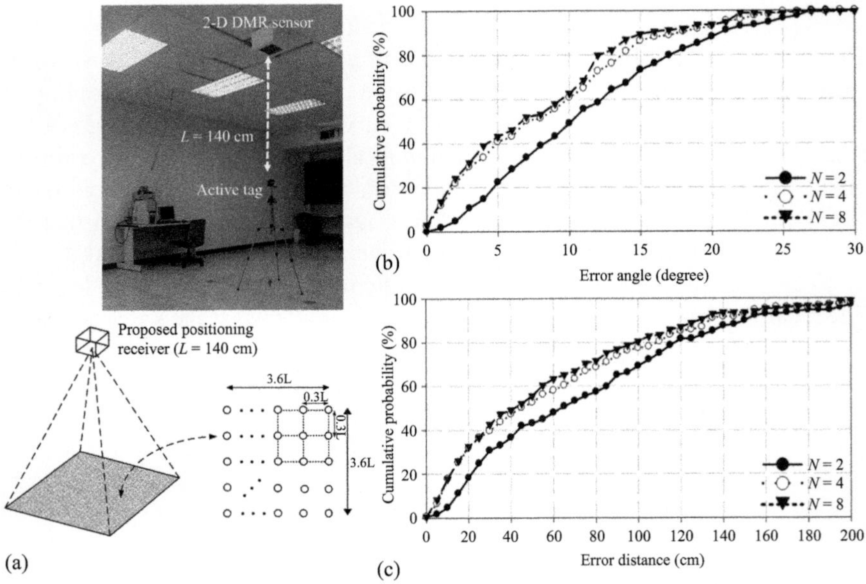

Figure 7.9 Demonstration of 2-D indoor positioning. (a) Experiment setup. (b) Cumulative probability of error angle. (c) Cumulative probability of error distance

radar. The data were collected within a 5.04 m × 5.04 m positioning area. The real experimental setup is shown in Figure 7.9(a).

In order to verify the accuracy improvement with the proposed selecting-and-averaging algorithm, the system with different numbers of switched beam N are measured and compared. Figure 7.9(b) shows the error angles between the estimated and actual positions with the different N values, whereas Figure 7.9(b) shows the average distance error. The experimental results reveal no significant improvement in error angles between the cases of $N=4$ and $N=8$, which suggests that $N=4$ may be an optimized choice between the estimation accuracy and positioning speed/update rate.

Once again, the performance has better accuracy around the center of the effective FOV, and is degraded when the test points are approaching the boundaries.

The multipath signals in this experimental space are unpredictable. However, the measurement data are similar with the previous analysis in Figure 7.7(c) under the case of $K=10$ dB. Since this laboratory space is empty, the multipath signals should be relatively weak comparing to the line-of-sight signal. Therefore, those results remain consistent.

7.3 Surgical instrument positioning system

The proposed 2-D DMR system is also applied for the short-range sensing. In this section, a surgical instrument tracking/positioning system is demonstrated. The

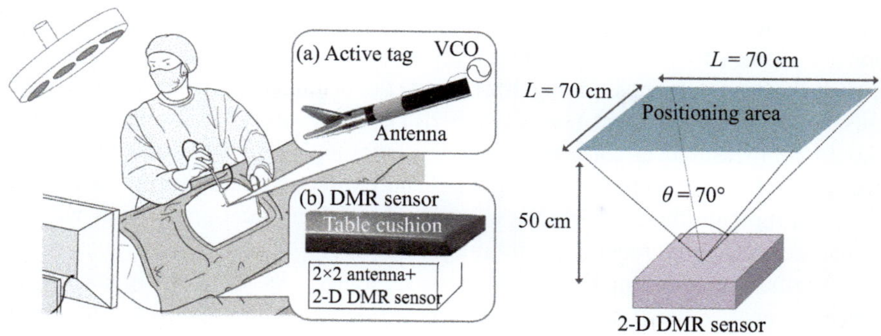

Figure 7.10 The scenario of this proposed system in the operating room. [2016] IEEE. Reprinted, with permission, from Ref. 39

motivation of this work is to design a distant wireless surgical positioning system that would guide the surgeons to the surgical site and perform the operation more precisely during the MIS procedure.

7.3.1 Design consideration

Figure 7.10 shows the proposed system scenario [39]. According to our application, this system is used in an operating room that is usually equipped with a large amount of stainless steel surgical instruments and facilities. How to distinguish the target from severe background clutters becomes a major task. There are several considerations to be taken into account before choosing an appropriate installation. For 2-D positioning, obviously, using one 2-D DMR sensor is more desirable than using two 1-D DMR sensors if considering system deployment. In order to prevent the signal obstruction as well as to save floor space, a positioning tag is attached to the surgical instrument serving as an RF transmitter, while the radar receiver is positioned underneath the surgical table cushions. Accordingly, the position region is upward instead of downward. This arrangement can also minimize the transmission distance, consequently reducing the propagation loss and increasing signal integrity. Accordingly, the distance between the positioning receiver and the transmitting antenna is approximately 0.5 m, and the estimated positioning area is 0.7 m × 0.7 m, which gives an angular coverage range of 70°. The tolerable error angle is 5.71° under 5-cm distance error. Meanwhile, the low power consumption and small circuit size are strictly confined.

7.3.2 Peak-tracking algorithm

Since this case is short-range sensing, the positioning accuracy should be further improved. As discussed in Section 7.2.5, the improvement in accuracy is very limited after N exceeds 4. Apparently, the selecting-and-averaging algorithm has its limitations and may not be suitable for this application.

There are several reasons for its performance to be limited. First of all, the premeasured Σ-Δ curves used in our work come from certain cuts of the beam patterns. More specifically, we only measure Σ-Δ curves on the *x-z* plane ($\phi = 0°$) and *y-z* plane ($\phi = 90°$). Once the target is not located on both planes, it could introduce the angular positioning errors since the beam shapes from different angle cuts can be very different (see Figure 7.11). To resolve this issue, Σ-Δ curves should be measured at different angle cuts thoroughly, not to mention that the beams need to be steered at each plane. The consequence is increasing the processing time significantly, which is not feasible for surgical application with the demand of real-time tracking/positioning.

Another issue is again the multipath inference. As seen in Figure 7.8, the measured peak values from the Σ-Δ curves are degraded in a real scenario, consequently reducing the slope of the power ratio change versus angle. With lower slope, angle estimation can be misleading due to the voltage frustration caused by the surrounding interference, which is difficult to recover simply using the selecting-and-averaging algorithm. Fortunately, the interference has less impact on the peak angle positions even the curve slope is degraded.

Based on those observations, we therefore proposed a peak-tracking algorithm for more accurate and stable angle estimation. As with the previous system, by progressively tuning the phase shifters, each switched beam can result in a corresponding Σ-Δ curve. With a target being placed at an angle of θ_T, a calibration curve specifically for this case can be reconstructed along phase tuning by plotting the power ratio readings from each Σ-Δ curve (Figure 7.12(a)) versus each curve's corresponding peak angle (Figure 7.12(b)). The same procedure is repeated everytime we change θ_T. Eventually, we can prebuild an entire set of calibration curves for all possible target angles. Later, when it comes to on-site measurement, we can progressively sweep the phase states to produce a curve for the target, and then find the best match within those prebuilt calibration curves through peak tracking. The angular information of this matched curve will be considered as the target's incident angle.

With more data from Σ-Δ curves, it will be easier to determine the peak from the calibration curves. Accordingly, we need highly distributed beam scanning. The finer the scanning angle, the higher the positioning accuracy. In this system, 321 beams are distributed within ~80° FOV by precisely controlling the phase shifters (with 20 mV per step). The consequent downside is that the system takes longer computational time to complete the scanning. Even so, the processing time of MCU

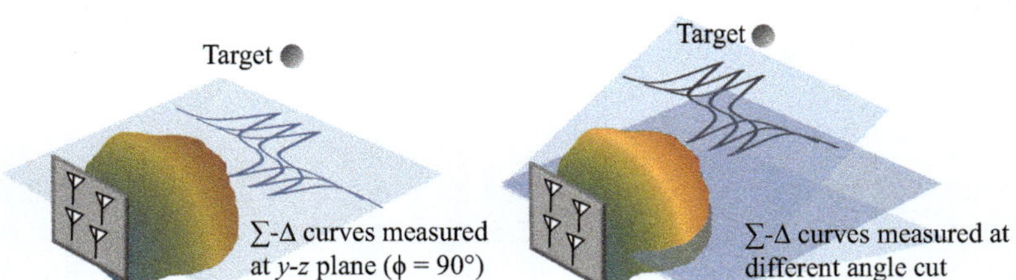

Figure 7.11 The premeasured Σ-Δ curves from the y-z plane (left), and from the other angle cut of the plane (right)

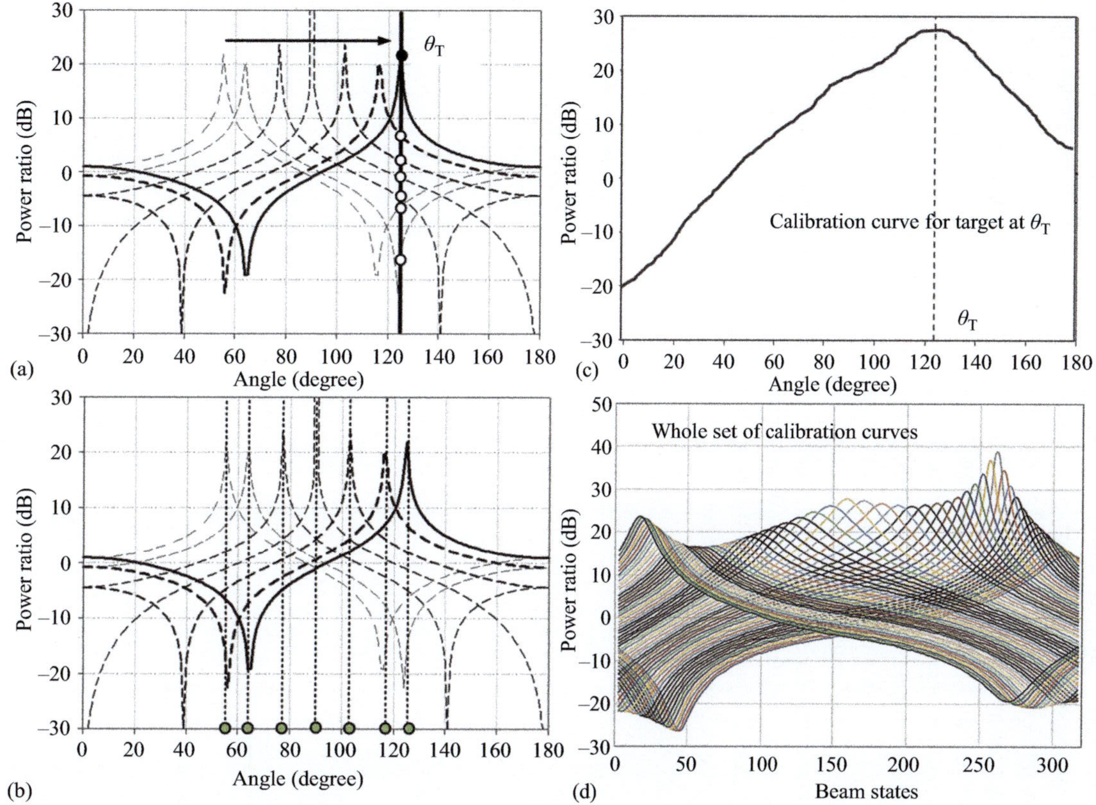

Figure 7.12 *(a) Power ratio readings from each Σ-Δ curve. (b) Each Σ-Δ curve has its corresponding peak angle. (c) Calibration curve for the target at θ_T. (d) Measured calibration curves from our implemented system. [2016] IEEE. Reprinted, with permission, from Ref. 39*

only takes milliseconds so that real-time positioning and tracking can be simultaneously achieved.

Figure 7.12(d) displays all measured peak-tracking curves using our implemented system. With both elevation and azimuth angles being determined, the target's location in a 2-D space can be obtained as well.

Figure 7.13(a) and (b) shows the angle errors measured in the anechoic chamber. At the 80th percentile of cumulative probability, both θ and ϕ have the maximum angle error less than 0.2°. Figure 7.13(c) compares the 1-D positioning performance in an interference-influence laboratory space using either selecting-and-averaging algorithm ($N = 8$) or this proposed peak-tracking algorithm. The measurement points correspond to the mobile tag transmitter being moved from 0.5 to 4.1 m in increments of 0.4 m along a straight line, whereas the spanning angle is from 55° to 125°. This will result in a total number of 50 measurements. As seen in Figure 7.13(c), the angle error is increased along the distance, but the packing-tracking algorithm (solid line) shows better performance than the selecting-and-averaging algorithm (dashed line). With error angle less than 3°, the cumulative percentile from the proposed peak-tracking algorithm is 82.5%, while the selecting-and-averaging algorithm only achieves 52.3%.

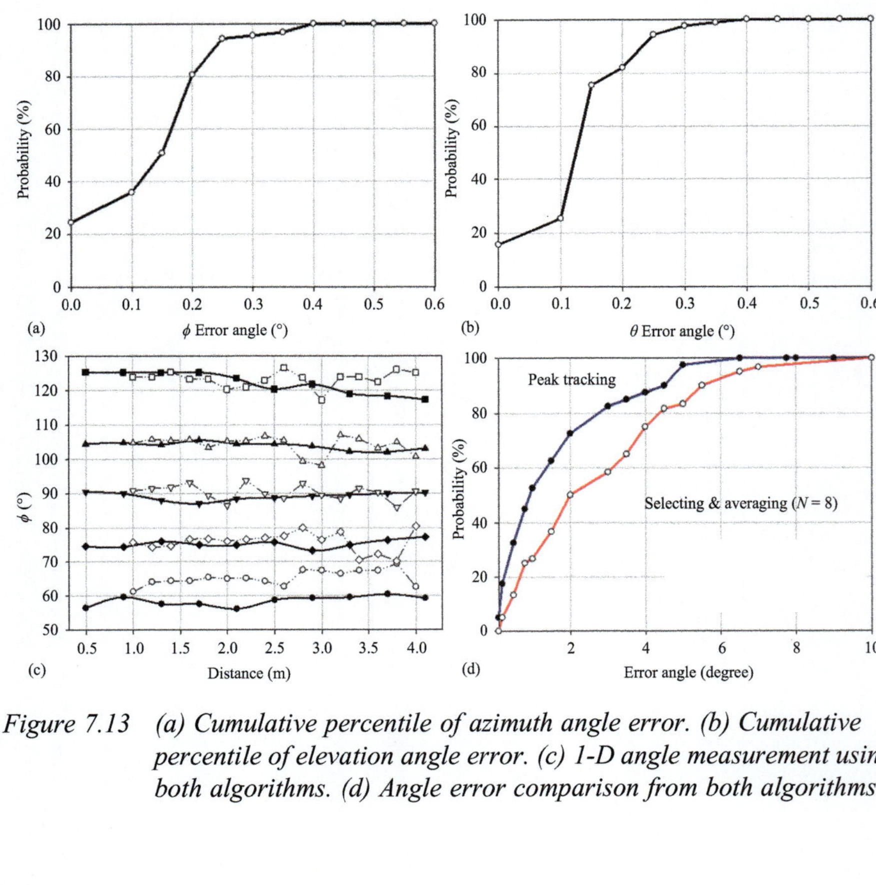

Figure 7.13 (a) Cumulative percentile of azimuth angle error. (b) Cumulative percentile of elevation angle error. (c) 1-D angle measurement using both algorithms. (d) Angle error comparison from both algorithms

7.3.3 Hardware design

In this application, we continue to use the 2.4-GHz 2-D DMR sensor from the previous indoor positioning. In order to track the surgical instrument, an active tag, which consists of a voltage-controlled oscillator (VCO) and an antenna, is designed and attached at the tip end of the tool serving as a transmitter. We choose this relatively simple circuit configuration not only to minimize the circuit size but also to reduce the power consumption. Such a low-cost, low-profile circuitry can be disposed after use.

However, several design challenges are encountered when it comes to the tag antenna design. First of all, the size of the antenna is highly restricted since the antenna will be curved and attached to the instrument surface. Therefore, the cylinder area should be less than 2 cm (height) $\times$ 0.5 cm (radius). In addition, the instrument makes contact with internal organs so that the loading effect caused by variety of tissues can result in an unexpected frequency shift. Accordingly, a certain amount of bandwidth is required for tolerance. Meanwhile, the omnidirectional radiation patterns are desired for better signal reception since the tool may be tilted or rotated during the manipulation.

Those design considerations are very similar to the antennas in the CE [38,40–47]. The reported CE literatures are basically divided into two types of antennas: embedded antenna and conformal antenna. The embedded antennas occupy the internal space of the CE, while the conformal antennas are curvedly attached to the surface of the CE without occupying internal space, which is more related to our design goal.

In this work, we started with a single-loop structure. By adding an additional arc for bandwidth enhancement and converting strips into slots using Babinet's principle, the resultant meander-slot loop is the final design, as shown in Figure 7 (a). In order to provide a certain degree of flexibility, a 10-mil Rogers RO3003 ($\varepsilon r = 3$, $\tan\delta = 0.001$) is chosen as substrate material. This meander-slot loop antenna can provide the quasi-omnidirectional radiation patterns we desire. Note that the tissue effect should be also included during the design procedure.

A VCO (HMC385LP4) is chosen as the signal source due to its extremely compact circuit size (~16 mm^2), which is a great benefit for packaging. This VCO can cover the frequency tuning range from 2.25 to 2.5 GHz with 4.5-dBm output power. In the practical use, multiple instruments can be assigned with different frequencies as identification. To prevent the direct heating on the tissues, this VCO is placed in the rear section of the instrument and connected to the tag antenna through an IPEX cable. Finally, we use a silicon membrane to seal the tag circuitry.

The tag circuitry have been assembled to the instrument and soaked into self-made tissue simulating liquid ($\varepsilon r \approx 54$@2.4 GHz). The measured operating frequencies from S11 are slightly up-shifted (~250 MHz) due to the uncertainty of the tissue liquid, but overall the data agree with the simulation predictions (see Figure 7.14).

7.3.4 Surgical instrument positioning demonstration

The system deployment has been illustrated in Figure 7.15. To simulate the surgical condition, the proposed 2-D DMR sensor is installed under a 10-cm thick foam

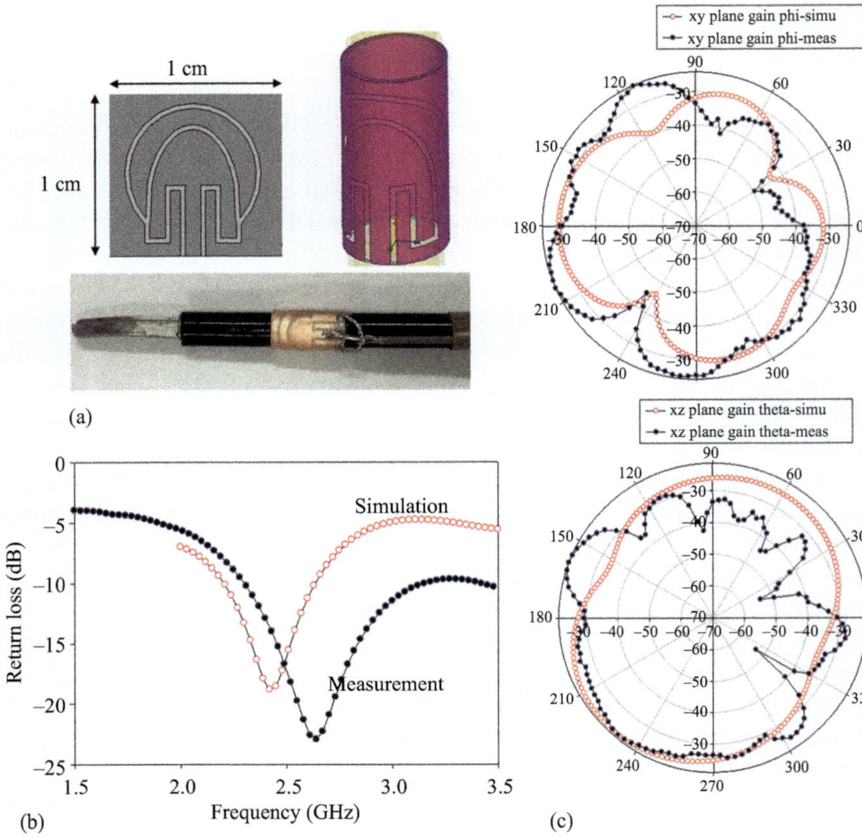

Figure 7.14 *(a) Implemented antenna assembled with surgical instrument. (b) Simulated and measured return loss. (c) Simulated and measured radiation patterns. [2016] IEEE. Reprinted, with permission, from Ref. 39*

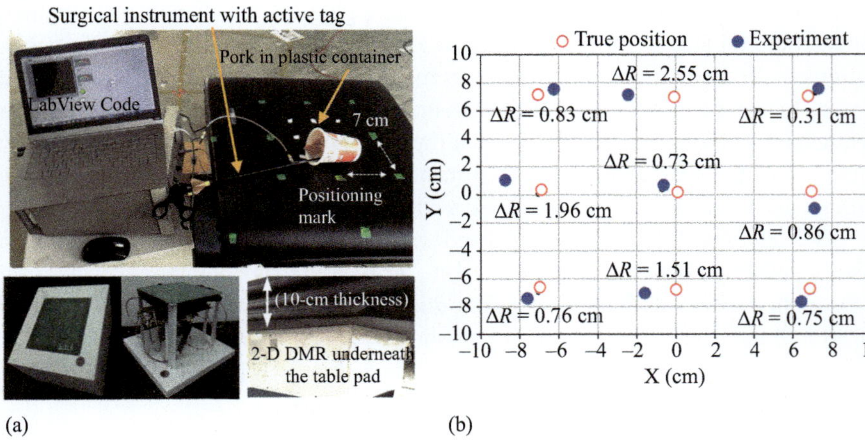

Figure 7.15 *Demonstration of surgical instrument positioning. (a) Experimental setup. (b) 2-D positioning measurement results. [2016] IEEE. Reprinted, with permission, from Ref. 39*

table cushion. Although the proposed 2-D DMR sensor is packaged into a metal box in our demonstration, it can be planarly assembled and embedded flatly underneath the table cushion. According to the application scenario, the surgical instrument will be surrounded by the body tissues, particularly in the abdominal cavity. In this experiment, the tip of the surgical instrument is inserted into a plastic container filled with pork. The data are collected at nine target locations within a 14 cm × 14 cm positioning area. Measurement results show that the minimum and maximum distance errors (ΔR) are 0.31 and 2.55 cm, respectively, and the overall mean error is 0.69 cm.

7.4 Conclusion

In this chapter, the theoretical concept and circuit implementation of the DMR sensor has been proposed. By electrically steering the radiation beams, we can generate numerous Σ-Δ curves in the desired FOV. Two positioning systems along with the accuracy enhancement algorithms are also introduced in this work. For indoor positioning application, the selecting-and-averaging algorithm was presented to moderate the impact from interference as well as to enhance the positioning accuracy. The great advantage is that only a single positioning node is needed for 2-D positioning. However, the positioning area is confined by the installation height and may not be suitable for large spaces.

The second application is to track surgical instruments during an operation. To further improve the positioning accuracy, the so-called "peak-tracking algorithm" has been introduced. A set of calibration curves are reconstructed from the precisely steered Σ-Δ curves, and the target angle can be estimated by finding the best match from those calibration curves via peak tracking. The experiments validate the significant improvement in angular accuracy. The tag antenna designed for surgical instruments poses a great challenge since its size is strictly confined. Although the preliminary results have demonstrated its potential in medical use, it still has a long way to go before this system can truly become practical in surgical navigation. This system is intended to provide a proof-of-concept study. Several issues are not fully addressed in this work yet, such as the instrument orientation sensing. Also, the regulation of medical devices should be further considered. At the current stage, this system can be used as a training platform, providing the surgeons with great assistance in surgical practice.

Acknowledgments

The authors would like to express their great appreciation to Prof. Mao-Ching Chiu for his valuable and constructive suggestions on multipath fading modeling. The authors would also like to thank the following former lab members for their contributions in developing the radar-based positioning systems through those years: Dr. Ting-Yueh Chin, Dr. Ruey-Hsuan Lee, Chia-Yi Jia, Chia-Tai Ho, Chia-Yang Hsu, Chih-Yuan Lu, Chung-Ting Lang, Zong-Da Tsai, and Shao-Hsuan Chang. It is truly a team effort to accomplish this work.

References

[1] M. D'Souza, B. Schoots, and M. Ros, "Indoor position tracking using received signal strength-based fingerprint context aware partitioning," IET Radar, Sonar & Navigation, vol. 10, no. 8, pp. 1347–1355, 2016.

[2] M. M. Atia, A. Noureldin, and M. J. Korenberg, "Dynamic online-calibrated radio maps for indoor positioning in wireless local area networks," IEEE Trans. on Mobile Computing, vol. 12, no. 9, pp. 1774–1787, 2013.

[3] M. Azizyan, I. Constandache, and R. Roy Choudhury, "Surround sense: mobile phone localization via ambience fingerprinting," in Proc. of MobiCom, pp. 261–272, 2009.

[4] C. Zhang, M. J. Kuhn, B. C. Merkl, A. E. Fathy, and M. R. Mahfouz, "Real-time noncoherent UWB positioning radar with millimeter range accuracy: theory and experiment," IEEE Trans. Microw. Theory Tech., vol. 58, no. 1, pp. 9–20, 2010.

[5] N. A. Alsindi, B. Alavi, and K. Pahlavan, "Measurement and modeling of ultrawideband TOA-based ranging in indoor multipath environments," IEEE Trans. Microw. Theory Tech., vol. 58, no. 3, pp. 1046–1058, 2009.

[6] C.-H. Lim, Y. Wan, B.-P. Ng, and C.-M. S. See, "A real-time indoor WiFi localization system utilizing smart antennas," IEEE Trans. Consumer Electronics, vol. 53, no. 2, pp. 618–622, 2007.

[7] A. Tennant, "Experimental two-element time-modulated direction finding array," IEEE Trans. Antennas Propag., vol. 58, no. 3, pp. 986–988, 2010.

[8] Y. S. Su, C. C. Chang, J. J. Guo, and S.F. Chang, "2-D wireless human subjects positioning system based on respiration detections," in Proc. of Int. Microw. Symp. (IMS), Montréal, Canada, 17–22, June 2012.

[9] R. H. Lee, J. C. Wu, S. H. Chang, S. F. Chang, C. C. Chang, and Y. M. Chen, "Radar design for wireless indoor positioning applications," in Proc. of the European Microw. Conf. (EuMC), Nuremberg, Germany, Oct. 6–11, 2013.

[10] A. S. Paul and E. A. Wan, "RSSI-based indoor localization and tracking using sigma-point Kalman smoothers," IEEE J. Sel. Topics Signal Process., vol. 3, no. 5, pp. 860–873, 2009.

[11] H.-S. Ahn and W. Yu, "Environmental-adaptive RSSI-based indoor localization," IEEE Trans. Automat. Sci. Eng., vol. 6, no. 4, pp. 626–633, 2009.

[12] P. W. Lu and R. S. Chen, "Infrared-based vehicular positioning with the automatic radiation-strength control," IET Intell. Transport System, vol. 8, no. 3, pp. 273–285, 2013.

[13] Y. S. Kuo, P. Pannuto, K.-J. Hsiao, and P. Dutta, "Luxapose: indoor positioning with mobile phones and visible light," in Proc. of MobiCom '14, New York, NY, USA, pp. 447–458, 2014.

[14] J. Krumm, S. Harris, B. Meyers, B. Brumitt, M. Hale, and S. Shafer, "Multi-camera multi-person tracking for easy living," in Proc. Third IEEE Int. Workshop on, Dublin, Ireland, pp. 3–10, 2000.

[15] W. Huang, Y. Xiong, X.-Y. Li, *et al.*, "Shake and walk: acoustic direction finding and fine-grained indoor localization using smartphones," in Proc. INFOCOM '14, Toronto, ON, Canada, pp. 370–378, 2014.

[16] S. Feldmann, K. Kyamakya, A. Zapater, and Z. Lue, "An indoor Bluetooth-based positioning system: concept, implementation and experimental evaluation," in ICWN '03, Las Vegas, Nevada, USA, vol. 272, 2003.
[17] M. Cypriani, F. Lassabe, P. Canalda, and F. Spies, "Open wireless positioning system: a Wi-Fi-based indoor positioning system," in VTC 2009-Fall, 2009 IEEE 70th, Anchorage, AK, USA, pp. 1–5, 2009.
[18] Y. Wang, X. Jia, H. K. Lee, and G. Y. Li, "An indoors wireless positioning system based on wireless local area network infrastructure," in SatNav, Melbourne, Australia, vol. 54, 2003.
[19] L. M. Ni, Y Liu, Y. C. Lau, and A. P. Patil, "LANDMARC: indoor location sensing using active RFID," in Proc. of IEEE Int. Conf. Pervasive Computing and Communication, pp. 407–415, 2003.
[20] M. Vossiek, R. Roskosch, and P. Heide, "Precise 3-D object position tracking using FMCW radar," in Proc. of 29th European Microw. Conf., Oct. 1999, vol. 1, pp. 234–237.
[21] R. H. Lee, Z. D. Tsai, C. T. Lang, C. C. Chang and S. F. Chang, "A switched-beam FMCW radar for wireless indoor positioning system," in Proc. of the European Radar Conf. (EuRAD), Manchester, UK, 9–14 Oct. 2011.
[22] Y. Xiao, J. Lin, O. Boric-Lubecke, and V. M. Lubecke, "Frequency tuning technique for remote detection of heartbeat and respiration using low-power double-sideband transmission in the Ka-band," IEEE Trans. Microw. Theory Tech., vol. 54, no. 5, pp. 2023–2032, 2006.
[23] F.-K. Wang, C.-J. Li, C.-H. Hsiao, *et al.*, "A novel vital-sign sensor based on a self-injection-locked oscillator," IEEE Trans. Microw. Theory Tech., vol. 58, no. 12, pp. 4112–4120, D2010.
[24] T. Y. Chin, K. Y. Lin, S. F. Chang, and C. C. Chang, "A fast clutter cancellation method in quadrature doppler radar for noncontact vital signal detection," in Proc. of 2010 IEEE Int. Microw. Symp. (IMS), Los Angeles, California, USA, 23–28 May 2010.
[25] H. S. Chang, H. C. Chu, P. T. Chen, C. C. Chang, and S. F. Chang "Human motion analysis based on multi-channel Doppler radar system," in Proc. of 2019 IEEE Int. Microw. Symp. (IMS), Boston, MA, USA, 2–7 June 2019.
[26] C. S. Lin, S. F. Chang, C. C. Chang, and C. C. Lin, "Microwave human vocal vibration signal detection based on Doppler radar technology," IEEE Tran. on Microw. Theory and Tech. vol. 58, no. 8, pp. 2299–2306, 2010.
[27] P. Molchanov, S. Gupta, K. Kim, and K. Pulli, "Short-range FMCW monopulse radar for hand-gesture sensing," In Proceedings of the 2015 IEEE Radar Conference (RadarCon), Arlington, VA, USA, 10–15 May 2015; IEEE: Piscataway, NJ, USA, 2015; pp. 1491–1496.
[28] P. R. Stauffer, D. B. Rodrigues, and P. F. Maccarini, "Utility of microwave radiometry for diagnostic and therapeutic applications of non-invasive temperature monitoring," 2014 IEEE Benjamin Franklin Symp. Microw. and Antenna Sub-systems for Radar, Telecommun., and Biomedical Applicat. (BenMAS), Philadelphia, PA, 2014, pp. 1–3.
[29] M. Bassi, M. Caruso, M. S. Khan, A. Bevilacqua, A. D. Capobianco, and A. Neviani, "An integrated microwave imaging radar with planar antennas

for breast cancer detection," IEEE Trans. Microw. Theory Tech., vol. 61, no. 5, pp. 2108–2118, 2013.
[30] U. Mezger, C. Jendrewski, and M. Bartels, "Navigation in surgery," Langenbeck's Archives of Surgery, vol. 398, no. 4, pp. 501–514, 2013.
[31] P. M. Novotny, D. T. Kettler, P. Jordan, P. E. Dupont, P. J. del Nido, and R. D. Howe, "Stereo display of 3D ultrasound images for surgical robot guidance," Int. Conf. IEEE Eng in Medicine and Biology Soc., New York, NY, 2006, pp. 1509–1512.
[32] Y. Nakajima, T. Dohi, T. Sasama, *et al.*, "Surgical tool alignment guidance by drawing two cross-sectional laser-beam planes," IEEE Trans. on Biomed. Eng., vol. 60, no. 6, pp. 1467–1476, 2013.
[33] N. Navab, S. M. Heining, and J. Traub, "Camera augmented mobile C-arm (CAMC): calibration, accuracy study, and clinical applications," IEEE Trans. Med. Imag., vol. 29, no. 7, pp. 1412–1423, 2010.
[34] Z. Sun, C. T. Soh, S. Udomsawaengsup, A. Shabbir and S. Foong, "Modular design of a real-time passive magnetic localization system for enhanced safety in nasogastric intubation," in Proc. 6th IEEE RAS/EMBS International Conference, 2016.
[35] T. D. Than, G. Alici, H. Zhou, and W. Li, "A review of localization systems for robotic endoscopic capsules," IEEE Trans. Biomed. Eng., vol. 59, no. 9, pp. 2387–2399, 2012.
[36] R. M. Page. "Accurate angle tracking by radar," NRL Report RA-3A-222A, December 28, 1944.
[37] J. C. Wu, C. C. Chang, T. Y. Chin, *et al.*, "Wireless indoor localization using dynamic monopulse receiver," in Proc. of 7th Radar Conference (EuRAD), pp. 69–72, Sept. 2010, Paris, France.
[38] J. Faerber, G. Cummins, and M. P. Y. Desmulliez, "Design of conformal wideband antennas for capsule endoscopy within a body tissue environment," 2016 46th European Microw. Conf. (EuMC), London, 2016, pp. 1223–1226.
[39] Y.-C. Chang, P. Y. Chen, C. T. Hsieh, C. C. Chang, and S.-F. Chang, "2-D wireless positioning on minimally invasive surgical instruments," in Proc. of Int. Microw. Symp. (IMS), San Francisco, CA, USA, 22–27 May, 2016.
[40] Z. Duan, L. Xu, S. Gao, and W. Geyi, "Integrated design of wideband omnidirectional antenna and electronic components for wireless capsule endoscopy systems," in IEEE Access, vol. 6, pp. 29626–29636, 2018.
[41] S. H. Lee, K. Chang, Y. J. Yoon, "A dual spiral antenna for wideband capsule endoscope system," in Proc. IEEE-APMC, Asia-Pacific Micro. Conf., Bangkok, Thailand, pp. 1–4, Dec. 2007.
[42] S. H. Lee, K. Chang, K. J. Kim, and Y. J. Yoon, "A conical spiral antenna for wideband capsule endoscope system," in IEEE AP-S. Symp., San Diego, California, pp. 1–4, July 2008.
[43] S. H. Lee and Y. J. Yoon, "Fat arm spiral antenna for wideband capsule endoscope systems," IEEE RWS., pp. 579–582, Jan. 2010.

[44] P. M. Izdebski, H. Rajagopalan, and Y. Rahmat-Samii, "Conformal ingestible capsule antenna: a novel chandelier meandered design," IEEE Trans. Antennas and Propag., vol. 57, no. 4, pp. 900–909, 2009.
[45] S. Yun, K. Kim, and S. Nam, "Outer-wall loop antenna for ultrawideband capsule endoscope system," IEEE Antennas Wireless Propag. Lett., vol. 9, pp. 1135–1138, 2010.
[46] Q. Wang, K. Wolf, and D. Plettemeier, "An UWB capsule endoscope antenna design for biomedical communications," in Proc. 3rd Int. Symp. ISABEL., Rome, Italy, pp. 1–6, Nov. 2010.
[47] X. Cheng, E. David, K. Cheolbok, and Y. K. Yoon, "A compact omnidirectional self-packaged patch antenna with complementary split-ring resonator loading for wireless endoscope applications," IEEE Antennas Wireless Propag. Lett., vol. 10, pp. 1532–1535, 2011.

Chapter 8

Noncontact healthy status sensing using low-power digital-IF Doppler radar

Hong Hong[1], Heng Zhao[1], Li Zhang[1], Chuanwei Ding[1], and Xiaohua Zhu[1]

Healthy status sensing is of great significance in the early disease prevention, clinical treatment and back-end home care. Vital signs can not only provide physiological information but also reflect various healthy status of human subject. Our emphasis is on discovering inner relationships between the Doppler radar-based noncontact vital sign detection and the healthy status of human subject. The custom-designed low-power digital intermediate frequency (digital-IF) continuous-wave (CW) Doppler radar has been designed to capture the vital signs with high accuracy and robustness. Then the compressed sensing (CS)-based method is proposed to enhance the resolution of the vital sign spectrum, and the synchrosqueezing transform (SST)-based algorithm is used to extract the instantaneous vital signs. Based on the digital-IF Doppler radar and the advanced signal processing algorithms, several healthy status sensing modules, including the breathing disorder recognition and sleep-stage estimation, have been realized by using advanced machine learning techniques. Laboratorial and clinical experiments demonstrate the effectiveness of noncontact healthy status sensing using the proposed radar system.

8.1 Digital-IF CW Doppler radar

A single carrier frequency CW vital sign detection radar based on digital-IF structure has been proposed. In particular, the digital-IF structure, based on the superheterodyne structure, directly samples IF signals and adopts digital processing from IF signal to baseband signal [1]. Compared with the traditional homodyne structure of the vital sign detection radar, the digital-IF structure has the advantages of high sensitivity, high precision, high integration and lower transmit power. In vital sign detection, the chest-wall displacement caused by cardiopulmonary motions (especially heart beat) is very small [2], so the digital-IF structure can effectively improve the successful detection rate and measurement accuracy of vital sign monitoring. Figure 8.1 shows the overall block diagram of the digital-IF

[1]Nanjing University of Science & Technology, China

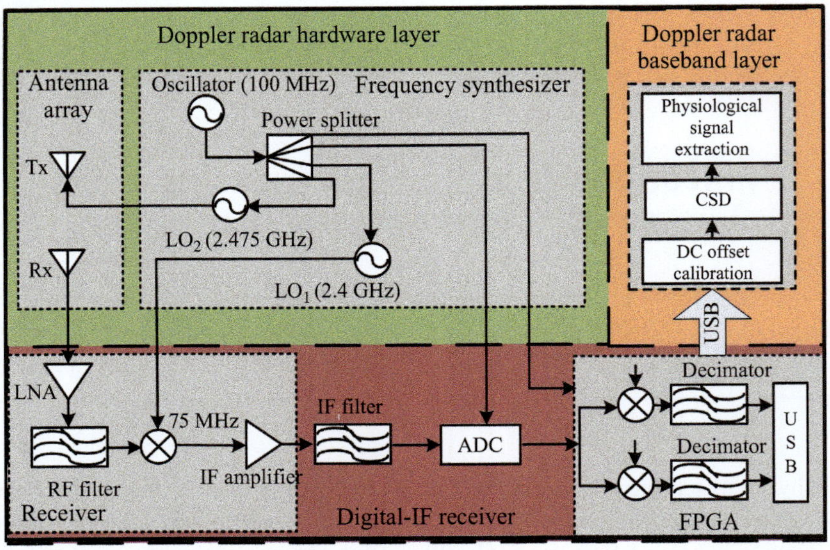

Figure 8.1 The block diagram of the digital-IF CW Doppler radar hardware layer [3]

CW Doppler radar. As shown in this figure, the proposed system consists of the radio frequency (RF) layer, the digital-IF layer and the baseband layer.

8.1.1 RF layer

The main function of the RF layer is to generate RF signals, realize the transmission and reception of RF signals and also perform the first-stage downconversion processing. As shown in Figure 8.1, the RF layer can be further divided into transmitting part and receiving part, including the antenna array, a receiver and a frequency synthesizer.

In the transmitting part, a microstrip antenna array is used as transmitting and receiving antenna. The size of a single antenna is 161 mm × 161 mm, the estimated beam width is approximately 40° and the maximum gain reaches 12.8 dB at 2.475 GHz. In the transmit branch, a custom-designed frequency synthesizer demonstrated in Figure 8.2(a) is designed to generate a 100-MHz clock signal, a 2.475-GHz transmitted signal and a 2.4-GHz reference signal. As shown in Figure 8.2(a), a temperature-compensated crystal oscillator (Tianma telecom TX3213B) is used to provide the 100-MHz reference clock signal. In order to guarantee the coherence of the Doppler radar, the 100-MHz clock signal is divided into four channels through a four-port power splitter (Mini-Circuits AD4PS-1+). The four channels are the clock signals for the analog-to-digital convertor (ADC), the field-programmable gate array (FPGA) and the two local oscillators (LO_1 and LO_2). Lastly, the 2.475-GHz transmitted signal and the 2.4-GHz reference signal are produced by two voltage-controlled oscillators with phase-locked loop (LO_1 and LO_2).

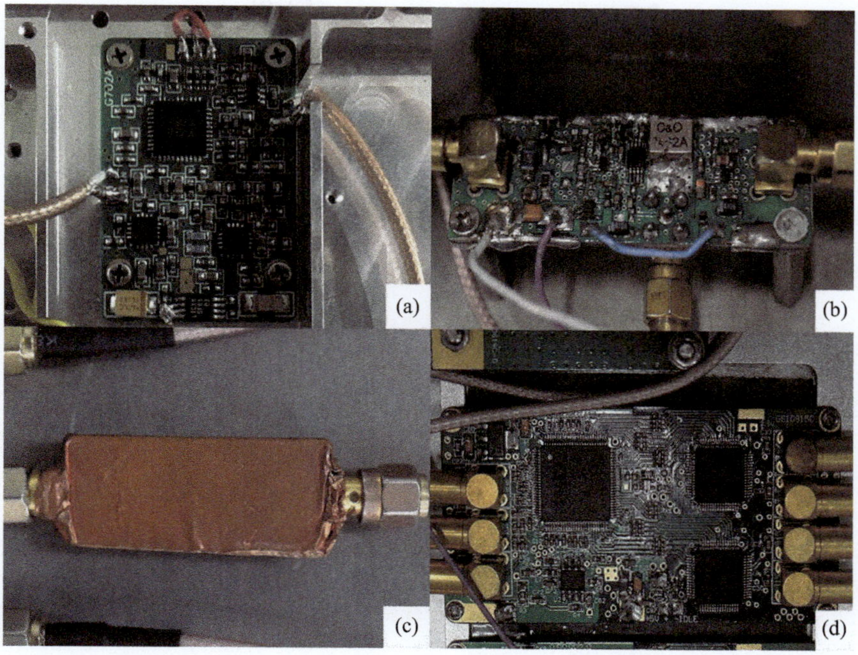

Figure 8.2 The photograph of the heterodyne CW Doppler radar [3]. (a) Custom-designed frequency synthesizer. (b) Custom-designed receiver. (c) Custom-designed IF filter. (d) 12-bit ADC

In the receiving part, a custom-designed digital-IF receiver shown in Figure 8.2(b) is designed for downconverting the received RF signal to the IF output. The echo signal is received by the receiving antenna and amplified by a low-noise amplifier (Mini-Circuits PSA-5451+). The amplified signal is then passed through an RF filter to filter out-of-band noise. The filtered signal is then mixed with the 2.4-GHz local oscillator signal to obtain an IF signal with a carrier frequency of 75 MHz. Then it is imported into the IF layer for subsequent signal processing.

8.1.2 IF layer

The main function of the IF layer is to implement analog-to-digital conversion, downconversion and IF signal processing. With the development of digital signal processing technology, digital IF receivers are gradually being applied to radar systems. Compared with the analog one, the radar receiver based on digital IF technology has the advantages of high processing precision, good flexibility, small device size, low power consumption and strong anti-interference ability.

In the IF layer, the 75-MHz IF signal generated by downconversion of the RF layer is first filtered by an IF filter to eliminate the out-of-band noise. The IF filter has a center frequency of 75 MHz with a 3-dB bandwidth from 71 to 79 MHz. Therefore, it can effectively filter out clutter and noise to improve the signal-to-noise ratio. Next, the filtered signal is amplified by an IF amplifier and converted to a differential signal by a single-ended-to-differential circuit. Then, the differential signal is input to a specially designed ADC sampling circuit for sampling.

Due to the low frequency of vital signs, the spectrum of the IF signal is concentrated around 75 MHz. According to the traditional Nyquist sampling theorem, the sampling frequency should be at least twice of the maximum component in the signal spectrum. Here, the bandpass sampling theorem is adopted to sample the IF signal [4]. According to the bandpass sampling theorem, when the sampling frequency satisfies:

$$\frac{2f_H}{n} \le f_s \le \frac{2f_L}{n-1} \tag{8.1}$$

where n is an integer, and satisfies the following relationship:

$$2 \le n \le int\left(\frac{f_H}{B}\right) \tag{8.2}$$

where $int(\bullet)$ indicates the largest integer, the sampled signal can be recovered without distortion. Due to the frequency ranges of the vital signs, it is assumed that the spectrum of the IF echo signal is concentrated during 74–76 MHz with a 2-MHz bandwidth. As a result, the range of n is $2 \le n \le 38$. When n is 2, the sampling frequency ranges from 76 to 148 MHz. Is this system, a 12-bit ADC (Analog Devices AD9430) demonstrated in Figure 8.2(d) digitalizes the analog IF signal with a 100-MHz sampling frequency.

After sampling, the digitized signal is sent to the FPGA (Xilinx XC6VLX130T) platform. Figure 8.3 shows the block diagram of the IF signal processing implemented in the FPGA, including digital downconversion and decimation filtering. In order to eliminate the null/optimal point problem of single-channel receiver, we use two quadrature digital local oscillator signals and IF digital echo signals for multiplication and mixing operations to achieve digital downconversion. The quadrature digital local oscillator signals are generated by direct digital frequency synthesis (DDS) techniques. DDS technology produces sinusoidal signals with a wide range of frequency variations with extremely high resolution. Since the baseband signal of vital sign signal does not require a sampling rate of 100 MHz, the two quadrature

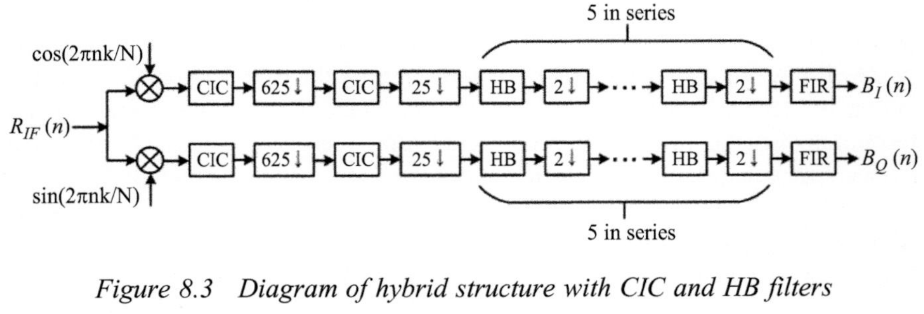

Figure 8.3 Diagram of hybrid structure with CIC and HB filters

signals after mixing need to be decimated to achieve a reduction in the sampling rate. A hybrid structure of a classic multistage cascaded integrator-comb (CIC) filter and a half-band (HB) filter is utilized to implement the decimation. The HB filter is suitable for realizing the decimation with a power of 2, which has high computational efficiency and strong real-time performance. Cooperated with the CIC filter, it can achieve integer times decimation. As shown in Figure 8.3, the sampling rate of each channel is reduced from 100 MHz to 100 Hz. The two quadrature signals are then sent to the baseband layer through a programmable Universal Serial Bus (USB) 3.0 peripheral.

8.1.3 Baseband layer

The baseband layer includes the DC offset calibration, the complex signal demodulation (CSD) and physiological signal extraction. For a digital-IF receiver, the amplitude imbalance can be neglectful, but the DC offset still exists in the quadrature channels [5]. As a result, the quadrature signals can be written as [6]:

$$B_I(n) = \cos\left[\theta + \frac{4\pi x(n)}{\lambda} + \Delta\Phi(n)\right] + DC_I \tag{8.3}$$

$$B_Q(t) = \sin\left[\theta + \frac{4\pi x(t)}{\lambda} + \Delta\Phi(n)\right] + DC_Q \tag{8.4}$$

where $\theta = 4\pi d_0/\lambda + \theta_0$ is the constant phase shift dependent on the nominal distance to the target d_0, $\Delta\Phi(n) = \Phi(n) - \Phi(n - 2d_0/c)$ is the total residual phase noise, $x(n)$ denotes the chest-wall displacement which consists of heartbeat signal $x_h(n)$ and respiration signal $x_r(n)$, and DC_I and DC_Q denote the DC offsets in the in-phase (I) and quadrature-phase (Q) channels, respectively.

As shown in [6], the physiological signal-induced chest-wall displacement is involved in $x(n)$. Since the phase shift θ and residual phase noise $\Delta\Phi(n)$ are nearly constant, the phase information can be considered as the respiration signal after subtracting these components. Since the DC offset reduces the linearity and demodulation accuracy, the l_p $(0 < p < 1)$ minimization-based method is used [5].

After removing the DC offset, the CSD is utilized to combine the quadrature channel outputs as [7]

$$\begin{aligned} S(t) &= B_I(t) + j \times B_Q(t) \\ &= \cos\left[\theta + \frac{4\pi x(t)}{\lambda} + \Delta\Phi(t)\right] + j \cdot \sin\left[\theta + \frac{4\pi x(t)}{\lambda} + \Delta\Phi(t)\right] \\ &= \exp\left\{j\left[\frac{4\pi x(t)}{\lambda} + \Delta\Phi(t)\right]\right\}. \end{aligned}$$

Finally, the demodulated signal $S(t)$ is filtered by two bandpass filters (0.1–0.6 and 0.8–2 Hz) to extract physiological signals including respiration signal $x_r(t)$ and heartbeat signal $x_h(t)$, respectively.

8.2 Advanced signal processing algorithms for physiological signal extraction

For the noncontact vital sign detection radar, the discrete Fourier transform (DFT) is the most widely used algorithm to extract the physiological signal frequencies [7]. However, due to the smearing and leakage problems caused by limited data length, the DFT suffers from significant performance reduction. To solve this problem, several algorithms have been proposed. A parametric and cyclic optimization algorithm, which is called the relaxation (RELAX) algorithm, can extract the heartbeat and respiratory components for a 12.015-s-long duration with a 20-Hz sampling frequency [8]. The multiple signal classification (MUSIC) algorithm has been demonstrated its feasibility to accurately obtain the heartbeat frequency in a 8–28-s time intervals [9]. The cyclostationary approach has been utilized [10]. This approach uses nonlinear transformation and Fourier analysis to obtain the periodicity of heartbeat and respiration information.

Even though these methods can eliminate the smearing and leakage problems, the required data length restricts their applications in processing the time-varying vital sign frequencies. In practice, the respiratory and heartbeat signals are nonstationary, which usually have the time-varying frequencies. The physiological dynamics hidden in the time-varying frequencies can reveal the health information of the cardiopulmonary system. In this section, the classical compressed sensing (CS) theory is first introduced to tackle the short-term vital sign frequencies [11]. The time window length can be reduced to 5 s. Afterwards, to remove the frequency grid in the classical CS method, the stepwise atomic norm minimization (ANM) method has been proposed [12]. The estimation accuracy of vital sign frequencies can be improved effectively. Furthermore, the concept of short-term vital sign has been extended to the instantaneous vital sign. A high-resolution time-frequency (TF) analysis algorithm, SST, has been used to extract the instantaneous vital signs [6]. Both simulated and laboratorial experiments have been carried out, and the results all demonstrate the effectiveness of the three methods.

8.2.1 CS and stepwise ANM

8.2.1.1 Sparse reconstruction

The phase information can be represented as a sum of sinusoids with their frequencies corresponding to the respiratory, heartbeat, harmonics and intermodulation components. Thus, the discrete expression can be written as

$$x(n) = \sum_{k=1}^{K} \alpha_k \cos(2\pi f_k n T_0 + \phi_k) + w(n) \tag{8.5}$$

where K is the number of sinusoids, T_0 is the sampling interval, $n = 1, 2, \cdots, N$ is the sampling points and $w(n)$ is the additive Gaussian noise. α_k and f_k are the amplitude and frequency corresponding to the kth sinusoid component, respectively. Dividing the spectral interval $[F_l, F_h]$ into P_s frequency girds $\{f_1, f_2, \cdots, f_{P_s}\}$, a dictionary can be constructed as

$$\mathrm{A} = [\mathrm{a}_1, \mathrm{a}_2, \cdots, \mathrm{a}_{P_s}] \in C^{N \times P_s} \tag{8.6}$$

where

$$\mathrm{a}_{P_s} = \left[1, e^{j2\pi f_{P_s} T_0}, e^{j2\pi f_{P_s} 2T_0}, \cdots, e^{j2\pi f_{P_s}(N-1)T_0}\right]' \tag{8.7}$$

Define a coefficient vector $\mathrm{s} = [s_1, s_2, \cdots, s_{P_s}]'$ where the entries yield:

$$s_{P_s} = \begin{cases} \alpha_k & f_{P_s} = f_k \\ 0 & f_{P_s} \neq f_k \end{cases} \tag{8.8}$$

we can reformulate the (8.5) as

$$\boldsymbol{x} = \mathrm{A}s + \boldsymbol{w} \tag{8.9}$$

where $\boldsymbol{x}$ and $\boldsymbol{w}$ are the $N \times 1$ measurement vector and noise vector, respectively. $\boldsymbol{x}$ is K-sparse under the dictionary A.

As we know, an increment on the grid number P_s leads to the improvement of spectral resolution of signal $\boldsymbol{x}$. When $P_s > N$, the problem becomes a super-resolution problem, which results in (8.9) undetermined. The dictionary A is assumed to be overcompleted. Thus, a unique solution of $\boldsymbol{x}$ can be determined by adding some sparse conditions.

In the practical use, s is sparse with only few nonzero elements. Thus, a sparse reconstruction problem can be set as: if an $N \times 1$ signal can be approximated by a linear combination of K $(K < N)$ atoms in an $N \times P_s(N < P_s)$ over-completed dictionary, it can be precisely reconstructed by solving the undetermined model with an l_0-norm minimization as follows:

$$\min \|\mathrm{s}\|_0 \qquad \text{s.t.} \quad \|\mathrm{x} - \mathrm{As}\| < \varepsilon \tag{8.10}$$

where $\|\cdot\|_0$ denotes the l_0-norm. As we know, the l_0-norm minimization is an non-deterministic polynomial (NP)-hard problem, which can be replaced by the l_1-norm minimization when the signal is sparse enough and the sensing matrices have sufficiently incoherent columns [13]. The l_1-norm minimization is the closest convex optimization, and it can be solved in a polynomial time.

8.2.1.2 MMV model of heartbeat detection

Since the amplitude of respiration is usually ten times of the heartbeat component, some enhancement of heartbeat component is necessary. The multiple measurement vectors (MMV) model is proposed to rearrange the demodulated signal [14]. The whole signal is divided into Q_s frames. Each frame is rearranged into L snapshots with a N_s-point sliding time window. Thus, (8.9) can be written as

$$\mathbf{X} = \mathbf{AS} + \mathbf{E} \tag{8.11}$$

where $\mathbf{X} = \left[\mathrm{x}^{(1)}, \mathrm{x}^{(2)}, \cdots, \mathrm{x}^{(L)}\right]$ is the L-snapshots measurement matrix, $\mathbf{S} = \left[\mathrm{s}^{(1)}, \mathrm{s}^{(2)}, \cdots, \mathrm{s}^{(L)}\right]$ is a row sparse matrix with nonzero rows corresponding to the dominant frequencies, $\mathbf{E}$ is the noise matrix. As a result, the MMV model of (8.10) is

$$\min_{\mathrm{S}} \|\mathrm{S}\|_0 \qquad \text{s.t.} \quad \|\mathrm{X} - \mathrm{AS}\| < \varepsilon_m \tag{8.12}$$

where $\|\mathrm{S}\| = |supp\,\mathrm{S}|$ is the promoted sparsity measure, $supp\,\mathrm{S} = \{1 \le j \le P : \mathrm{s}^j \neq 0\}$ and s^j denotes the jth row of $\boldsymbol{S}$.

Theocratically, the estimation performance can be improved with more snapshots. However, more snapshots will reduce the ability of short-time processing. Based on experiments, the sliding step is set within one period of the heartbeat.

The RMFOCUSS algorithm is used to solve the MMV model in (8.12) [14]. The key step of the regularized multi-vector focal undetermined system solver (RMFOCUSS) is using the iterations to asymptotically eliminate the row vectors of estimator $\|\hat{\mathrm{S}}\|$ according to the sparsity measure as

$$J^{(q)}(\mathrm{S}) = \sum_{n=1}^{P_s} (\|\mathrm{s}[n]\|)^q = \sum_{n=1}^{P_s} \left(\sum_{l=1}^{L} |\mathrm{s}^{(l)[n]}|^2 \right)^{q/2} \tag{8.13}$$

When q is close to zero, the value of the sparsity measure represents the number of nonzero rows in $\mathbf{S}$. A regularization parameter λ can be used to balance the sparsity and the recovery error. Since the third or fourth harmonics of the respiration are very close to the heartbeat component in the spectrum. The heartbeat component is estimated within a certain frequency scope such as 0.8–2.2 Hz.

8.2.1.3 Stepwise ANM

The ANM can separate the spectral lines with a low sampling rate and short duration only if the spectrum is "sufficient separation" [15]. However, as the minimum frequency separation between heartbeat and respiration harmonics is very small, the ANM cannot be directly used to noncontact vital sign detection. The stepwise ANM (StANM) is proposed to eliminate the limitation [12]. The proposed algorithm first extracts the respiratory frequency to estimate the locations of its harmonics. Then the location of heartbeat can be obtained by solving a modified ANM. The details of stANM can be summarized as follows.

The first step is the conventional ANM procedure, the time-domain signal $r \in C^{N\times 1}$ is recovered from the noisy measurement by solving the semidefinite program as

$$\min_{\boldsymbol{r},\boldsymbol{T},t} \frac{\mu}{2N} tr(\boldsymbol{T}) + \frac{\mu}{2} t + \frac{1}{2} \|\boldsymbol{y} - \boldsymbol{r}\| \; subject\ to \begin{bmatrix} \boldsymbol{T} & \boldsymbol{r} \\ \boldsymbol{r}^* & t \end{bmatrix} \ge 0 \tag{8.14}$$

where $\mathbf{T}$ is a Toepliz matrix, and $tr()$ and $(\cdot)^*$ denote the trace and conjugate transpose, respectively. N is the signal length. Then the solution v can be obtained as

$$\max_{v,Q} Re(\boldsymbol{v}^*\hat{\boldsymbol{r}})\ subject\ to \begin{bmatrix} \boldsymbol{T} & \boldsymbol{r} \\ \boldsymbol{r}^* & t \end{bmatrix} \geq 0$$

$$\sum_{i=1}^{N-1} \boldsymbol{Q}_{i,i+j} = \begin{cases} 1, & j=0 \\ 0, & j \in [1, N-1] \end{cases} \tag{8.15}$$

$\boldsymbol{Q}$ *is Hermitian*

The frequencies can be obtained when the dual polynomial reaches μ, which is

$$\left| \sum_{n=0}^{N-1} ve^{-j2\pi nf} \right| = \mu \tag{8.16}$$

At last, the corresponding amplitudes can be solved by a least squares problem.

Assuming the component with the maximum amplitude is the respiratory component, the harmonics of respiratory signal are defined as $\hat{f}_k = k\hat{f}_1$, $k = 2, 3, \cdots, K-1$.

The last step is solving a modified a semidefinite program as

$$\min_{r, r_{res}, T, t, \beta_k} \frac{\mu}{2N} tr(\boldsymbol{T}) + \frac{\mu}{2} t + \frac{1}{2} \|\boldsymbol{y} - \boldsymbol{r}\|$$

$$subject\ to \begin{bmatrix} \boldsymbol{T} & \boldsymbol{r}_{res} \\ \boldsymbol{r}^* & t \end{bmatrix} \geq 0 \tag{8.17}$$

$$\mathbf{r} = \boldsymbol{r}_{res} + \sum_k \beta_k \boldsymbol{a}_k, \quad k \in P$$

where P is the set of indices of the estimated frequencies $\hat{f}_k$, β_k is the amplitude and $\boldsymbol{a}_k = \left[1 \quad e^{j2\pi \hat{f}_k} \ldots \quad e^{j2\pi(N-1)\hat{f}_k}\right]^T$ is the atom of the kth frequency. The dual optimal solution $\hat{\boldsymbol{v}}_{res}$ is calculated with the residual optimal $\hat{\boldsymbol{r}}_{res}$ and the heartbeat frequency is defined as the one with dominant amplitude.

8.2.1.4 Laboratorial experiments

The first experiment is the comparison of fast Fourier transform (FFT), MUSIC and RMFOCUSS for noncontact vital sign detection. The sampling rate is 100 Hz. During the experiment, the human subject sat 2 m away from the radar antenna. The contact monitor (JP2000-09) was used as the reference. The total signal length is 80.94 s. In the MMV model, each frame length is 5.12 s. The interval between two successive frames is 1 s. Thus, Q_s is 55 and L is 100 in each frame. For the RMFOCUSS algorithm, q is 0.8, γ is 0.01 and λ is 5. For MUSIC algorithm, the signal and noise are divided into 64 subspaces.

To evaluate the ability of the short-time processing, the TF spectrums of MUSIC and RMFOCUSS algorithms are calculated for the whole signal. Figure 8.1 shows the TF spectrums in a frequency scope ranged from 0.8 to 1.4 Hz. In Figure 8.4(a), only

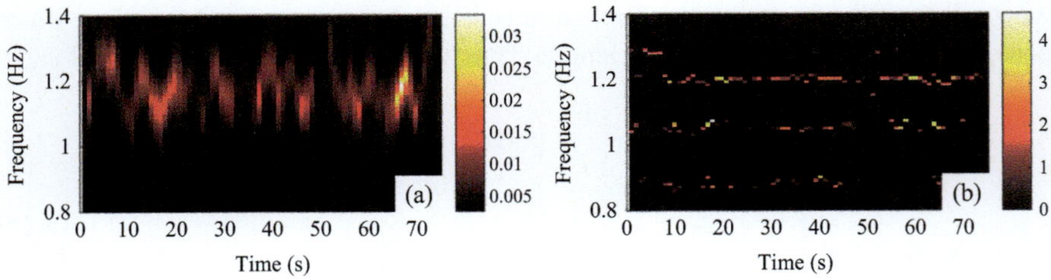

Figure 8.4 The TF spectrums derived from (a) the MUSIC algorithm and (b) the RMFOCUSS algorithm [11]

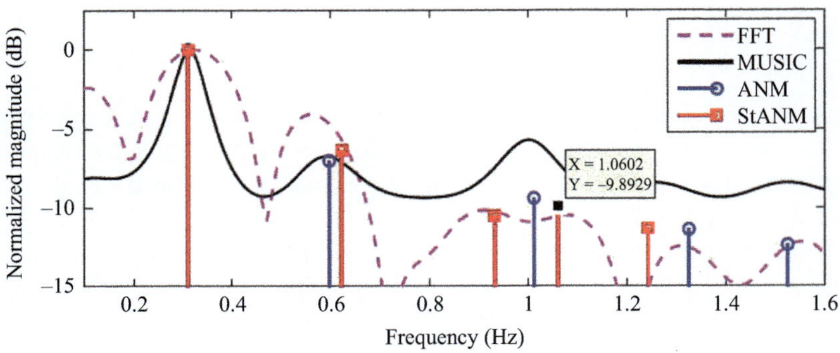

Figure 8.5 The vital sign estimation results derived from FFT, MUSIC, ANM and stANM [12]

one swing track can be seen, which is combined with the heartbeat and fourth harmonic of respiration. However, in Figure 8.4(b), it is obvious that three clear tracks can be found, which are corresponding to the third harmonic of respiration, heartbeat and fourth harmonic of respiration. Therefore, the result demonstrates the superiority of RMFOCUSS in short-time noncontact vital sign.

Furthermore, Figure 8.5 demonstrates a comparison of spectral estimation performance of FFT, MUSIC, ANM and stANM. In the experiment, a 6.5-s-long signal frame was randomly chosen. The human subject was seated 2 m away from the antenna. As shown in the figure, all the algorithms perform well in estimating the respiratory component. However, only the StANM can distinguish the heartbeat component from the harmonics. The heart rate estimated from the StANM is 64 beats per min (BPM;1.06 Hz), which is close to the result from JP2000-09.

8.2.2 SST for instantaneous vital sign detection

8.2.2.1 Synchrosqueezing transform

The SST is a high-resolution TF algorithm [16,17]. Due to the energy spread of continuous wavelet transform (CWT) on the time-scale plane, it will be unavoidably smeared by the wavelet function. The SST reassigns the energy distribution of CWT so that a high-resolution TF representation can be obtained. In particular, it allows reconstruction operation in the selected TF region.

The vital sign signals can be reformulated as a series of amplitude-demodulated–frequency-demodulated (AM–FM) signals as

$$s(t) = \sum_{k=1}^{K} s_k(t) \tag{8.18}$$

where $s_k(t) = A_k(t)\cos[\varphi_k(t)]$ is the AM–FM signal with instantaneous amplitude $A_k(t)$ and frequency $\varphi_k'(t)$ satisfying $\varphi_{k+1}'(t) > \varphi_k'(t)$ for all t. Based on the AM–FM model, the SST algorithm can be described in detail as follows.

First, the CWT $W_f(a, b)$ of the signal $f(t)$ is calculated as

$$W_f(a, b) = \int f(t)a^{-1/2}\overline{\nu(\frac{t-b}{a})}dt \tag{8.19}$$

where $\nu(t)$ is the mother wavelet and $\overline{\nu(t)}$ denotes the complex conjugate. For any points on time-scale plane satisfying $W_f(a, b) \neq 0$, the candidate instantaneous frequency can be obtained as

$$\omega_f(a, b) = -i\left[W_f(a, b)\right]^{-1}\frac{\partial W_f(a, b)}{\partial b} \tag{8.20}$$

Due to the existence of noise, a threshold $\gamma > 0$ can be set to ignore any points where $|W_f(a, b)| < \gamma$. The SST is defined as

$$T_f(\omega, b) = \int_{A(b)} W_f(a, b)a^{-3/2}\delta(\omega_f(a, b) - \omega)d\omega \tag{8.21}$$

where $A =: \{a; W_f(a, b) \geq \gamma\}$.

According to the ranges of respiratory and heartbeat frequencies, $f_r(t)$ and $f_h(t)$, the instantaneous respiratory frequency $if_r(t)$ and heartbeat frequency $if_h(t)$ are obtained by extracting the peaks within the corresponding ranges from the TF spectrum, which can be represented as

$$if_r(t) = \max_{w\in[00.7]Hz} |T_f(\omega, b)|^2 \tag{8.22}$$

$$if_h(t) = \max_{w\in[0.73]Hz} |T_f(\omega, b)|^2 \tag{8.23}$$

To eliminate the random fluctuations induced by noise and computational errors, the extracted curves are smoothed by a simple moving average. Let l_r and l_h be the indices of a small frequency band around $if_r(t)$ and $if_h(t)$ respectively, the instantaneous time-domain respiratory and heartbeat signals can be reconstructed as

$$x_r(b) = 2\mathcal{R}_\nu^{-1}\Re\mathfrak{e}\left[\int_{l_r} T_f(\omega, b)d\omega\right] \tag{8.24}$$

$$x_h(b) = 2\mathcal{R}_\nu^{-1}\Re\mathfrak{e}\left[\int_{l_h} T_f(\omega, b)d\omega\right] \tag{8.25}$$

where

$$\mathcal{R}_v = \int_0^\infty \overline{\hat{v}(\xi)} \frac{d\xi}{\xi} \tag{8.26}$$

here $\hat{v}(\xi)$ is the Fourier transform of $v(t)$ and $\mathcal{R}_v^{-1}$ denotes the reciprocal of $\mathcal{R}_v$.

When the mother wavelet satisfies $\xi_0 \leq \frac{1-\Delta}{1+\Delta}$, where the support in the frequency domain is $[1-\Delta, 1+\Delta]$ and ξ_0 is the ratio of the two frequencies, two signals can be separated by SST. It is known that the ratio of heartbeat and respiration is

$$\xi_0 = \frac{f_r(t)}{f_h(t)} \approx 0.3 \tag{8.27}$$

As a result, the support of $v(t)$ should be smaller than $(\frac{6}{13}, \frac{20}{13})$. Thus, the Morlet wavelet is chosen in the experiments due to the proper frequency support.

8.2.2.2 Laboratorial experiments

To validate the effectiveness of the SST algorithm for noncontact instantaneous vital sign detection, the post-exercise instantaneous vital sign rate was measured. First, the human subject was asked to do some physical exercise for about 5 min to increase his vital sign rates. Once he stopped exercising, the subject sat on a chair while reclining against the back immediately to recover. Then a 120-s vital sign signal was recorded during his recovery.

In this experiment, the TF distributions calculated by short-time Fourier transform (STFT), CWT and SST have been calculated. In Figure 8.6 [6], it is obvious that two trajectories can be seen in all figures, which correspond to the instantaneous respiratory and heartbeat rates. Note that the result from CWT is a time-scale distribution, the y-axis is inversely proportional to the frequency. The instantaneous heart rate shows a decreasing monoexponential fashion from 120 to

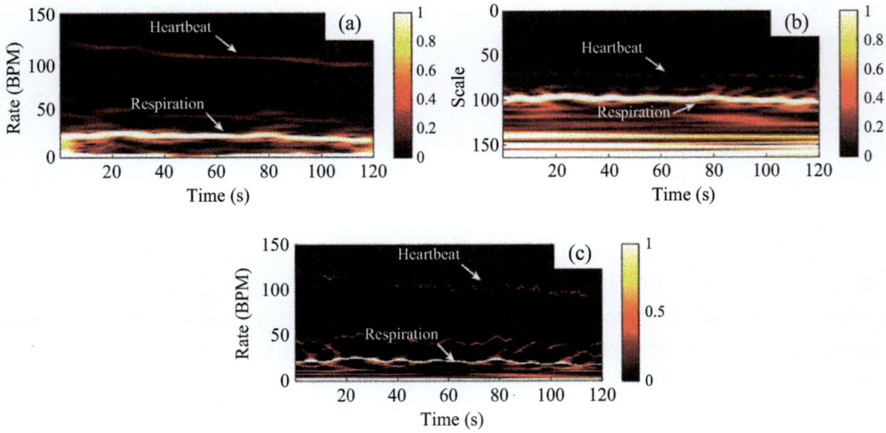

Figure 8.6 The TF spectrums derived from (a) the STFT algorithm, (b) the CWT algorithm and (c) the SST algorithm [6]

97 BPM. However, the instantaneous breathing rate decreases from 21 to 16 BPM. Comparing the three figures, the SST yields the most concentrated TF representation. Moreover, some fluctuations in the heart rate curve can only be observed in the SST result. The instability of the instantaneous HR during recovery may be induced by the cardiovascular parasympathetic system. Also, the second harmonic of the respiratory signal can be identified clearly. This experiment demonstrates that the SST algorithm has the capability of significantly variational instantaneous vital sign rate measurement and long-term monitoring.

8.3 Noncontact healthy status sensing

8.3.1 Breathing disorder recognition

Breathing is one of the important human vital signs which strongly correlates to health conditions. Unlike the regular parameters such as breathing rate and tidal volume, the breathing disorder is difficult to be accurately captured but provides significant clues for disease diagnosis and prognosis. The breathing disorder is also characterized as irregular breathing pattern. It can be induced by several diseases such as stroke and heart failure [18]. As a result, breathing disorder recognition is of great importance for health monitoring and disease prognosis.

Conventional methods for monitoring breathing disorder are costly and need medical personnel to operate. Fortunately, it has been proved that the biomedical radar can capture the time-domain waveform accurately. Therefore, it is considered to be a good candidate to acquire the breathing disorder.

Based on the proposed digital-IF Doppler radar sensor, A noncontact breathing disorder recognition using the digital-IF Doppler radar has been proposed [19]. Figure 8.7 shows the block diagram of the system. The proposed system consists of the digital-IF Doppler radar sensor and the breathing disorder recognition module. The details of the Doppler radar sensor can be found in Section 8.1. The breathing disorder recognition module includes the breathing activity segmentation, feature extraction and breathing disorder classification. The details are shown in the next section.

8.3.1.1 Breathing disorder recognition module

In the breathing disorder recognition module, six kinds of breathing patterns can be identified, which include the normal breathing, Cheyne–Stokes breathing, Cheyne–Stokes variant breathing, dysrhythmic breathing, Biot's breathing and the Kussmaul's

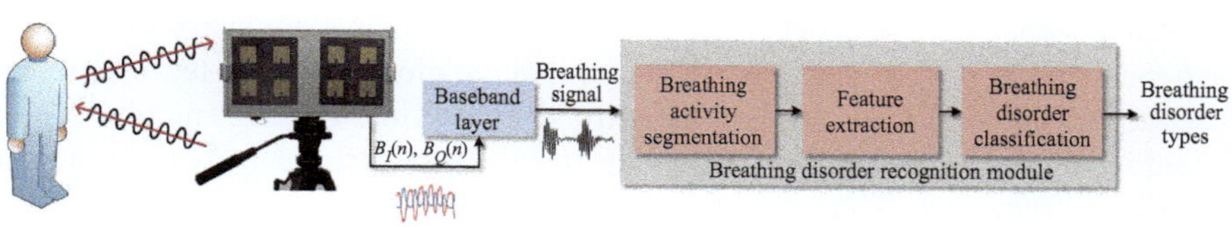

Figure 8.7 The block diagram of the noncontact breathing disorder recognition system

Table 8.1 Thirteen features used in the classification [19]

Categories	Indexes	Features
Time-domain features	1	Peak amplitude variability (σ_a)
	2	Average peak amplitude ($\bar{a}$)
	3	Maximum peak amplitude ($\max_a$)
	4	Number of peaks (N_a)
Short-term domain features	5	Variability (σ_E)
	6	Maximum ($\max_E$)
	7	Minimum ($\min_E$)
	8	Range (R_E)
TF domain features	9	Variability (σ_f)
	10	Mean ($\bar{f}$)
	11	Maximum ($\max_f$)
	12	Minimum ($\min_f$)
	13	Range (R_f)

breathing. A database was constructed first from ten healthy human subjects including seven males and three females. They were instructed to imitate the patterns, and each pattern was repeated six times and lasted for 30 s.

The breathing signal is segmented to separate the breathing activity and apnea. An adaptive segmentation algorithm, the Vertical Chart Box (V-Box), is utilized. The V-Box algorithm is a nonparametric detection rule which relies on the concept of a moving and vertically trimmed box [20]. The construction does not use a priori knowledge on underlying probability distribution. The V-Box algorithm uses a box to detect the breathing activities. When the observations locate in the box, the number is counted. When the counted number exceeds the given threshold, the interval in the box is defined as breathing. Conversely, that out of the box is an apnea. Due to the amplitude difference of the six breathing patterns, the normalization is carried out before using the V-Box algorithm. The parameters are selected empirically. Considering the nature of apnea, if the interval between two breathing activities is less than 3 s, the two breathing activities can be merged as one.

Then, considering the feasibility and computational complexity, 13 distinct features including 4 time-domain features, 4 short-term energy features and 5 TF features are extracted to characterize the breathing patterns. Table 8.1 shows the 13 features. These features are extracted from each 30-s period. The time-domain waveform contains crucial information of breathing patterns. For the radar-detected breathing signal, the peaks denote the transition between inspiration and expiration, and the absolute peak amplitude reflects the tidal volume. The short-term energy is used to represent the time-varying intensity since the time-varying energies of the breathing disorders are quite different. The instantaneous frequency corresponds to the time-varying breathing rate, which is also called rhythm. It is important to discriminate the breathing patterns.

In order to optimize the classifier, five algorithms including *k*-nearest neighbor (KNN), decision tree (DT), ensemble learning, discriminant analysis and support vector machine (SVM) are evaluated. First, 13 features are used to search the best configuration of each classifier. Then with the best configurations, the feature selection is fulfilled by using the ReliefF algorithm with the incremental number of features [21]. Seven out of 13 features are selected. Using the selected features, the overall accuracies of the SVM, LDA, KNN, bagged tree (BT) and DT are 94.7%, 92.8%, 94.4%, 90% and 94.2%, respectively. Therefore, the optimized SVM with seven features is chosen.

8.3.1.2 Experiments

The experiment was carried out with real patient in the Nanjing Chest Hospital to assess the overall system performance. A patient diagnosed with nocturnal breathing disorders were involved in the experiment. The radar sensor is mounted under the ceiling with its radiation beam facing the patient's chest wall. The distance between each other is 2 m. The experiment was approved by the institutional review board of the Nanjing Chest Hospital.

Figure 8.8 illustrates the subject's hourly distribution of Cheyne–Stokes breathing and Cheyne–Stokes variant breathing during sleep. These two breathing disorders occur mostly in the sixth, seventh and eighth hours. The references were manually labeled by a respiratory physician. The time-varying number shows the relationship with sleep stage. The error in each hour is no more than four except the first hour. In the first hour, the patient just fell asleep, the random body movements distorted the breathing signal, and thus the result is not accurate.

Furthermore, the recognition of six breathing patterns was carried out. After removing the body movements, 832 valid epoches can be used for testing. Figure 8.9 presents the classification results of the subject's 8-h breathing signal. In this figure, NSVM means the correctly predicted numbers by the SVM classifier, and NHAND is the hand-labeled numbers. As a result, the accuracy is defined as the ratio between NSVM and NHAND. During his sleep, five breathing patterns were found except the Biot's breathing. The average accuracy for recognizing them is 86.5%.

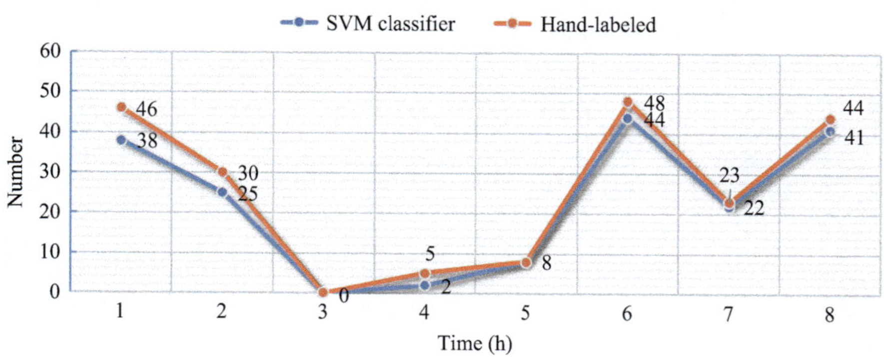

Figure 8.8 The hourly distribution of the Cheyne–Stokes breathing and the Cheyne–Stokes variant breathing during sleep [19]

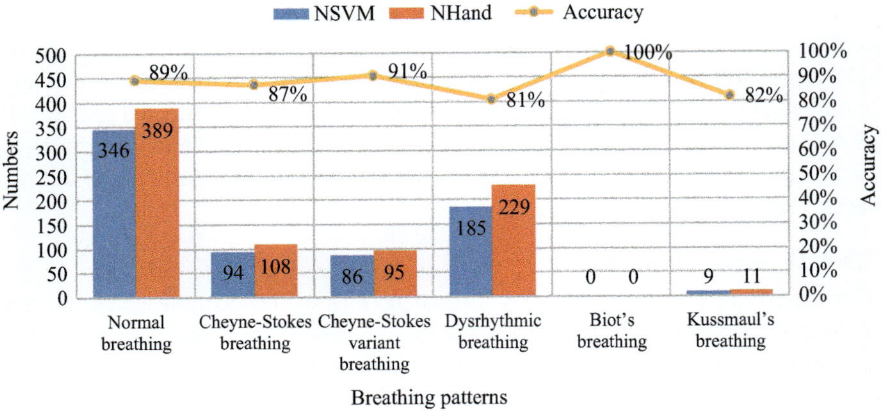

Figure 8.9 The classification results of the six breathing patterns during sleep [19]

8.3.2 Sleep-stage estimation

With the accelerated pace of life, the sleep health issues of modern people have received much attention under the influence of large work pressures and irregular life schedules. Quality and effective sleep is not only related to the health of the individual but also directly affects the spirit and efficiency of work. Sleep medicine has become an important part of modern medicine. The observation and monitoring of sleep and exploring to improve sleep quality and efficiency are important topics in sleep medicine.

The polysomnography (PSG) is a gold standard for assessing sleep quality and diagnosing sleep disorders in sleep medicine. PSG can measure the patient's sleep time, quality, efficiency and sleep stages by monitoring the physiological parameters during whole night using the electroencephalography (EEG), electrocardiography (ECG), electrooculography (EOG), and electromyography (EMG). However, there are some limitations of this device, which are the following:

1. The patient needs to be directly connected to a complete set of electrodes and other sensors which limits patient activity.
2. The monitoring results are required to be manually analyzed by professional, and only certified hospitals can carry out.
3. The contact between the electrode and the sensor directly affects the quality of sleep and the physiological parameters of the patient.
4. It's expensive, and not suitable for long-term monitoring.

To overcome the above limitations and make sleep monitoring more close to nature state, and as shown in previous sections, biomedical radar can obtain physiological information from chest-wall displacement which has attracted wide attention in this field. It frees the subject from equipment or electrodes and does not infringe on the subject's privacy. Thus a noncontact sleep-stage estimation system based on biomedical radar have been proposed. Figure 8.10 shows the system architecture of the noncontact sleep-stage estimation system.

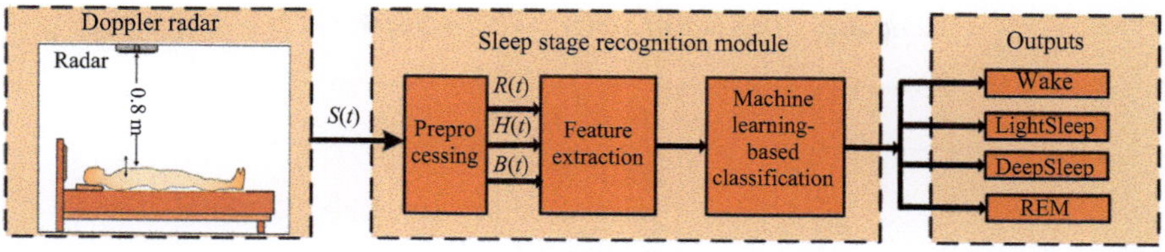

Figure 8.10 The overall architecture of the sleep-stage estimation system [3]

8.3.2.1 Sleep-stage recognition module

In the sleep disorder recognition module, a combination of respiration, heartbeat and body information is used to determine four kinds of sleep stages including Wake, LightSleep, DeepSleep and Rapid Eye Movement (REM) stages.

First, the respiratory and heartbeat signals can be extracted from the phase information. And we can divide the radar signal $S(t)$ into continuous nonoverlapping epochs S_k of 60 s. In the kth epoch, the Riemann integral $A_k(i)$ of S_k with respect to the 10-s period is defined as

$$A_k(i) = \sum_{t=t_k-i*1,000}^{t_k-(i-1)*1,000} |S_k|dt \tag{8.28}$$

where t_k denotes the time of S_k. $A_k(i)$ ($i = 1,2, ..., 6$) is calculated every 10 s. The algorithm reports a body movement if there is a change of more than an order of magnitude $A_k(i)$. The detected movements are validated against the posture information provided by the PSG system.

Douglas *et al.* [23] showed that different sleep stages have different vital signs patterns. For example, the breathing pattern was predominantly regular in non-REM sleep except for a few minutes after arousal, when irregularity occurred in all subjects and for much longer periods during the first half of the night. And irregular breathing in REM stage on the other hand occurred throughout all REM periods in both men and women. Based on this, a new feature set was extracted from the sleep-related physiological signals in this chapter. Eleven features including a body movement feature, five respiration features, four heartbeat features and a time feature have been extracted, which are shown in Table 8.2.

- **Body movement index (BMI):** During sleep, with the occurrence of body movement, body movement monitoring is essential for determining the sleep stage [22]. Body movement usually appears in the Wake and LightSleep stages due to the changes of sleep posture occurs every 5–10 min. Conversely, lack of body movement can indicate the DeepSleep and REM stages. Thus, we use $BI(k)$ to represent the amplitude of body movement, which is defined as

$$BI(k) = \sum_{i=1}^{6} A_k(i) - A_k(i)_{\min} \tag{8.29}$$

where $A_k(i)_{\min}$ is the minimum value of $A_k(i)$ at the 60-s interval S_k.

Table 8.2 The list of sleep features [3]

No.	Features list
1	BMI [22]
2	Respiration per min (RPM) [23,24]
3	Variance of RPM [25]
4	ADA of respiratory signal
5	REM parameter [26]
6	Sample entropy [27]
7	Heartbeat per min (HPM)
8	Variance of HPM
9	ADA of heartbeat signal
10	Deep parameter [26]
11	Time feature [28,29]

- **Amplitude difference accumulation (ADA) of respiration/heartbeat:** Respiration/heartbeat varies significantly in both amplitude and frequency during the REM stage. Besides the frequency feature, we can utilize ADA of respiration/heartbeat to distinguish between REM and NREM stages. Furthermore, we can use this information to determine whether a subject is in DeepSleep stage. Here, the ADA in every minute is defined as

$$ADA(k) = \sum_{i=1}^{n} \left| P_{i+1}^{k} - P_{i}^{k} \right| \tag{8.30}$$

where P_i^k denotes the ith peak point in the kth minute and n is the total number of peak points in the kth minute.

Today, machine learning algorithms attract more and more attention in classification applications. However, the sleep-stage classification is an imbalanced task, which means the number of LightSleep periods far exceeds other sleep stages. Therefore, the single machine learning algorithm like SVM, discriminant analysis and KNN has usually imperfect results of classification. In statistics and machine learning, ensemble methods use multiple learning algorithms to obtain better predictive performance that could be obtained from any of the constituent learning algorithms alone [30]. Unlike a statistical ensemble in statistical mechanics, which is usually infinite, a machine learning ensemble only refers to a concrete finite set of alternative models, but typically allows for much more flexible structure to exist among those alternatives. In our experiment, we have chosen the BT [30], boosted trees [30] and subspace KNN [31] as the ensemble algorithms.

8.3.2.2 Experiments

The data for this example come from a study in [3]. The Doppler radar located on the top of healthy subjects to record physiological signals from 11:30 PM to 5:30 AM. A PSG (SOMNOCheck 2 RK) was set as standard.

Figure 8.11(a), (b) and (c) shows a 6-h sleep-stage estimation experiment including respiratory, heartbeat and body movement signals from a male subject.

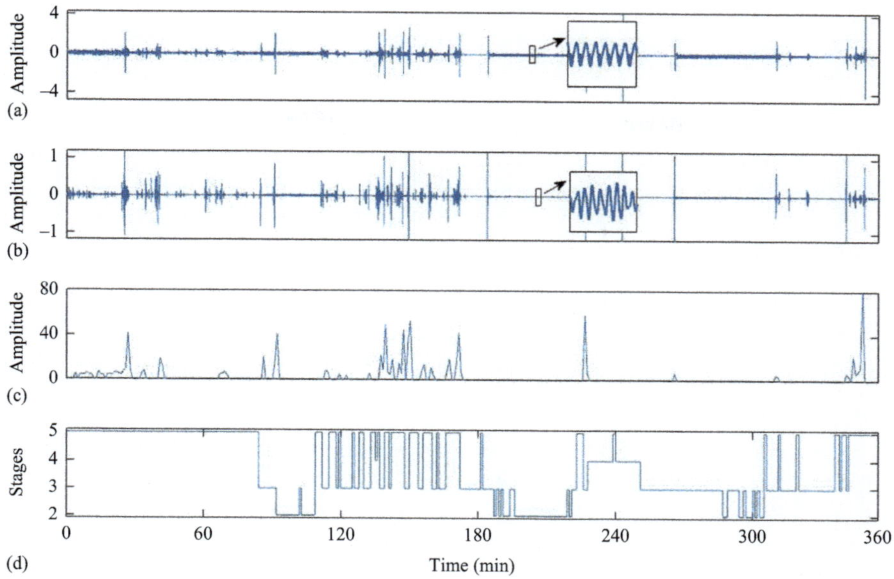

Figure 8.11 *(a) 6-h respiratory signal. (b) 6-h heartbeat signal. (c) 6-h body movement signal. (d) Reference sleep stages provided by SOMNOCheck 2 RK device [3]*

The sleep-stage estimation framework divides the 6-h signals into successive epochs with each epoch being 60-s long. Figure 8.11(d) shows the reference sleep stages provided by SOMNOCheck2 RK device. Here, we define the stages 5, 4, 3 and 2 to be Wake, REM, LightSleep and DeepSleep, respectively. And Figure 8.12 demonstrates the final sleep stages determined using the ensemble learning classifiers including boosted trees, BT and subspace KNN. As shown in the figure, all the estimated sleep stages and the reference sleep stages show substantially similar results. For the boosted trees algorithm, the classification accuracy of the stages "DeepSleep," "LightSleep," "REM" and "Wake" are 74.5%, 81.1%, 82.6% and 82.7%, respectively. For the BT algorithm, the classification accuracy of the four stages are 80.0%, 78.8%, 87.0% and 80.6%, respectively. Meanwhile, for the subspace KNN algorithm, the classification accuracy of the four stages are 89.1%, 75.8%, 87.0% and 82.0%, respectively.

To further validate the efficiency of sleep-stage recognition module, an experiment was conducted on the data 1 which collected from ten subjects, including one female and nine males. As shown in Figure 8.13, the subspace KNN algorithm performs better than both the boosted trees and the BT in 9 out of 11 valid data. More specifically, the accuracy of the subspace KNN algorithm ranges from 76.5% to 86.6%, with the average value of 81.0%. On the contrary, for the boosted trees, the range of accuracy is from 70.9% to 86.4%, and the BT ranges from 72.7% to 85.7%. Compared with other two ensemble algorithms, we can find that the subspace KNN performs better especially in the lower bound of accuracy.

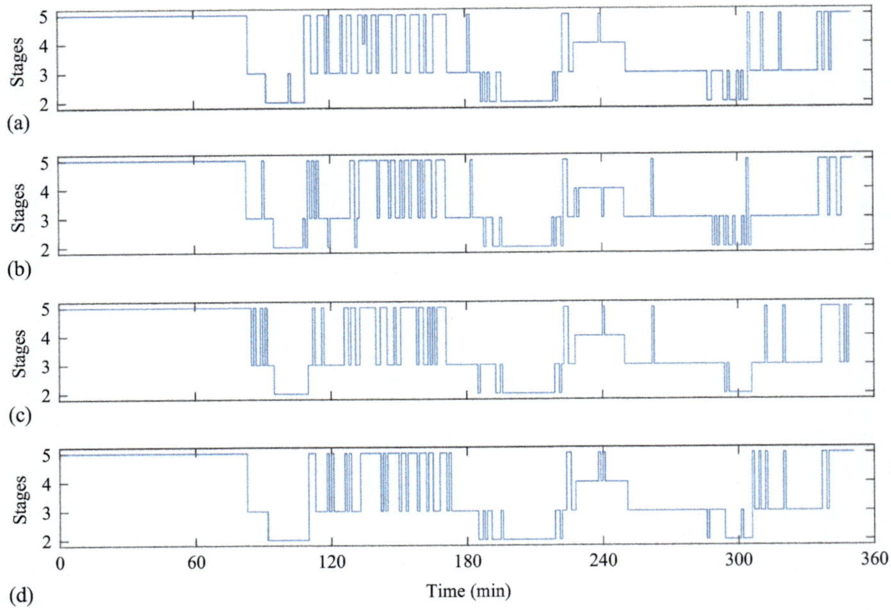

Figure 8.12 *(a) Reference sleep stages provided by the SOMNOCheck 2 RK device. (b) New sleep stages estimated through the boosted trees algorithm. (c) New sleep stages estimated through the BT algorithm. (d) New sleep stages estimated through the subspace KNN algorithm [3]*

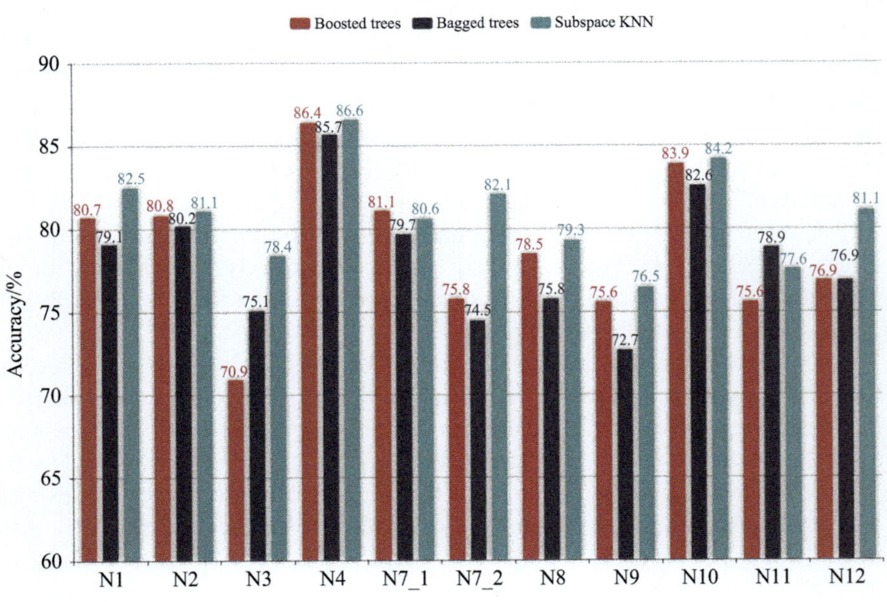

Figure 8.13 *Comparison of different classifiers in terms of accuracy [3]*

References

[1] Wu Y, and Li J. The design of digital radar receivers. IEEE Aerospace and Electronic Systems Magazine. 1998;13(1):35–41.

[2] Li C, Cummings J, Lam J, *et al.* Radar remote monitoring of vital signs. IEEE Microwave Magazine. 2009;10(1):47–56.

[3] Hong H, Zhang L, Gu C, *et al.* Noncontact sleep stage estimation using a CW Doppler radar. IEEE Journal on Emerging and Selected Topics in Circuits and Systems. 2018;8(2):260–270.

[4] Vaughan RG, Scott NL, and White DR. The theory of bandpass sampling. IEEE Transactions on Signal Processing. 1991;39(9):1973–1984.

[5] Zhao H, Hong H, Sun L, *et al.* Accurate DC offset calibration of Doppler radar via non-convex optimisation. Electronics Letters. 2015;51(16):1282–1284.

[6] Zhao H, Hong H, Sun L, *et al.* Noncontact physiological dynamics detection using low-power digital-IF Doppler radar. IEEE Transactions on Instrumentation and Measurement. 2017;66(7):1780–1788.

[7] Li C, Lubecke VM, Boric-Lubecke O, *et al.* A review on recent advances in Doppler radar sensors for noncontact healthcare monitoring. IEEE Transactions on Microwave Theory and Techniques. 2013;61(5):2046–2060.

[8] Li C, Ling J, Li J, *et al.* Accurate Doppler radar noncontact vital sign detection using the RELAX algorithm. IEEE Transactions on Instrumentation and Measurement. 2010;59(3):687–695.

[9] Bechet P, Mitran R, and Munteanu M. A non-contact method based on multiple signal classification algorithm to reduce the measurement time for accurately heart rate detection. Review of Scientific Instruments. 2013; 84(8):927–934.

[10] Kazemi S, Ghorbani A, and Amindavar H. Cyclostationary modelling of amplitude and frequency modulated signals in heart and respiration monitoring Doppler radar systems. IET Radar Sonar and Navigation. 2015;9(2):116–124.

[11] Sun L, Li Y, Hong H, *et al.* Super-resolution spectral estimation in short-time non-contact vital sign measurement. Review of Scientific Instruments. 2015;86(4):105–133.

[12] Sun L, Hong H, Li Y, *et al.* Noncontact vital sign detection based on stepwise atomic norm minimization. IEEE Signal Processing Letters. 2015; 22(12):2479–2483.

[13] Candes EJ, and Wakin MB. An introduction to compressive sampling. IEEE Signal Processing Magazine. 2008;25(2):21–30.

[14] Cotter SF, Rao BD, Engan K, *et al.* Sparse solutions to linear inverse problems with multiple measurement vectors. IEEE Transactions on Signal Processing. 2005;53(7):2477–2488.

[15] Tang G, Bhaskar BN, Shah P, *et al.* Compressed sensing off the grid. IEEE Transactions on Information Theory. 2013;59(11):7465–7490.

[16] Daubechies I, Lu J, and Wu HT. Synchrosqueezed wavelet transforms: an empirical mode decomposition-like tool. Applied and Computational Harmonic Analysis. 2011;30(2):243–261.

[17] Auger F, Flandrin P, Lin YT, *et al.* Time-frequency reassignment and synchrosqueezing: an overview. IEEE Signal Processing Magazine. 2013; 30(6):32–41.
[18] Lanfranchi PA, Braghiroli A, Bosimini E, *et al.* Prognostic value of nocturnal Cheyne-Stokes respiration in chronic heart failure. Circulation. 1999;99(11):1435.
[19] Zhao H, Hong H, Miao D, *et al.* A noncontact breathing disorder recognition system using 2.4-GHz digital-IF Doppler radar. IEEE Journal of Biomedical and Health Informatics. 2019;23(1):208–217.
[20] Rafajlowicz E, Pawlak M, and Steland A. Nonparametric sequential change-point detection by a vertically trimmed box method. IEEE Transactions on Information Theory. 2009;56(7):3621–3634.
[21] Robnik, Ikonja M, and Kononenko I. Theoretical and empirical analysis of ReliefF and RReliefF. Machine Learning. 2003;53(1–2):23–69.
[22] Kagawa M, Sasaki N, Suzumura K, *et al.* Sleep stage classification by body movement index and respiratory interval indices using multiple radar sensors. In: Engineering in Medicine and Biology Society (EMBC), 2015 37th Annual International Conference of the IEEE. IEEE; 2015. pp. 7606–7609.
[23] Douglas N, White D, Pickett CK, *et al.* Respiration during sleep in normal man. Thorax. 1982;37(11):840–844.
[24] Lin F, Zhuang Y, Song C, *et al.* SleepSense: a noncontact and cost-effective sleep monitoring system. IEEE Transactions on Biomedical Circuits and Systems. 2017;11(1):189–202.
[25] Samy L, Huang MC, Liu JJ, *et al.* Unobtrusive sleep stage identification using a pressure-sensitive bed sheet. IEEE Sensors Journal. 2014;14(7): 2092–2101.
[26] Kurihara Y, and Watanabe K. Sleep-stage decision algorithm by using heartbeat and body-movement signals. IEEE Transactions on Systems, Man, and Cybernetics-Part A: Systems and Humans. 2012;42(6):1450–1459.
[27] See AR, and Liang CK. A study on sleep EEG using sample entropy and power spectrum analysis. In: Defense Science Research Conference and Expo (DSR), 2011. IEEE; 2011. pp. 1–4.
[28] Watanabe T, and Watanabe K. Noncontact method for sleep stage estimation. IEEE Transactions on Biomedical Engineering. 2004;51(10):1735–1748.
[29] Lavie P, and Zomer J. Ultrashort sleep-waking schedule. II. Relationship between ultradian rhythms in sleepability and the REM-non-REM cycles and effects of the circadian phase. Electroencephalography and Clinical Neurophysiology. 1984;57(1):35–42.
[30] Zhou ZH. Ensemble Methods: Foundations and Algorithms. London: Chapman and Hall/CRC; 2012.
[31] Ho TK. Nearest neighbors in random subspaces. In: Joint IAPR International Workshops on Statistical Techniques in Pattern Recognition (SPR) and Structural and Syntactic Pattern Recognition (SSPR). Heidelberg Germany: Springer; 1998. pp. 640–648.

Chapter 9

Radar measurement of the angular velocity of moving objects

Eric Klinefelter[1] *and Jeffrey A. Nanzer*[1]

9.1 Radar measurements

A radar system transmits electromagnetic energy in a way that permits the extraction of information from the signals reflected off an object [1]. After a detection of the received signal has been made, it is generally of interest to determine the precise location of the object in three-dimensional (3D) space, requiring the measurement of its range as well as its angle (both azimuth and elevation). If tracking is desired, or if it is of interest to predict the future location of the object, the trajectory or time rate of change of the position in both range and angle must be measured. Ideally, radar measurements should form a complete set in $\mathbb{R}^3$ for both position and velocity, meaning that any target position or velocity can be represented as a linear combination of direct measurements. Clearly, measurements of range, azimuth angle, and elevation angle form an orthogonal basis for position. Similarly, direct measurements of radial velocity and angular velocity (both in azimuth and elevation) are required to fully represent any possible target trajectory.

In this chapter, a direct measurement is defined as one which is performed in radar fast time, which means that the time to take the measurement is on the order of the time it takes for the electromagnetic wave to propagate to the target and back. Typical radar systems measure range, radial velocity, and angle in fast time. This is different from a typical measurement of angle rate, in which the radar performs a fast-time measurement at successive angles to determine the angular rate of change of a target. Before coherent Doppler processing became commonplace, range rate was similarly calculated by taking successive range measurements and dividing by the time between samples. Measurement of angular rates of change will be described using terminology matching that from traditional radar textbooks (e.g. [2]), where range rate is used to describe successive repeated measurements of range, and radial velocity or Doppler velocity is used to describe the direct measurement through the coherent processing of the Doppler shift. In this manner, angle rate describes typical scanning of the scene to determine angular motion,

[1]Electrical and Computing Engineering, Michigan State University, USA

while the direct measurement technique outlined in this chapter is described as measuring angular velocity.

This chapter reviews a recently developed technique for directly measuring the angular velocity of moving objects using a distributed radar aperture with interferometric processing. Interferometry has been studied in both optical and radio systems for many years, and can generally be described as extracting information from the interference caused by electromagnetic waves. The wave nature of light was demonstrated by Young in 1803 through interferometry [3], and Michelson and Morley disproved the theory of a luminiferous aether in 1887 using optical interferometry [4]. It was Fizeau and later Michelson who suggested using optical interferometry to measure the angular dimensions of astronomical objects [5]. The angular resolution of an aperture was given by Rayleigh as

$$\theta = 1.220\frac{\lambda}{D} \tag{9.1}$$

where λ is the free-space wavelength and D is the aperture diameter. At the time it was known from optics that for increased angular resolution a filled aperture of size D was not required, only two small slits or two receiving lenses separated by a distance D could be used [6]. With knowledge that astronomical objects were radio sources, this same technique was applied using antennas instead of optical devices, which led to Ryle and Vonberg publishing the first stellar observations using an interferometric radio telescope in 1946 [7]. Separating the receiving elements creates a fringe pattern that has ambiguous absolute angle but narrow resolution. Described in more detail later, Figure 9.1 shows the fringe pattern created by an interferometer on the Earth and the response generated by distant radio sources as the pattern scans past.

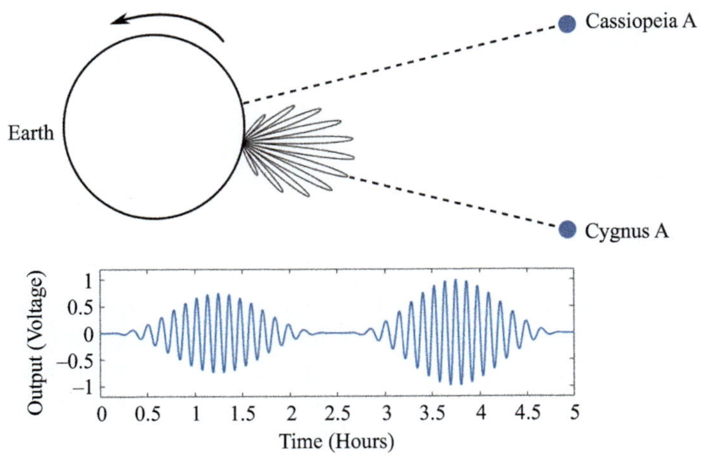

Figure 9.1 The fringe pattern of a two-element interferometric array has lobes of ambiguous absolute angle but increased angular resolution. The diagram shows the response of the two point sources as the interferometric fringe pattern scans the sky due to the rotation of the Earth

When the radio source is within the field of view the interferometer output is proportional to the fringe pattern. The absolute position is not known, yet when the source enters and exits the field of view accurate angular resolution can be obtained. This is how Michelson and others could resolve double stars and measure star diameters, which was technologically impossible before. More recently, radio interferometry has been used to synthesize large electromagnetic antenna apertures using collections of smaller antennas [7], such as the Karl G. Jansky Very Large Array (VLA) located in New Mexico. Today, interferometric processing is used where increased angular resolution is required, for example, with the Minitrack network which tracked satellite trajectories. In radar imaging, interferometric synthetic aperture radar (InSAR) has become a standard method for obtaining high-resolution topographic maps [8].

9.2 Interferometric measurement of angular velocity

The interferometric measurement of the angular velocity of moving objects is a relatively new technique that can provide direct angular velocity measurements in a manner that is similar to the measurement of radial velocity via the Doppler shift [9]. This method is implemented in the processing of a distributed receiver, and as such the technique can be implemented on either a passive or active system: the received signal can be either transmitted and reflected off the object or intrinsically radiated by the object. The signals received at the two antennas are correlated and integrated, forming a correlation interferometer, the output of which generates a response whose frequency is proportional to the angular velocity of the target. This section reviews the theory behind this measurement technique, which is based on developments from radio astronomy [7]. Although the derivation of the correlation interferometer response has been known, using the response to measure angular velocity has not been discussed until relatively recently [10], in part because previously the moving platform (such as the Earth, or a plane) is typically used to detect a static target and in these instances the angular velocity is expressly known. The novelty of the method described here comes by way of a static system detecting a moving target and the fact that angular velocity manifests as a simple frequency response on the radar output.

A two-element correlation interferometer consists of two receivers separated by a baseline D, the outputs of which are multiplied and integrated. Figure 9.2 shows a simplified diagram of an interferometer with a complex correlator producing the in-phase (I) and quadrature (Q) components of the received signal. If the interferometer is viewing a monochromatic response from a point source in free space, the voltage responses to a plane wave at incident angle θ at the inputs to the in-phase multiplier are

$$\begin{aligned} V_1(t) &= a_1\sqrt{A_1(\theta)}\cos(2\pi ft) + n_1(t) \\ V_2(t) &= a_2\sqrt{A_2(\theta)}\cos\big(2\pi f\big(t-\tau_g\big)\big) + n_2(t) \end{aligned} \tag{9.2}$$

where a_i are the amplitudes of the received signals with $i = 1, 2$, A_i are the antenna power patterns, f is the frequency of the signal, and τ_g is the geometrical time delay

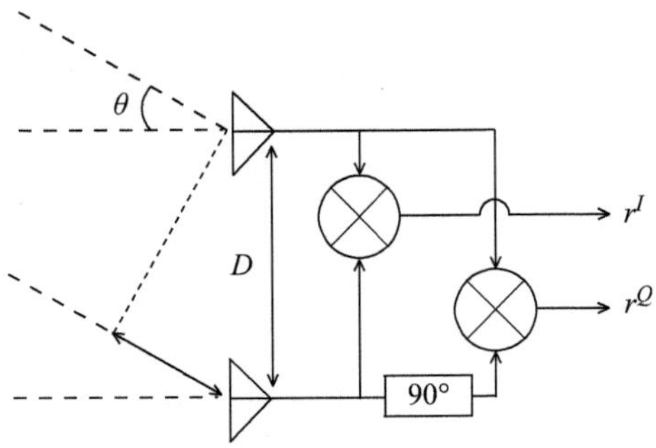

Figure 9.2 Diagram of a complex correlation interferometer [10]

in reception of the signal between the two antennas. This delay is given in terms of the antenna separation or baseline D by

$$\tau_g = \frac{D}{c}\sin(\theta) \tag{9.3}$$

The cosine terms of (9.2) represent the signal components and the n_i terms represent noise components. Recall that the equations above and those that follow are for a monochromatic plane wave; thus, the target or radiation source must be in the far field of the antenna array. This value is known as the Fraunhofer distance $d_f = 2D^2/\lambda$, where D is now the antenna baseline instead of the individual antenna size. It is likely that for short-range motion sensing the target will be near field to the array (but not necessarily the individual antenna elements); however, experimental measurements discussed later indicate that the above approximation is nonetheless a good predictive model. It can be reasonably assumed that the individual antenna patterns are identical or very nearly so; thus, the output of the in-phase multiplier becomes:

$$\begin{aligned} r^I(\tau_g) &= \langle V_1(t)V_2(t)\rangle \\ &= \langle a_1a_2A(\theta)\cos(2\pi ft)\cos\left[2\pi f(t-\tau_g)\right]\rangle \\ &= \left\langle \frac{a_1a_2}{2}A(\theta)\left(\cos\left[2\pi f(2t+\tau_g)\right]+\cos(2\pi f\tau_g)\right)\right\rangle \\ &= \frac{a_1a_2}{2}A(\theta)\cos(2\pi f\tau_g) \end{aligned} \tag{9.4}$$

where $\langle\cdot\rangle$ represents the time-averaging process by which higher frequency terms depending on $2f_c$ have been filtered out. Since the noise signals are uncorrelated, their time average tends to zero. Similarly, the signal at the quadrature output is given as

$$\begin{aligned} r^Q(\tau_g) &= \langle a_1a_2A(\theta)\cos(2\pi ft)\cos\left[2\pi f(t-\tau_g)+90^\circ\right]\rangle \\ &= \frac{a_1a_2}{2}A(\theta)\sin(2\pi f\tau_g) \end{aligned} \tag{9.5}$$

The complex output of the correlator can be found by combining the in-phase and quadrature signals. In terms of angle, the output becomes:

$$r(\theta) = \frac{a_1 a_2}{2} A(\theta) e^{j2\pi D_\lambda \sin(\theta)} \tag{9.6}$$

where $D_\lambda = D/\lambda$ is the antenna baseline measured in wavelengths. The variation over θ in (9.6) produces an interference pattern called the fringe pattern. Figure 9.3 shows examples of the magnitude of the in-phase fringe pattern for multiple baselines. As the baseline increases, the frequency of the spatial oscillation increases.

The maxima and minima in the fringe pattern are the angles where the difference in the distance from the source to the two receivers causes a phase difference which results in constructive and destructive interference. As an object passes through the system pattern (a function of the fringe pattern), the voltage output of the interferometer oscillates in conjunction with the maxima and minima. The output of the system is stated in terms of angle θ, yet the angular velocity is given as the time rate change of angle as

$$\omega = \frac{d\theta}{dt} \Rightarrow \theta = \omega t + \theta_0 \tag{9.7}$$

where θ_0 represents the initial angle which, without loss of generality, is assumed to be zero. The output can be expressed in terms of angular velocity by substituting (9.7) into (9.6), resulting in

$$r(\omega t) = \frac{a_1 a_2}{2} A(\omega t) e^{j2\pi D_\lambda \sin(\omega t)} \tag{9.8}$$

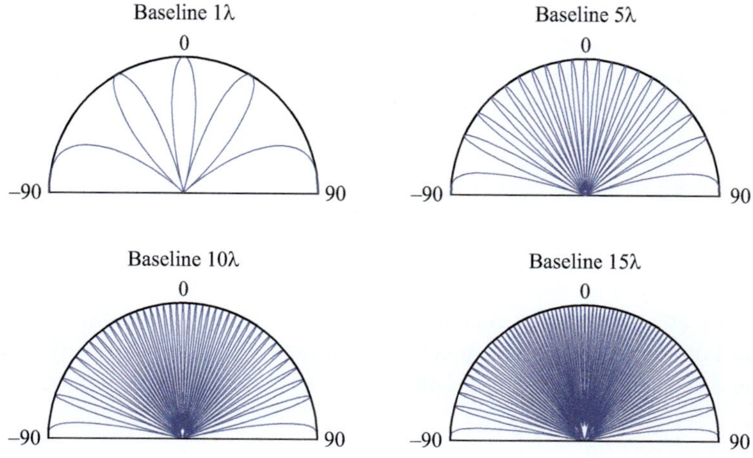

Figure 9.3 Examples of the magnitude of the in-phase signal fringe pattern for multiple antenna baselines [10]

The frequency shift is the instantaneous frequency (IF) of the interferometer response, which is the time derivative of the phase term of (9.8), given by $2\pi f_s = d\phi/dt$, which results in

$$f_s = \omega D_\lambda \cos(\omega t) \tag{9.9}$$

Thus the frequency shift at the output of the correlation interferometer is directly proportional to the angular velocity of the far-field point source. If the antenna pattern is sufficiently directive and near broadside, the cosine term in (9.9) is approximately one; this is the narrow-beam approximation. The sign of the change in angle in (9.7) depends on angular direction of the object and can be either positive or negative, resulting in positive or negative frequencies. Such directional measurements require both the in-phase and quadrature parts of the output signal and then produce a response similar to the Doppler frequency, where the frequency shift relative to the carrier frequency indicates approaching or receding trajectories.

Orthogonal velocity measurements in multiple dimensions are required to fully describe target trajectories. For example, consider the 2D case where the total velocity of target is given as $v = v_r\hat{r} + v_\phi\hat{\phi}$, where v_r is the radial velocity component and v_ϕ is the angular velocity component. From the equation for Doppler shift, the radial component is given as

$$v_r = \frac{f_D \lambda}{2} \tag{9.10}$$

where f_D is the Doppler shift. Similarly, with the relationship that $\omega = v/R$, and assuming the narrow-beam approximation the angular velocity component is

$$v_\phi = \frac{f_s \lambda R}{D} \tag{9.11}$$

Thus a system which employs both Doppler and interferometric processing is theoretically capable of fully resolving velocity trajectories for a point source. This example is readily extended to 3D with the addition of a second antenna pair with a baseline oriented orthogonal to the first, where one antenna may be reused for both baselines, thus requiring only three antennas.

9.3 Measurement resolution and accuracy

The analysis above shows the closed-form solution for the response of a correlation interferometer to a monochromatic plane wave. This section discusses the resolution of the angular velocity measurement, or the ability of the system to resolve multiple targets which are close together with respect to angular velocity. Following this, the theoretical accuracy of the angular velocity measurement will be examined.

9.3.1 Resolution

It was shown that the angular velocity of a target is directly proportional to the frequency shift of the interferometer response, and as such a measurement of the frequency shift is equivalent to a scaled measurement of angular velocity. This implies that the resolution of the frequency shift is proportional to the resolution of the angular velocity. Given a signal in the presence of noise, the resolution of a measurement is generally defined to be the half-power bandwidth of the response of a point source. If the responses are separated by less than this amount, they will overlap each other in the frequency domain and the system will be unable to resolve them as individual targets. Thus to determine the frequency resolution of the system, the spectral response of the interferometer is needed.

The spatial filtering effect of the antenna pattern was shown in the equations above, but in general, there is an additional filtering effect created by the bandwidth pattern. If the incident radiation is not monochromatic, such as in the case of a passive system detecting thermal radiation, the interferometer response must be integrated over the bandwidth of the receiver [11]. Assuming a rectangular pass-band with bandwidth Δf, the bandwidth pattern is given by

$$F(\theta, \Delta f, D) = \mathrm{sinc}\left(\pi \Delta f \frac{D}{c} \sin(\theta)\right) \tag{9.12}$$

Thus if the system has a wide bandwidth, the spatial filtering response of (9.6) becomes increasingly narrow.

Since the number of fringes in the interferometer radiation pattern depends on λ, the use of a wideband signal will cause the response to have more fringes at the highest frequency contained in the signal and fewer fringes at the lowest frequency. The center fringe remains stationary in angle, although its width will change; however, the more prominent effect is evident at the outer edges where the fringes move into or out of the field of view as the frequency changes. However, for an active, continuous wave system, the signal is essentially monochromatic, and thus the bandwidth pattern is nearly constant in angle. The antenna pattern and bandwidth pattern can be combined into a single function called the system beam pattern, which is given as

$$K_B(t, \Delta f, D) = A(\theta) F(\theta, \Delta f, D) \tag{9.13}$$

As described above, this has a spatial filtering effect; thus the normalized complex interferometer response becomes:

$$r_n(t) = K_B(t, \Delta f, D) e^{j2\pi D_\lambda \sin(\omega t)} \tag{9.14}$$

Determining the spectral response is simplified by the narrow-beam approximation, or assuming that the system beamwidth is small such that $\theta \ll 1$ rad, and thus $\omega t \ll 1$, which results in

$$\begin{aligned} \cos(\omega t) &\approx 1 \\ \sin(\omega t) &\approx \omega t \end{aligned} \tag{9.15}$$

Now the Fourier response of (9.13) becomes:

$$R_n(f) = \hat{K}_B(f - f_s, \Delta t, D) \tag{9.16}$$

where $\hat{K}_B$ is the Fourier transform of the system beam pattern and f_s is the frequency shift (9.11).

A simple way of demonstrating the spectral response is by assuming the system beam pattern takes on a Gaussian shape such as

$$K_B(\theta) = e^{-\frac{\theta^2}{2\sigma^2}} \tag{9.17}$$

where σ is the standard deviation. The half-power beam width of this pattern is

$$\theta_{BW} \approx 2.355\sigma \tag{9.18}$$

Now the complex interferometer response becomes:

$$r_n(t) = e^{-\frac{\omega^2}{2\sigma^2}t^2} e^{j2\pi D_\lambda \sin(\omega t)} \tag{9.19}$$

of which the Fourier transform, assuming the narrow-beam approximation, is

$$R_n(f) = \sqrt{\frac{2\pi\sigma^2}{\omega^2}} e^{-\frac{2\pi^2\sigma^2}{\omega^2}(f-f_s)^2} \tag{9.20}$$

which is simply a Gaussian centered on the frequency shift. Using (9.18), the resolution is the half-power bandwidth of (9.20), given as

$$\delta f = 2.355 \frac{\omega}{\pi\theta_{BW}}. \tag{9.21}$$

Thus better resolution (smaller δf) occurs with wide system beam patterns and slow-moving targets. Effectively, either slow-moving targets or wider beam patterns results in longer temporal responses at the output of the interferometer. It is well known that the resolution of a frequency measurement is dependent on the temporal duration of the signal; thus, in this case extending the temporal response yields a longer signal and thus better resolution on the frequency shift. The variability in resolution with increasing frequency is shown in Figure 9.4. A further implication is that objects with time-varying angular velocity are measured with time-varying resolution.

Additionally, because the angular velocity is equal to the object's linear velocity divided by its range, an object moving with the same linear velocity at different ranges is also measured with different resolution. The effect of the system beam width for multiple antenna baselines is shown in Figure 9.5. There is an inverse relationship between system beam pattern and resolution, which is the physical result of a narrower system beam pattern having fewer fringes and thus a shorter signal response with respect to time.

The desire for high resolution (small δf) is critical for micro-motion sensing, for example, if you consider the various fingers of a hand each moving with different angular velocities with respect to the receiver. If the resolution of the system

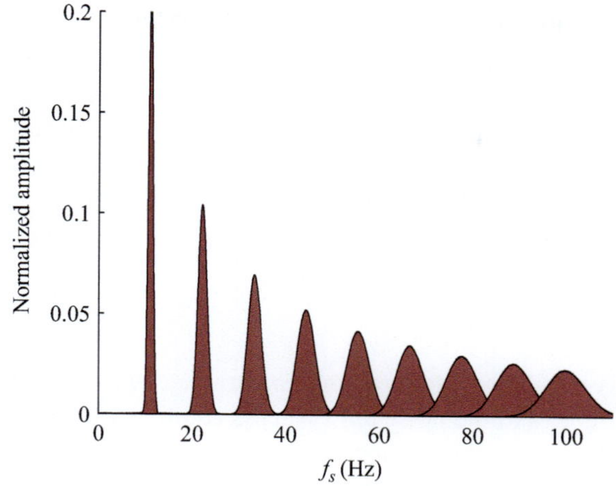

Figure 9.4 Spectral response of an interferometer with a Gaussian system beam pattern: the spectral amplitude is inversely proportional to f_s while the bandwidth is proportional to f_s [11]

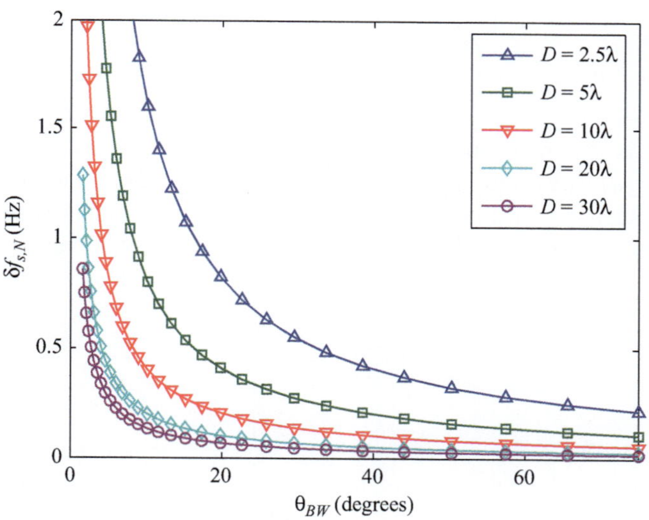

Figure 9.5 Normalized frequency resolution of a system with a Gaussian system beam pattern [11]

is too low, the individual micro-motions will occlude one another and thus appear as a single response in the time-frequency plot. It is critical then to design a system which has a small enough δf to give the desired resolution in the expected frequency ranges of interest.

9.3.2 Accuracy

Whereas resolution refers to the ability to determine the presence of multiple objects, the accuracy of a given measurement is the error between the estimate and the true value of the parameter being estimated. While the fundamental limitations on the accuracy of radar measurements are dependent on the level of noise present, the form of the signal also plays an important role [2]. From estimation theory, a radar measurement is typically modeled as a signal with unknown parameters in noise [12]. This can be given in general by

$$s_r(x) = \alpha s(x; u) + w(x) \tag{9.22}$$

where $s(x; u)$ is a function of the unknown parameter u to be estimated, α is the complex amplitude, and the noise term $w(x)$ is additive white Gaussian noise. An estimator that minimizes the variance between the measurement and the unknown parameter u is desired. The minimum theoretical variance of any unbiased estimator is lower bounded by the Cramer–Rao bound (CRB) [13]. Although the CRB may not always be achievable, the form of the CRB can provide insight into the optimal design of the signal to be used such that the variance is reduced and the measurement accuracy is improved. For signals of the form of (9.22), the CRB can be expressed as [12]:

$$\mathrm{var}(\hat{u} - u) \geq \frac{N_0}{2|\alpha|^2}\left(\int \left|\frac{\partial s(x; u)}{\partial u}\right|^2 dx - \frac{1}{E_s}\left|\frac{\partial s(x; u)}{\partial u}^{*} s(x; u)dx\right|^2\right)^{-1} \tag{9.23}$$

where N_0 is the noise power per unit bandwidth, α is the amplitude, and $E_s \triangleq \int | s(x; u) |^2 dx$ is the signal energy. In [14], it was shown that this relationship can be expressed in a general form to bound the four fundamental radar measurement types: range, radial velocity, angle, and angular velocity. The CRB simplifies to

$$\mathrm{var}(\hat{u} - u) \geq \frac{N_o}{2|\alpha|^2\left(\zeta^2 - \frac{1}{E_s}\mu^2\right)} \tag{9.24}$$

where μ is the first moment (or mean) and ζ^2 is the second moment for the relevant quantity associated with one of the four measurements [14]. The first and second moments for a given radar measurement are described in the Fourier transform domain. For example, when estimating range (which is accomplished by measuring the time delay), the first and second moments are the mean bandwidth and mean-square bandwidth; when estimating Doppler frequency, the moments are the mean and mean-square time durations of the signal. In measuring the angular velocity, the exact form of the moments depends on the receiver used. For the case of an interferometer using a correlation receiver, the accuracy of the measurement of angular velocity is dependent on the spatial distribution of the radiation intensity (i.e. the antenna pattern). For signals which display symmetry, such as in the case of the antenna pattern of an interferometer using a correlation receiver, the mean μ is zero and thus the second term in the denominator can be ignored, resulting in

$$\mathrm{var}(\hat{u} - u) \geq \frac{N_o}{2|\alpha|^2\zeta^2} \tag{9.25}$$

The second moment is given by

$$\zeta^2 = \int (2\pi\theta)^2 \, |A_s(\theta)|^2 \, d\theta \tag{9.26}$$

where $A_s(\theta)$ is the system pattern.

Since the accuracy is inversely proportional to the mean-square antenna pattern, for a given signal-to-noise ratio (SNR) it is desirable to maximize ζ to minimize the variance of the estimate. Theoretically, the second moment is maximized when the power is concentrated into two infinitesimally narrow beams separated by as wide of an angle as possible. Such an antenna pattern is not practical to implement, although it can be approximated using, for example, digital arrays or multi-feed reflector antennas. This implies that the object whose angular velocity is being measured must pass through both antenna beams during the measurement, and furthermore that the angular velocity does not change when doing so. These constraints may not always be feasible.

A more practical solution is the use of the two-element correlation interferometer as described above. While not achieving optimal accuracy, it enables fast and direct measurement of the angular velocity since the fringe pattern extends over all angles within the element beamwidth. The system pattern for the interferometer is the product of the individual antenna pattern, the bandwidth pattern, and the fringe pattern. As mentioned earlier, for an active system the bandwidth pattern is negligible; thus the system pattern is

$$A_s(\theta) = A(\theta)\cos\left(2\pi f \tau_g\right) \tag{9.27}$$

and the expression for the variance of the frequency measurement becomes:

$$\mathrm{var}\left(\hat{f}_s - f_s\right) \geq \left[2\ \mathrm{SNR} \int (2\pi\theta)^2 |A(\theta)\cos\left(2\pi f \tau_g\right)|^2 d\theta\right]^{-1} \tag{9.28}$$

where the SNR is $|\alpha|^2/N_0$. Thus to lower the minimum variance of the angular velocity measurement, the SNR and the area under the system pattern should be maximized. This implies that for maximum accuracy, a wide individual antenna pattern should be used. The baseline and frequency parameters are embedded in the τ_g term, but overall have little effect on the area under the system pattern.

Figure 9.6 compares the antenna patterns for a dual-beam antenna compared to an interferometer with the same electrical size. The antenna pattern is modulated by a $\cos^2\theta$ function approximating a wide-beam element pattern. The separation of the dual-beam pattern highlights the idea that the object must pass through both beams over an appreciable angle with the same angular velocity to ensure a correct

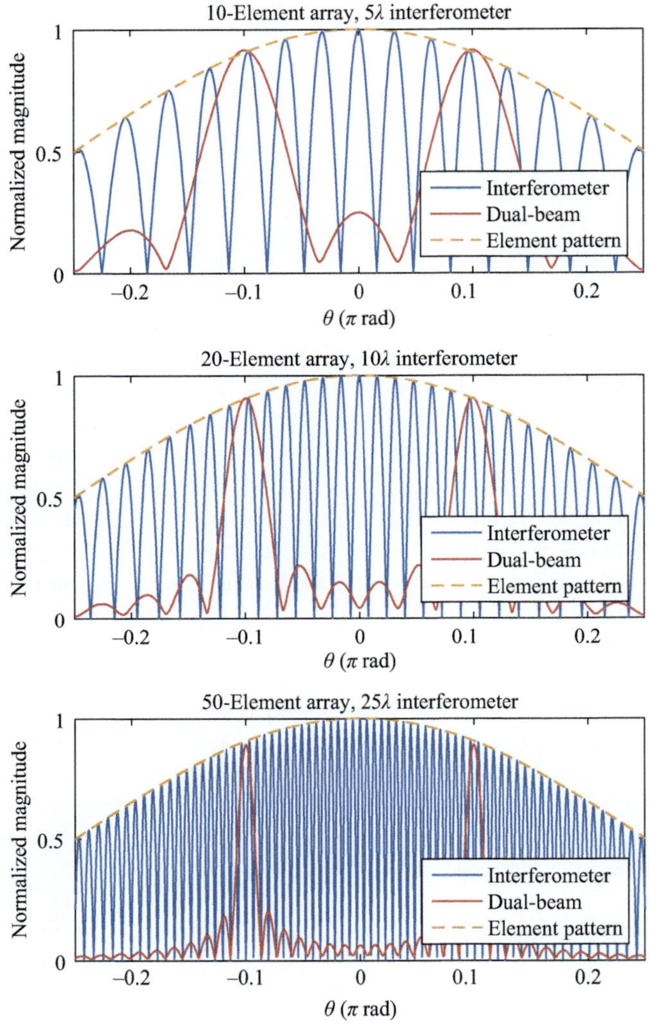

Figure 9.6 Comparison of the antenna patterns of a dual-beam system and a two-element interferometer. Both arrays are modulated by a $\cos^2(\theta)$ *element pattern. The separation of the beams in the dual-beam array is* $\pi/10$ *[15]*

measurement of the angular velocity. In contrast, the interferometric pattern clearly imparts the fringe pattern over the field of view, indicating that the angular velocity can be measured when the object is at any angle.

The relative accuracy of the interferometer compared to a dual-beam implementation with a 20-element array spaced at $\lambda/2$ increments is shown in Figure 9.7 for various angular separations of the two beams. While there are clearly instances where the dual-beam outperforms the interferometer, there are also instances where

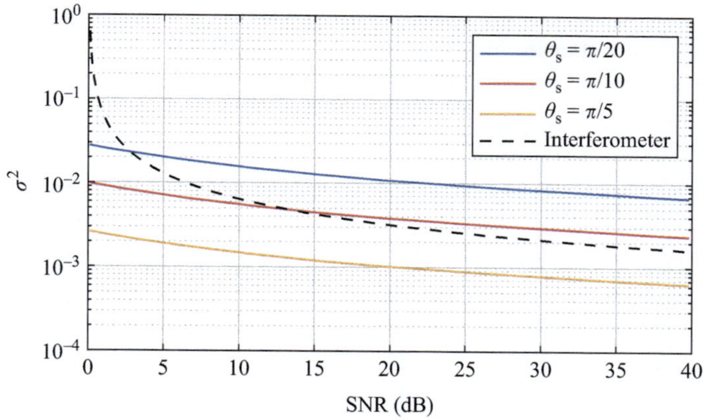

Figure 9.7 Accuracy versus SNR for a 20 element dual-beam radiation pattern with beam separations of $\pi/20$, $\pi/10$, and $\pi/5$, and for a two-element interferometer [15]

the interferometer actually achieves better accuracy. The reason is that the dual-beam response, despite more closely aligning to the optimal case of two infinitesimally small beams, still contains power outside of the beams, reducing the accuracy. Thus, the interferometric implementation can achieve good accuracy along with the benefit of being simpler to implement.

9.4 Nonlinear distortion and mitigation

The analysis above was given for a single monochromatic point source; however, in practice, there may be multiple targets or a single nonrigid target that has various angular velocities, such as the moving fingers of a hand or the swinging arms and legs of a person. The result is that the received signals will be composed of multiple frequency components from the various Doppler shifts due to relative motion. One consequence of the use of a correlation interferometer is the generation of nonlinear distortion arising from the multiplication of the received signals when multiple frequency components are present. While this problem is mitigated in the case of angular-only motion, most practical cases will involve some nonnegligible radial and angular motion. In this section, the nonlinear distortion is characterized, and two system-level approaches to mitigating the effects of the distortion are presented.

Consider a moving target with two separate scatterers moving at different radial velocities. The received signals at the two antennas at the input to the correlator are then given by

$$\begin{aligned} s_1(t) &= e^{j2\pi f_1 t_1} + e^{j2\pi f_2 t_2} \\ s_2(t) &= e^{j2\pi f_1 \left(t_1 - \tau_{g,1}\right)} + e^{j2\pi f_2 \left(t_2 - \tau_{g,2}\right)} \end{aligned} \tag{9.29}$$

where $f_n = f_c + f_{D,n}$ is the frequency generated by the nth scatterer, which is dependent on the transmitted carrier frequency f_c plus the Doppler shift $f_{D,n}$. The normalized output at the correlator is then

$$r(t) = e^{j2\pi f_1 \tau_{g,1}} + e^{j2\pi f_2 \tau_{g,2}} + \langle e^{j2\pi(f_2 t_2 - f_1 t_1 + f_1 \tau_{g,1})} \rangle + \langle e^{j2\pi(f_1 t_1 - f_2 t_2 + f_2 \tau_{g,2})} \rangle \quad (9.30)$$

The desired terms are the first two terms on the right-hand side of (9.30), where the IF is proportional to the angular velocities of the two objects, whereas the second two terms are the intermodulation products which distort the desired signal.

In practice, more than two scattering centers are likely to be present. The effect of the distortion compounds as the number of scatterers increases, as can be seen in Figure 9.8. This shows the ideal simulated response of a walking person without nonlinear distortion (left); notice the oscillations due to the movement of the arms and legs. Shown on the right is the simulated response with distortion along with a measurement conducted with a 30-GHz active interferometric radar transmitting a continuous-wave signal and received by two elements separated by 90λ. The bulk motion of the person is clearly evident in both distorted cases; however, the fine details of the arm swing and leg swing are lost in the distortion.

Two system-level methods for reducing the distortion, using long-wavelength signals and using short pulse-length signals [17], are described in the following. Both methods successfully reduce the nonlinear distortion though at the cost of a large physical baseline or a wide-bandwidth system. The first method involves reducing the relative difference between the Doppler shifts from various parts of the target. The Doppler shift imparted on scattered signals is given in terms of the scatterer velocity v_n by $f_{D,n} = 2v_n/\lambda$; thus by increasing the wavelength, the relative Doppler shifts between scatterers are reduced. As indicated in (9.30), as the Doppler shift decreases, $f_1 \approx f_2$, and the terms in brackets after time averaging reduce to copies of the first two terms. The distortion will begin to be suppressed when the differential Doppler frequency shift, the difference between the Doppler

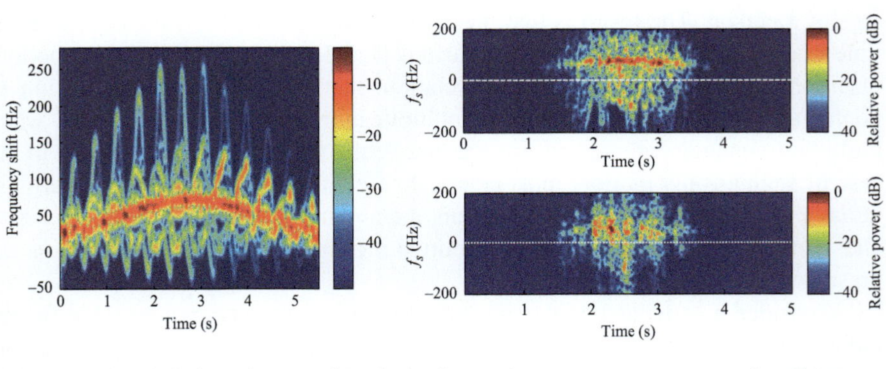

Figure 9.8 At left is the simulated ideal interferometric signature of walking human at 5 m with baseline $D_\lambda = 200$, top right is simulated return including intermodulation products, and bottom right is a measurement, both with $D_\lambda = 90$ [16,17]

shifts Δf_D of the various objects moving with differential velocity Δv, becomes lower than the interferometric frequency shift:

$$\Delta f_D = \frac{2\Delta v}{\lambda} < \frac{\omega D}{\lambda} = f_s \tag{9.31}$$

With $D/\lambda = 90$ and typical velocities of arm and leg swing of a human, the requirement on system wavelength to suppress the interference is 210 MHz. Figure 9.9 shows the simulated signal of a walking person using a 90λ interferometer, with wavelength increasing from 10 to 100 mm. The interference clearly reduces despite λ being well under the requirement above.

At such low frequencies, long electrical baselines may be difficult to implement for short-range applications. For example, a $\lambda = 100$ mm system with

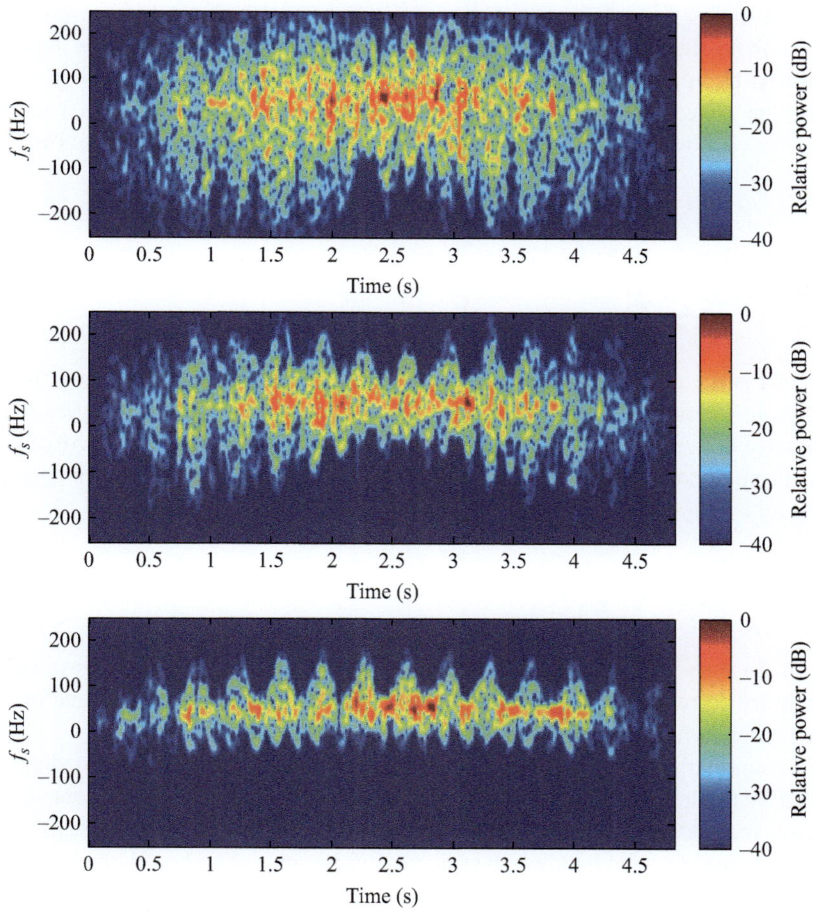

Figure 9.9 Simulation of walking human past 90λ baseline interferometer with signal wavelengths of 10, 30, and 100 mm (top to bottom) [17]

$D = 90\lambda$ requires a baseline of 9 m. A second issue with this method is that the system can no longer measure Doppler velocity, thus losing the ability to track velocity in the radial dimension.

The second method described here for distortion mitigation involves using short-pulse signals which reduce the temporal overlap between returns from different parts of the target. When multiple objects are observed at the same range, overlap of received signals will be unavoidable. However, pulsed signals with sufficiently short duration can minimize the overlap in cases when the objects are at different distances. Furthermore, although the previous analyses focused on continuous-wave operation, the same signal response manifests with pulsed waveforms. Without overlap from multiple objects, the inputs to the correlator contain only a single-frequency component, and the output of the correlator contains only the desired term and no intermodulation products.

An example is the separation of the signals scattered off the arms and torso of a person walking tangentially past the interferometer (i.e. the radar is viewing the person from the side). The surfaces of the arms and torso facing the radar are separated by approximately 10 cm, yielding a temporal difference of $\tau = 670$ ps. Using the typical radar range resolution equation $\Delta R = c/(2\delta f)$ yields a required waveform bandwidth of $\delta f = 1.5$ GHz. The signals at the correlator inputs become:

$$
\begin{aligned}
s_1(t) &= \sum_n \Pi\left(\frac{t - \tau_n}{\tau}\right) e^{j2\pi f_n (t-\tau_n)} \\
s_2(t) &= \sum_n \Pi\left(\frac{t - \tau_n - \tau_{g,n}}{\tau}\right) e^{j2\pi f_n (t-\tau_n-\tau_{g,n})}
\end{aligned}
\tag{9.32}
$$

where Π is the rectangular pulse function, τ_n is the delay caused by the difference in distance between scatterers, and $\tau_{g,n}$ is the delay caused by the geometrical difference between the two antennas. As long as the pulse width τ is less than all τ_n, the intermodulation distortion will be avoided, yielding a correlator response given by

$$
r(t) \propto \sum_n \Pi\left(\frac{t - \tau_n}{\tau}\right) e^{j2\pi f_n \tau_{g,n}}
\tag{9.33}
$$

In the above, the nonlinear distortion is not present since only single-frequency signals are correlated. This is illustrated in Figure 9.10, where pulsed waveforms of varying bandwidths are simulated.

In summary, two system-level methods for reducing the nonlinear distortion caused by the intermodulation products of multiple scatterers were presented. Both of these methods successfully reduce this distortion, though with the added cost of large baselines or wide bandwidth systems. Further methods should be explored for solving this issue, especially in the case for micro-motion sensing when it is critical to resolve multiple independent scatterers which are spatially close together.

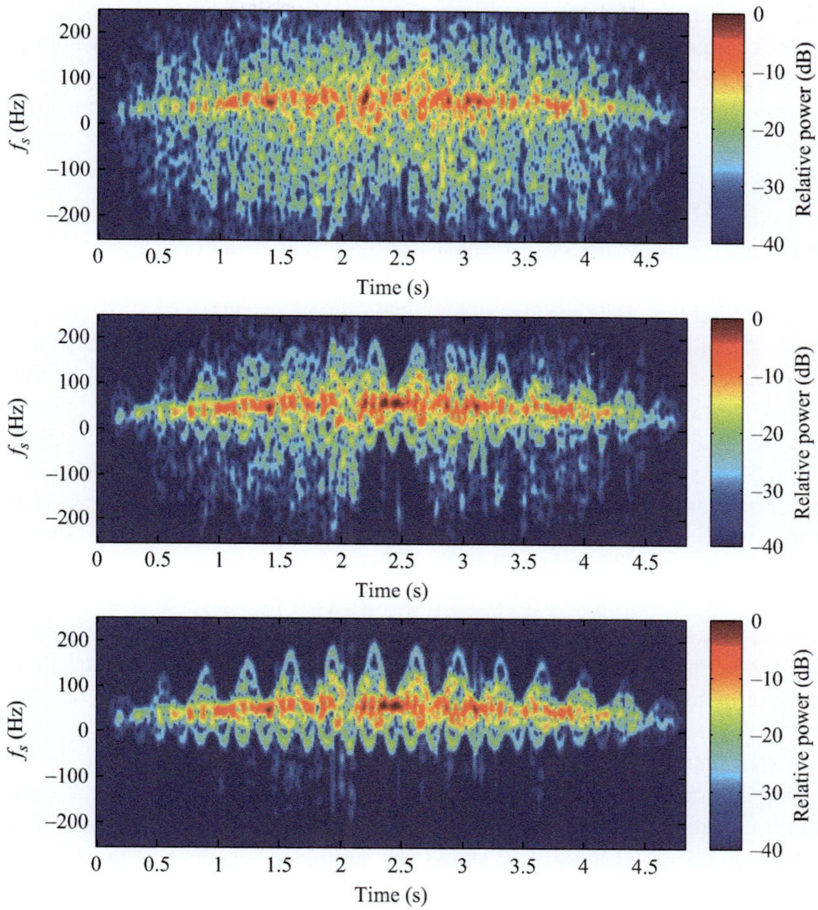

Figure 9.10 Simulation of person walking past 90λ baseline interferometer with pulse lengths of 100, 10, and 1 mm (top to bottom) [17]

9.5 Experimental system examples

In this section, two correlation interferometer systems are reviewed along with their experimental results. As noted earlier, the interferometric angular velocity measurement technique is a receiver-side implementation. As such, it can be implemented using an active system that transmits a signal and measures the reflected radiation, or it can be implemented with a passive interferometric receiver that receives signals that are generated by another source and scattered off the object or signals that are intrinsically generated by the object such as thermal radiation. The first system example demonstrates the latter case, implemented in a K_a-band passive correlation interferometer that detects the thermally generated electromagnetic radiation from the human body to measure the angular velocity of a walking person. The second system is an active K_a-band interferometric radar that transmits a continuous-wave signal and detects the scattered signals from a walking person.

9.5.1 Passive 27.4-GHz correlation interferometer system

Passive millimeter-wave systems measuring thermal radiation, called radiometers, generally require very high-sensitivity receivers since the radiation power levels emitted by humans and the surrounding background are on the order of picowatts or femtowatts at microwave and millimeter-wave frequencies. In addition, the radiation is noise-like, meaning that coherent detection techniques such as matched filtering are not viable methods to increase the signal gain. As such, the receivers typically require high gain amplifiers with low noise.

In [10], a passive 27.4-GHz correlation interferometer was implemented to measure the angular velocity of walking people by detecting the thermal radiation emitted by the human body. The system block diagram is shown in Figure 9.11. The antennas were Gaussian optical lens antennas with 32.7-dBi gain, followed by low-noise amplifiers (LNAs) with 32-dB gain and 3.2-dB noise figure. The 27.4-GHz received signals were downconverted to an intermediate frequency of 1.5 GHz before being input to the correlator, which was implemented in analog hardware using a mixer and a low-pass filter with bandwidth of 2 kHz. The sensitivity of the correlation radiometer was measured to be 270 mK.

Experiments were conducted for three antenna baselines with the person walking in straight line at a distance of 5.75 $\pm$ 0.25 m from the receivers. The antennas were separated by 14λ, 23λ, and 30λ while the person walked with a velocity of 1.5 m/s. Figure 9.12 shows the simulated response of the passive correlation interferometer with additive white Gaussian noise (AWGN) on the left, with measured data on the right showing a good match to the theoretical model. It is clear that the frequency increases as the antenna baseline increases, giving a direct measure of the bulk angular velocity of the walking person.

Because the thermal radiation emitted from a person is of such low power, antennas with high gain were necessary to achieve sufficient signal detection. As a result, the bulk velocity is easy to discern; however, the resolution of the signal response is coarse since the total signal duration is low due to the short duration

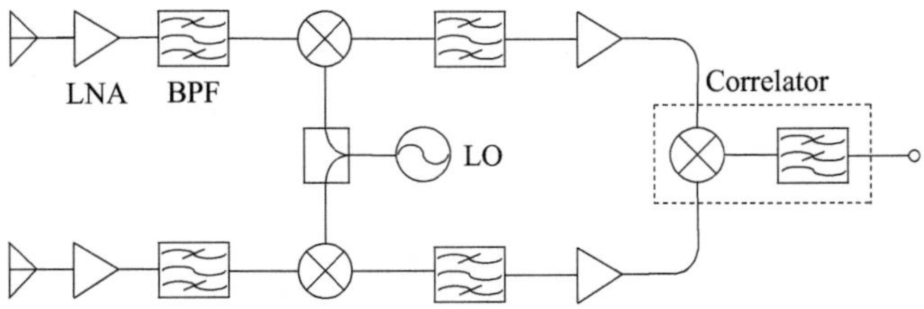

Figure 9.11 Block diagram of the two-stage, passive correlation interferometer [10]

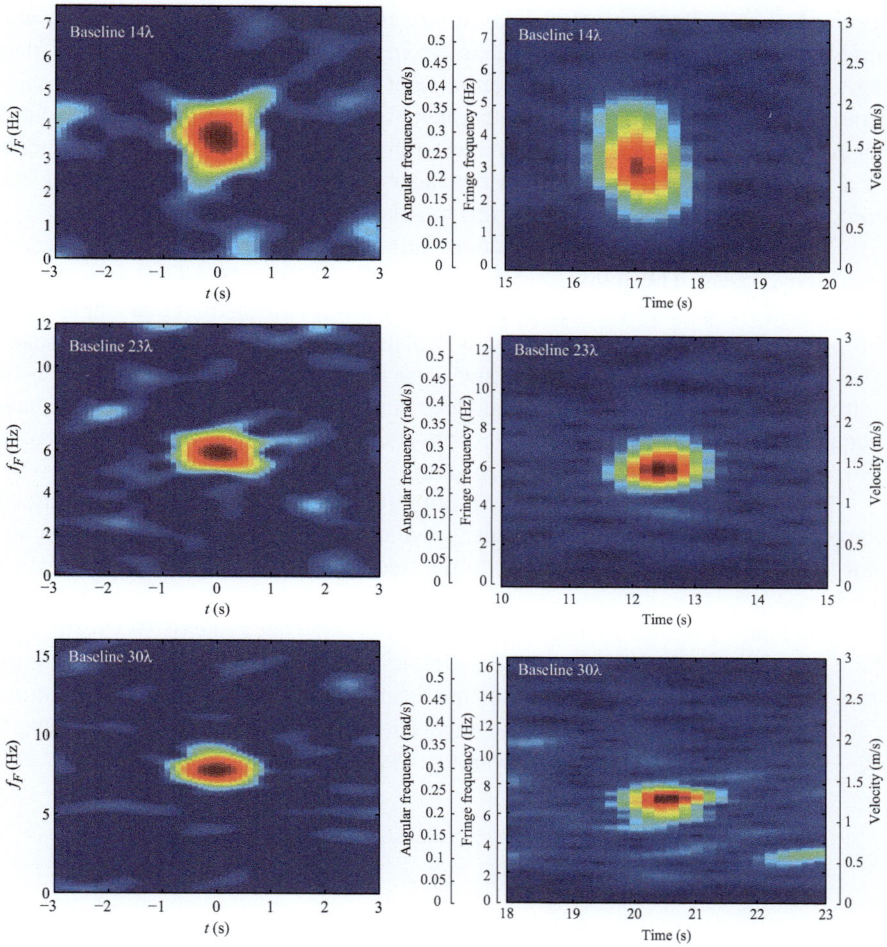

Figure 9.12 At left are simulated data of a person walking tangentially past the interferometer at a velocity of 1.5 m/s at a range of 5.75 m at three different baselines. AWGN has been added to approximate the noise in the experimental system. At right are experimental data examples of the same parameters above [10]

within which the person is within the beam, as such, any fine features from the motion of the arms or legs are not present. Narrow-beam systems can thus provide quite accurate measurements of the bulk angular velocity of close-range objects, even with nonrigid objects such as humans where multiple moving parts are present.

9.5.2 Active 29.5-GHz dual interferometric-Doppler system

While passive receivers can measure angular velocity, they lack the ability to measure Doppler which requires coherent processing of the transmitted and received signal. To demonstrate the ability to directly measure both radial and angular velocity, a K_a-band dual interferometric-Doppler radar system was demonstrated in [18]. The system consisted of a continuous-wave transmitter centered between the two receivers, the outputs of which were digitized and processed for both Doppler frequency shift and interferometric frequency shift. Measurements of a person walking tangentially to the radar were conducted, demonstrating the complementary nature of the radial and angular velocity measurements.

The interferometric radar system, shown in Figure 9.13, was a quasi-monostatic radar. The transmitter consisted of a 5-dBm, 29.5-GHz continuous-wave signal transmitted through a horn antenna with 15-dBi gain. Each antenna had a $\theta_{HPBW} = 30°$ in both the E-plane and H-plane. The two receivers each consisted of a 15-dBi horn antenna feeding a low-noise amplifier with 40-dB gain and 3-dB noise figure. The signals were then downconverted to a 2.7-GHz intermediate frequency through a mixer with 5-dB conversion loss. The IF signal was boosted with a 22-dB gain amplifier before quadrature downconversion to baseband.

The quadrature mixer generated both the in-phase and quadrature signal components of the received signals centered at DC; thus, the only frequency content in the baseband signals was due to the frequency shifts from the movement of the person.

The I and Q signals for each receiver were digitized at a rate of 10 kS/s using a National Instruments USB-6210 analog-to-digital converter (ADC). Time-frequency analysis is performed offline using MATLAB®, using a short-time Fourier transform with a 1,000 sample Hamming window and an FFT length of 2,048 samples. The same processing was used on both the Doppler and

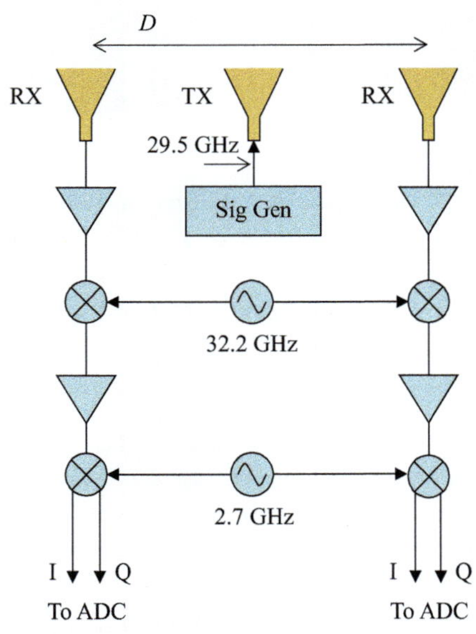

Figure 9.13 Schematic of the dual-mode millimeter-wave active interferometric system [18]

interferometric signals. The Doppler frequency shift was obtained by calculating the spectrogram of the complex signal of either one of the two receivers while the interferometric signal was calculated by first correlating and then low-pass filtering the two complex signals before calculating the spectrogram.

Figure 9.14 shows the results of the dual measurements of the radial and angular velocity of a walking person at various trajectories relative to the radar. The trajectories are shown in the insets. For instance, the first plot on the top left indicates motion directly toward the radar. The time-frequency response of the

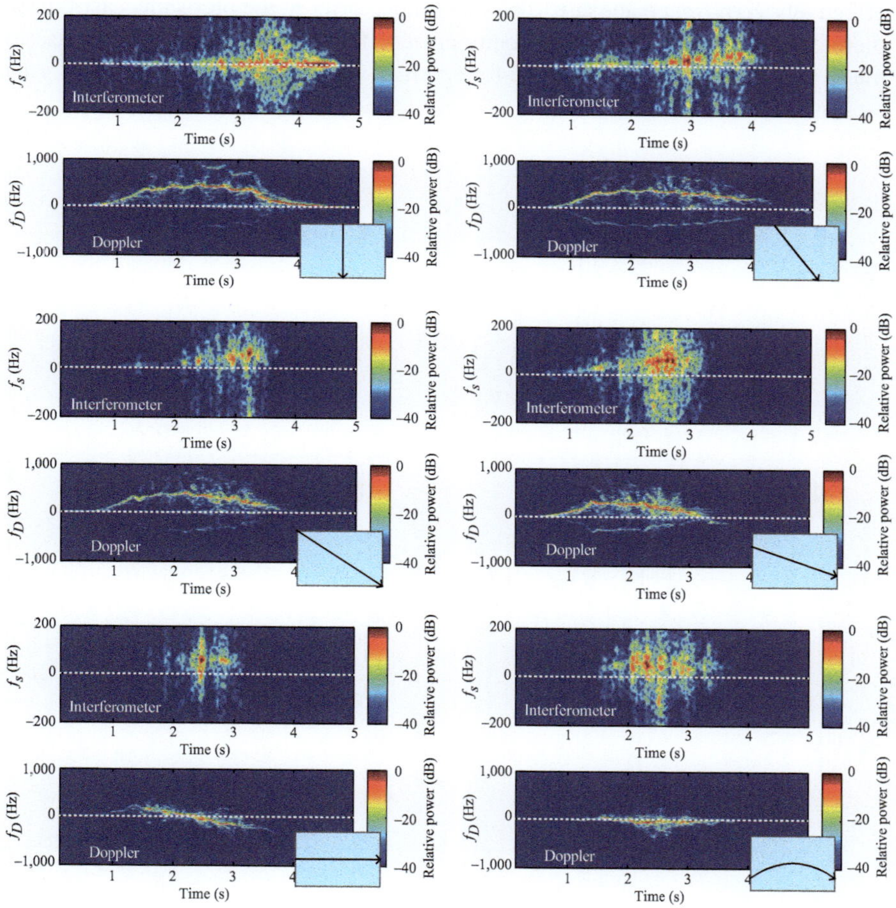

Figure 9.14 Measurements results for six trajectories (shown in the inset). The top plot of each spectrogram shows the interferometric time-frequency response while the bottom plot shows the Doppler time-frequency response. While the Doppler frequency shift decreases as the trajectory becomes more angular, the interferometric frequency shift increases [18]

interferometric signal is shown on the top of each plot, and the time-frequency response of the Doppler signal is shown on the bottom of each plot. When the trajectory is directly toward the radar, the Doppler frequency shift increases as expected due to the large radial velocity of the person. The interferometric response shows a bulk frequency of 0 Hz, indicating that no bulk angular velocity is detected. The nonlinear distortion on the signal is present due to the presence of some angular motion of the various parts of the body. At the other extreme is the plot on the lower right, showing the results of a trajectory that is only angular, with little to no radial velocity. In this case, the bulk interferometric shift displays a positive frequency indicating clockwise angular velocity while the Doppler shift shows effectively zero frequency shift. The other trajectories in the plot show cases where there is both radial and angular velocity present. Figure 9.15 shows the responses of the two modes at one time instance for a person walking directly radial, at 45°, and tangentially to the radar.

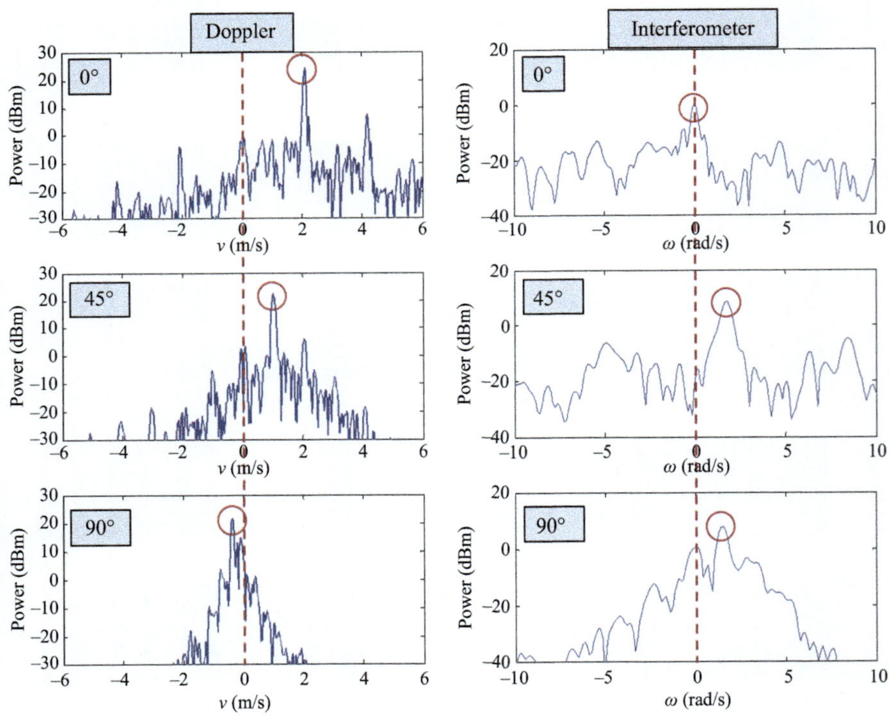

Figure 9.15 Measurements for a single time instance of the Doppler and interferometric signal responses for three trajectories of motion: toward the radar, at a 45° angle, and tangential to the radar, demonstrating the complementary nature of the interferometric and Doppler measurements

9.6 Conclusions

The use of a distributed interferometric radar provides a direct measurement method for the angular velocity of moving objects, a quantity which can traditionally only be inferred over multiple measurements. Radar systems which are able to directly measure range, radial velocity, angle (azimuth and elevation), and angular velocity form a basis in $\mathbb{R}^3$ and thus can fully describe 3D position and velocity without relying on successive measurements. Currently range, radial velocity, and angle can be directly measured, yet angular velocity typically requires beam scanning or complex signal processing algorithms with antenna arrays. Recently, a new technique for estimating target angle using a single interferometric array with a chirped pulse waveform was described in [19]; along with the angular velocity measurement described in this chapter, the angle-estimation method provides a path toward simultaneous high-accuracy angle and angular velocity measurements.

The resolution of angular velocity measurements was discussed and it was shown that the resolution varies with system design parameters (baseline and beamwidth) and target parameters (velocity and radial distance). It was shown that system designs which incorporate wide antenna patterns with large system baselines are better able to resolve multiple targets. Likewise, slowly moving targets are more easily resolved, which may also depend on the radial distance to the target. When designing an interferometric system for micro-motion sensing, an analysis should be performed, which computes minimum required baseline and beamwidth for a given desired frequency resolution based on expected target ranges and velocities. The accuracy of the angular velocity measurements was also discussed and it was found that the accuracy, or variance of the frequency measurement, is inversely proportional to the SNR and the mean-square antenna pattern. This implies that systems with high SNR or long integration times and wide antenna patterns provide a lower theoretical variance on the angular velocity measurement and thus are capable of more accurate estimates. Two physical methods were described to mitigate the nonlinear distortion due to the intermodulation products of the correlator. Both methods performed well, though with the added cost of either a large system baseline and the inability to detect Doppler velocity, or a wide-bandwidth pulsed system. For micro-motion sensing where detection of radial velocity is critical, using a pulsed system for distortion mitigation is a suitable option.

Combined interferometric-Doppler measurements in a single system were demonstrated, providing a starting point toward full trajectory radar measurements in $\mathbb{R}^3$. Combining measurements of range, radial velocity, azimuth angle, and elevation angle with two measurements of angular velocity in azimuth and elevation, full 3D position and velocity information are possible to obtain, which may lead to improvements in radar tracking and classification in future remote sensing applications. Notably, the signal form generated by the correlation interferometer is mathematically very similar to that of the Doppler radar response. This indicates that signal processing methods developed for Doppler processing may be easily applied to angular velocity processing as well, potentially minimizing the need for additional software design associated with the new measurements.

References

[1] Richards MA, Scheer JA, and Holm WA. Principles of Modern Radar: Basic Principles. Institution of Engineering and Technology. Raleigh, NC: SciTech Publishing, Inc.; 2010.
[2] Skolnik MI. Introduction to Radar Systems. New York: McGraw-Hill; 2001.
[3] Crew H. The Wave Theory of Light – Memoirs by Huygens, Young and Fresnel. Read Books. New York: American Book Company; 2009. Available from: https://books.google.com/books?id=ftlipVx47cUC.
[4] Michelson AA, and Morley EW. On the Relative Motion of the Earth and the Luminiferous Ether. American Journal of Science (1880–1910). 1887; 34(203):333.
[5] Born M, and Wolf E. Principles of Optics: Electromagnetic Theory of Propagation, Interference and Diffraction of Light. Vol. 6. Oxford: Pergamon Press; 1984.
[6] Wohlleben R, Mattes H, and Krickbaum T. Interferometry in Radioastronomy and Radar Techniques. Dordrecht, Netherlands: Kluwer Academic; 1991.
[7] Ryle M, and Vonberg DD. Solar Radiation on 175 Mc./s. In: Classics in Radio Astronomy. Studies in the History of Modern Science. Vol 10. Springer, Dordrecht; 1946.
[8] Zebker HA, and Goldstein RM. Topographic Mapping From Interferometric Synthetic Aperture Radar Observations. Journal of Geophysical Research. 1986;91:4993–4999.
[9] Nanzer JA. Microwave and Millimeter-Wave Remote Sensing for Security Applications. Norwood, MA: Artech House; 2012.
[10] Nanzer JA. Millimeter-Wave Interferometric Angular Velocity Detection. IEEE Transactions on Microwave Theory and Techniques. 2010;58(12): 4128–4136.
[11] Nanzer JA. On the Resolution of the Interferometric Measurement of the Angular Velocity of Moving Objects. IEEE Transactions on Antennas and Propagation. 2012;60(11):5356–5363.
[12] Van Trees HL, Bell KL, and Tian Z. Detection, Estimation, and Modulation Theory. Vol. 1, 3. Hoboken, New Jersey: Wiley-Interscience; 2001.
[13] Kay SM. Fundamentals of Statistical Signal Processing. Upper Saddle River, New Jersey: Prentice Hall; 2010.
[14] Nanzer JA, and Sharp MD. On the Estimation of Angle Rate in Radar. IEEE Transactions on Antennas and Propagation. 2017;65(3):1339–1348.

[15] Sharp MD, and Nanzer JA. Accuracy of Angle Rate Measurements Using a Distributed Radar With a Correlation Receiver. IEEE Antennas and Wireless Propagation Letters. 2018;17(2):209–212.
[16] Nanzer JA. Simulations of the Millimeter-Wave Interferometric Signature of Walking Humans. In: Proceedings of the 2012 IEEE International Symposium on Antennas and Propagation; 2012. pp. 1–2.
[17] Nanzer JA, and Zilevu KS. Distortion Mitigation in Interferometric Angular Velocity Measurements. Electronics Letters. 2014;50(18):1316–1318.
[18] Nanzer JA, and Zilevu KS. Dual Interferometric-Doppler Measurements of the Radial and Angular Velocity of Humans. IEEE Transactions on Antennas and Propagation. 2014;62(3):1513–1517.
[19] Vakalis S, Klinefelter E, and Nanzer JA. Angle Estimation Using Wideband Frequency Modulation and an Active Distributed Array. IEEE Microwave and Wireless Components Letters. 2018;28(11):1059–1061.

Chapter 10

Continuous-wave radar sensor for structural displacement monitoring

Shanyue Guan[1] and Jennifer A. Bridge[2]

Infrastructure systems are tied closely to the function of human society in many aspects. Effective approaches to maintaining infrastructure are critical for public safety. The assessment of structural conditions, diagnosis of structural damage, and infrastructure safety protection may be achieved by monitoring and analyzing structural responses. Many approaches have been pursued to monitor different types of structural responses; displacement monitoring is one of the commonly used approaches. Various types of sensors have been developed to monitor structural displacement with a range of cost, ease-of-use, and effectiveness. Radar systems provide a family of high-accuracy displacement measurements. Specifically, continuous-wave (CW) radar is a promising technology for structural applications. CW radar transmits the microwave to the target continuously and captures the signal reflected. By analyzing the difference between the transmitted and reflected microwaves, the radar can measure the relative displacement between the target and the radar. Some specific types of CW radars have been developed to monitor structural displacement and have demonstrated promising performance. Radar-based structural monitoring applications are presented and specific types of CW radars developed for structural monitoring are introduced. The operating mechanism for conducting displacement measurement with CW radars is presented. The hardware designs of the CW radars and the signal processing from radar signals to final displacement information are described. The structural displacement monitoring using the CW radar is quantified in both laboratory and field validation experiments. Challenges associated with applying the CW radar to monitoring structural displacement are summarized, and future work to improve the performance of CW radar is suggested.

[1]Department of Engineering, East Carolina University, USA
[2]Department of Civil and Coastal Engineering, University of Florida, USA

10.1 Introduction

Modern society relies on the consistent and stable functionality of its civil infrastructure. However, after structures are in service for many years, their performance may degrade, resulting in functionality and safety problems. In the United States, about eight decades after a period of increased construction during the 1930s, the condition of many bridges has declined, yielding maintenance concerns and even failures [1], such as the I-35W Mississippi River Bridge collapse in 2007. In 2017, the American Society of Civil Engineers [2] gave a grade D+ to the state of infrastructure in the United States. Aging infrastructure is a worldwide concern. For example, according to the Japan Transport Ministry, in Japan, the country has about 140,000 bridges with 15 m or longer and approximately 20% of the bridges were 50 years or older in 2016. The number will increase to 47% in less than 10 years.

Assessing and quantifying the condition of a structure is critical to maintaining its performance. In addition to degrading performance caused by long-term loading or environmental effects, structures are vulnerable to natural catastrophes such as earthquakes, hurricanes, and tsunamis. For instance, many structures were damaged or collapsed, after the Northridge (1994) [3] and Kobe (1995) earthquakes [4]. Monitoring and protecting structure's pre- and post-disaster performance are important for the assurance of public safety and the allocation of repair and maintenance resources.

The necessity of monitoring structural conditions has led to the development of structural health monitoring (SHM) technologies over the last two decades. SHM provides a range of approaches to capture structural conditions for various purposes. First, vibration-based monitoring is capable of identifying structural models from measured structural response under different load cases such as wind loads. Second, long-term monitoring systems can provide structural responses before and after severe natural or manmade catastrophes. This information can be analyzed to assess the damage and integrity of structures before investing money to repair or reconstruct them. In addition, field tests can be conducted on structures to determine whether the as-built structures have a similar response to that expected of the original design, which may also provide a better prediction of structural response under extreme loads.

Vibration-based monitoring methods are commonly used to measure response and assess structural conditions. These methods, using displacement, velocity, and acceleration measurements, have been widely applied on various infrastructures around the world. In addition to dynamic measurements, designers and owners often select static deflections as an important approach to evaluate structural design, especially for bridges. Besides monitoring structural responses, some sensors have been implemented to monitor environmental conditions such as wind speed, temperature, or humidity.

There are several commonly used sensors to measure or extract displacement of structures. Some sensors such as accelerometers lack either accuracy or practicality for realizing displacement measurements. Other technologies such as Global Position Systems do not provide adequate sampling rate or accuracy of measurements, unless expensive units are used [5]. Laser Doppler vibrometers have better accuracy, but their high price does not allow implementing a sensor network for practical applications.

The technology of CW radar is a promising approach to measuring structural displacement. CW radars transmit and receive signals that are reflected by a target. The phase difference between the transmitted and received signal is a function of the relative distance between the radar and the target. Using a CW radar provides a lower cost and more convenient approach to measuring structural displacement than currently available technologies.

10.2 Background

10.2.1 Structural health monitoring

SHM technologies aim to monitor, evaluate, and maintain structural integrity and safety. SHM has developed rapidly since the 1950s and consists of many instrumentations and methodologies applied to observe, measure, assess, and diagnose the conditions of engineering structures in civil engineering [6], mechanical engineering [7], aerospace engineering [8], and other fields. SHM development requires interdisciplinary knowledge. Early SHM technologies were applied mostly on machinery and aircraft [9,10]; however, in the last three decades, SHM technologies have also been applied widely to civil engineering structures.

For civil engineering structures, the primary cause of structural defects is deterioration due to aging problem. In addition, extreme loading events, such as hurricanes, earthquakes, or overloading, may cause damage to structures [3,4]. Developing fast, efficient, and low-cost approaches for evaluating structural condition due to deterioration and extreme events is a challenging task for civil engineers. Another role for structural monitoring is to evaluate the performance of the as-built structure compared to its intended design. There are often some deviations between the final constructed structures and the original design; comparing the structural performance to the design calculations can be carried out through analysis of SHM data.

Many aspects of SHM have been developed to address the above challenges. Researchers have developed many types of sensors to capture the structural response accurately and created new damage detection techniques to quantify the damage conditions of structures. Also, many full-scale experiments have been conducted on different structures to validate whether the response of practical structures is similar to the theoretical calculations from the design [11,12]. Furthermore, newer, high-profile structures are often designed with integrated, real-time monitoring systems installed during construction. Real-time monitoring systems are intended to monitor the structures response under normal and extreme cases which could provide understanding of the structural conditions in real time and protect structures subjected to extreme events [13,14].

As SHM technologies have developed, they have been implemented on various structures. SHM has been implemented broadly on bridges [5,15], buildings

[16,17], dams [18,19], pipelines [20,21], tunnels [22,23], and other types of engineering structures [24,25]. SHM technologies have been developed in a range of areas including novel sensor development, signal-processing techniques, structural control, real-time monitoring, signal transmission, data management, damage detection and diagnosis, and full-scale experiments.

10.2.2 Existing displacement sensing technologies

In general, vibration monitoring and static deflection monitoring are commonly adopted approaches to measure structural motion, and the results may be used for structural conditions evaluation [26] and damage detection [27]. As previously mentioned, accelerometers are used to measure vibration. However, to detect the displacement (both static and dynamic) directly with a high accuracy is still challenging.

Modern sensing technologies tend to be more convenient, more accurate, and less expensive, making them better suited to long-term applications. Based on different operating mechanisms, displacement sensors can be classified into seven major categories: electromechanical, electromagnetic, electroacoustic, laser/high-power energy, imaging, surveying technology, and radar techniques.

To select the most appropriate sensor for the application, factors including accuracy, cost, and convenience must be balanced and compared comprehensively. The accuracy of a sensor is determined by its resolution, sensitivity, noise level, measurement range (amplitude, frequency, distance), and repeatability. In addition to accuracy, the cost of the sensor and the cost and ease of its installation are also important factors in sensor selection.

10.2.3 Radar techniques

Radar techniques are a large group of systems to detect and locate objects. In general, the radar system operates by emitting the radar wave (pulse, sinusoidal, sawtooth, and square wave) from a transmitting antenna. Then the radar wave reaches the target and the reflected radar wave is captured by the receiving antenna. By analyzing the reflected signal, the location or velocity of the target may be obtained. The radar's signal is often modulated by frequency or amplitude to carry some information in the transmission. After receiving the signal, the transmitted information is obtained by demodulation from the radar signal which is the inverse process of modulation. Via this approach, the information can be transmitted over long distance without too much signal loss. Radar has been applied widely for military [28], aerospace [29], and communication applications [30]. More recently, radar has also been adopted for civil engineering applications [31,32], including the detection of debonding in concrete structures [33], and for monitoring deteriorating concrete dams [34]. Applications of radars for SHM, including ground penetrating radar (GPR), remote sensing radar, and distance measurement radars, are introduced in this section.

10.2.3.1 Ground penetrating radar

GPR transmits the radar pulse which has very strong penetration performance through some materials such as ground surface to image the subsurface. By comparing the transmitted signal with the received signal, the subsurface conditions can be obtained. The radar wave frequency stays in the range from 10 MHz to 2.6 GHz. A GPR transmitter emits electromagnetic energy into the ground. When the energy encounters a buried object or a boundary between materials with different permittivities, it may be reflected to the surface. A receiving antenna can then record the variations in the return signal. The depth to a target is determined based on the amount of time it takes for the radar signal to reflect to the unit's antenna. Radar signals travel at different velocities through different types of materials. It is possible to use the depth to a known object to determine a specific velocity and then calibrate the depth calculations.

GPR is a nondestructive testing (NDT) approach which can be used to scan a large area if it is carried by a mobile vehicle or airborne. Because of its high penetrating performance, it may be used to monitor many types of materials. The largest limitation of using the GPR is in high-conductivity material such as clay soils that are salt contaminated and for which the reflected signal strength is weak. Performance is also limited by signal scattering in heterogeneous conditions (rocky soils).

GPR has been applied for NDT on concrete structures [35]. Mobile GPRs have been used to detect the pavement [36] and bridge deck [37] conditions. The resolution of mobile-based radar systems depends on the radar's performance and the speed of the transportation. The speed at which a radar signal travels is dependent on the composition of the material being penetrated. Sometimes, to get higher quality measurements, more cycles of the measurements may need to be conducted.

10.2.3.2 Remote sensing radar

The radars can be installed on the satellites to conduct the sensing remotely. The radar transmits the microwave over large areas and captures the reflected microwave from the target areas. Remote sensing is an example where the time delay between emission and return is measured, from which the location, speed, and direction of an object can be calculated.

Remote sensing makes it possible to collect data of dangerous or inaccessible areas such as post-disaster area. Remote sensing is more focused on monitoring conditions over a large area with side length up to 1,000 m while the detailed local condition is not detected.

Satellite radar has been used widely for remote sensing for archaeological sites [38] and dams [39], and to assess changes of large areas after disasters such as earthquakes [40,41] or tornados [42]. However, the image quality from the satellite radar is easily affected by atmospheric conditions. Frequency of image updating is low (approximately one image for several hours), which is challenging to develop a real-time monitoring system.

10.2.3.3 Distance measurement radar

Besides detecting the structural conditions using penetrating performance or conducting remote sensing to monitor large areas, there are many types of radars that have been used for displacement or distance measurements. To measure the distance between the radar and the target, there are two general approaches: time-of-flight radar and interferometric radar. The theory of the time-of-flight radar is similar to a laser or ultrasonic system. By calculating the traveling time between transmitting the radar wave and receiving the radar wave, the distance between the radar and the target may be measured as the radar wave travels with the speed of light. The interferometric radar compares the difference between the transmitted signal and received signal on signal amplitude, frequency, or phase difference to calculate the distance between the radar and the target.

Time-of-flight laser radar
The mechanism of the time-of-flight radar is the transmission of a pulse signal. By calculating the time difference between transmitting and receiving the laser pulse signal, as the speed of the pulse transmission is known, the distance between the radar and the target can be calculated.

The advantage of using this type of radar is that, due to its straightforward operating theory, it is convenient to deploy and conduct measurement. Usually, this type of radar is installed at a stationary location and keeps transmitting the pulse to the target over long distance. However, the time difference measurement between transmitting and receiving measurement is not very accurate, and the distance obtained from this type of radar has some considerable errors.

The radar has been used to measure a target with a periodic motion successfully [43]. The radar has also been applied to measure the spatial motion of a target [44].

Pulse radar
Pulse radar is a system that transmits short wavelength and high-energy pulses over long distance. Compared with the time-of-flight laser pulse radar, the pulse signal energy of pulse radar is higher. It is usually designed for long distance measurements and transmits a relatively high-power pulse. For each transmission, only one pulse will be transmitted. The pulse radar transmits short pulses which are reflected by the target. By measuring the traveling time between sending a pulse of radar wave and receiving the echo of the object, as the radar wave travels at the speed of light, the distance between the radar and the target may be calculated. Besides the distance measurements, the speed of the target may also be calculated based on the Doppler effect by calculating the difference between the transmitted and received signal's frequency.

Pulse radar is useful for measuring over long distance and may output both the distance and velocity. However, pulse radar systems are large, which may make them impractical for SHM applications.

Pulse radar has been used for aircraft range detections [45] and wind speed measurement [46]. It has also been applied in human health care, such as fall risk assessment and fall detection [47].

FMCW radar
Due to the errors of the measuring the time of flight, interferometric radars are used more frequently to measure distance or displacement. Frequency-modulated CW (FMCW) radar transmits continuous wave, and the frequency of signals varies over

a stable period. The received signal reflected from the target is mixed with the transmitted signal to produce the interference (beat signal) since the frequencies of the transmitted signal and received signal are similar. The beat signal provides the distance between the target and the radar after demodulation [48]. A variety of modulations are possible: sine wave, sawtooth wave, triangle wave, and square wave.

FMCW radar is a short-range measuring radar capable of determining distance. It provides distance measurement along with speed measurement, which is essential when there is more than one source of reflection arriving at the radar antenna. Another advantage of FMCW is that radar operates at relatively low frequencies, which makes hardware design easier to achieve. Conversely, the FMCW radar transmission range is shorter and the power consumption is lower compared with the pulse radar. FMCW radar has been used to measure the distance between a metallic target and the radar [49]. It has also been tested in-door to measure the speed of a moving bicycle and the distance between the bicycle and the radar [50].

Step frequency interferometric radar

The step frequency interferometric radar transmits the microwave with the signal frequency increasing in linear steps. The general working principle of using the step frequency interferometric radar to measure the distance is to compare the transmitted signal and the reflected signal. The radar's received signal includes both the amplitude and phase information of the signal reflected by the target. The peak of the amplitude signal corresponds to the reflected signal which has the highest strength. The phase difference between the reflected radar wave and the transmitted radar wave may be used to calculate the distance between the radar and the target.

The step frequency interferometric radar has a very good accuracy over many other radar systems for measuring displacement. The accuracy may be close to submillimeter level [51]. However, the radar system is large and bulky, and is thus inconvenient to transport. Also the cost of the interferometric radar (up to several thousand US dollars) is much higher than most displacement sensors.

The interferometric CW step frequency radar has been validated for displacement monitoring on many types of structures, including vibrating stable cables [52], bridges [53–57], and buildings [51]. Another type of interferometric radar has been used for SHM application is the interferometric synthetic aperture radar (InSAR). The InSAR has also been used for monitoring the concrete structure deformation [39,58] and it has also been used to monitor subsidence and structural stability [59]. Interferometric radar has also been applied in human health care, such as fall risk assessment and fall detection [60].

CW radar

CW radar is a promising technology to measure structural displacement. The CW radar transmits a known stable-frequency continuous microwave and then receives the reflected from the target. The motion information of the target is modulated into the radar's signal by comparing the phase difference between the transmitted and received signal. Through the demodulation of the radar's signal, the displacement of the target can be measured. Different from the pulse radar, the CW radar is transmitting the microwave continuously.

The size of the CW radar may be very compact, making it easy to install and transport. The cost of the CW radar is reasonable compared with other sensors. In addition, the CW radar has a sub-millimeter accuracy which is very promising for SHM applications [61].

CW radar has been applied to conduct noncontact vital sign detections [62–65] and cancer radiotherapy [66,67]. It has been applied on some preliminary SHM applications to monitor the structural displacement [61,68–70].

To apply the CW radar sensor for bridge monitoring, the passive backscattering strategy may be used. To enable bridge dynamic and static displacement monitoring, one or more radars may be mounted to the underside of the bridge. For the passive backscattering mode, each sensor transmits a continuous microwave to a stationary target surface (ground or other reflective materials such as metallic surface which is sensitive to the microwave) under the bridge and then in turn receives the reflected signal [71], as illustrated in Figure 10.1. The received signal is modulated with the relative motion between each sensor and the stationary target surface. The displacement of bridge at each sensor location is obtained through appropriate signal processing and demodulation.

The passive backscattering strategy is convenient to setup for measuring displacement. However, the largest challenge of the backscattering strategy is that the signal strength attenuates quickly as the target distance increases. For the case of a single transmitter and receiver pair, a free space microwave propagation model is used to illustrate the signal attenuation:

$$P_r(d) = \frac{P_t G_t G_r \lambda^2}{(4\pi)^2 d^2 L} \tag{10.1}$$

where P_r is the received power, P_t is the transmitted power, G_t is the transmitter antenna gain, G_r is the receiver antenna gain, d is the distance between the transmitter and receiver, L is the system loss factor not related to propagation, and λ is the microwave wavelength [72]. For the case when the transmitter also acts as the receiver, such as when the radar operates in the passive backscattering mode, the

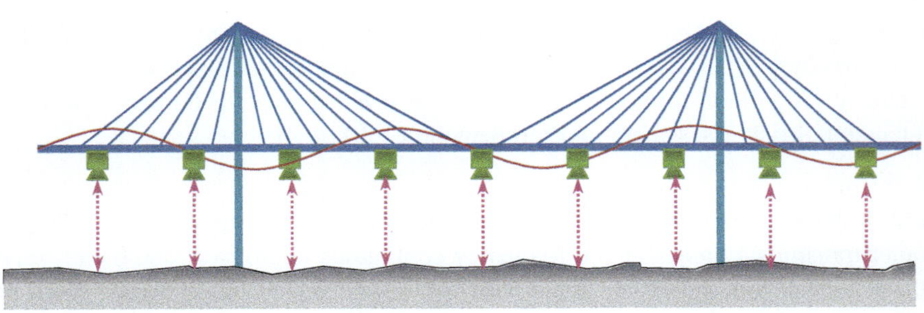

Figure 10.1 Diagram of passive backscattering configuration

signal is reflected by the target surface and travels round trip before it is received. According to the above equation, assuming there is no energy loss during the signal reflection at the target surface, the received signal power of the radar in the back-scattering mode will be [73] the following:

$$P_{\mathrm{r}}(d) = \frac{P_{\mathrm{t}}G_{\mathrm{t}}^{2}G_{\mathrm{r}}^{2}\lambda^{2}\sigma}{(4\pi)^{4}d^{4}} \tag{10.2}$$

where d is the distance between the radar and the target surface and σ is the cross-sectional area of the radar. Equation (10.2) demonstrates that the received signal power attenuates rapidly with increases in the target distance. For many applications, since there is a considerable distance between the sensor and the target, the reflected signal strength is too low for further analysis.

The baseband signal of the CW radar is original signal which will be transferred to be the radar's carrier signal and transmitted via the antenna. The frequency of the baseband signal is from 0 Hz to some cutoff frequency which is usually smaller than the radar's carrier signal frequency. The baseband signal is a complex signal which has the real part (I) and the imaginary part (Q). The baseband (I/Q) components generated by the radar is given by

$$B_I(t) = A_I \cos[\theta + 4\pi x(t)/\lambda + \Delta\varphi] + DC_I \tag{10.3}$$

$$B_Q(t) = A_Q \sin[\theta + 4\pi x(t)/\lambda + \Delta\varphi] + DC_Q \tag{10.4}$$

where A_I and A_Q are the amplitudes of I and Q signals, θ is a constant phase shift due to the transmission path and target surface, $\Delta\varphi$ is the total residual phase noise of the radar, and DC_I and DC_Q are the DC offsets in I and Q baseband signals, respectively. The $4\pi x(t)/\lambda$ term is the phase information corresponding to the target movement, $x(t)$, and λ is the carrier wavelength.

DC offset exists in the baseband signals from the radar. It is caused by the reflection from the environment and the radar signal transmission. To obtain the accurate displacement information from the radar sensor, the DC offset should be carefully removed.

Most of the DC offset can be canceled by sophisticated hardware design of the radar sensor. However, some remaining DC offset components generated by other stationary reflection surfaces in the environment must be removed by DC offset calibration to collect the correct displacement measurements.

Calibrating DC offset values begins by plotting the original baseband signals together to create a trajectory in a constellation diagram. The DC offset and amplitude values of the baseband signals are adjusted manually until the plotted I/Q trajectory fits a unit circle with the center at the origin. Mathematically, the raw baseband data is subtracted the DC offset values and then scaled them by the amplitude values until the sum of the corrected baseband data square equals to one. This process is demonstrated in Figure 10.2(a) for the measurement of a 5-mm amplitude sinusoidal motion with a frequency of 1.75 Hz. Figure 10.2(c) and (d) shows the constellation graph before and after DC offset calibration. Figure 10.2(b) shows the final displacement results from both correct and incorrect DC offset

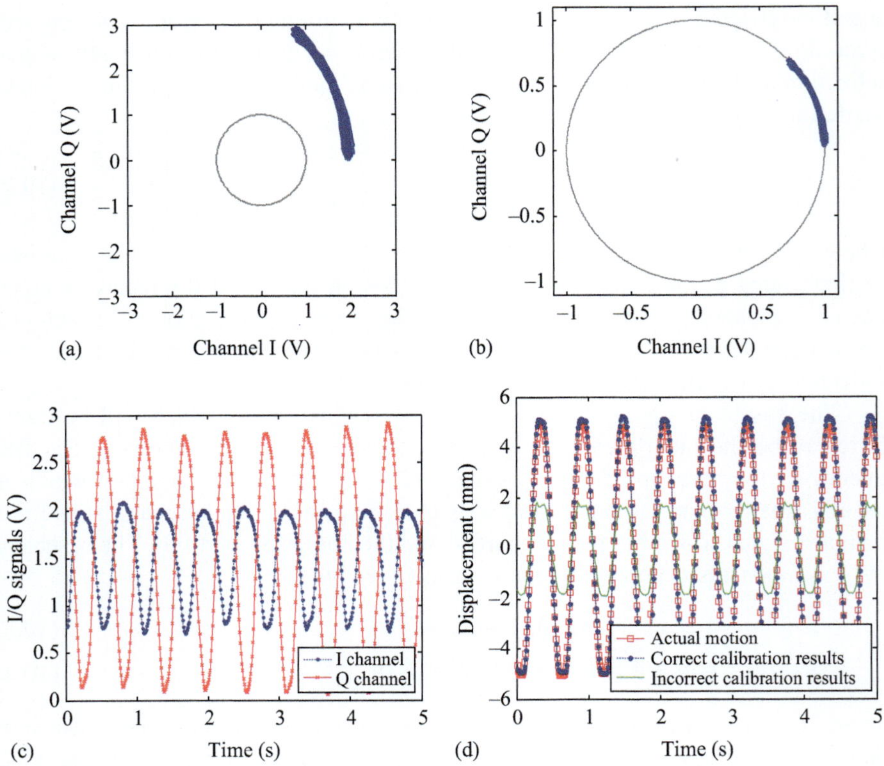

Figure 10.2 DC offset calibration process: (a) original I/Q time histories; (b) actual motion displacement and measured displacement (w/o correct DC offset calibration); (c) initial signals before DC offset calibration in complex plot; (d) signals after DC offset calibration in complex plot

calibration compared with the actual motion, demonstrating that the DC offset calibration step is very important to the final results.

The baseband signals after DC offset calibrations are

$$B_{I,corr}(t) = \cos[\theta + 4\pi x(t)/\lambda + \Delta\varphi] \tag{10.5}$$

$$B_{Q,corr}(t) = \sin[\theta + 4\pi x(t)/\lambda + \Delta\varphi] \tag{10.6}$$

By appropriately demodulating the corrected baseband signals, the displacement information can be obtained.

10.3 Continuous radar sensor hardware

Using CW radar sensing in medical applications, such as vital sign detection, involves a specific set of system design parameters [63]. The radar must accurately capture heart rate (frequency) up to several hundred hertz; however, the measured amplitude of motion is not critical. The system is mostly designed for indoor applications and the distance between the radar and the target (human body) is within a small range (less than ~1 m).

In contrast to the controlled environment of medical applications, applying the radar to SHM requires meeting a different set of design parameters. First, the radar measurements must capture both the amplitude and the frequency information of structural displacement accurately, as both are important aspects of structural response. Second, an advanced radar system for SHM applications should have the ability to measure low-frequency displacement and even static deflection, in addition to higher frequency motion. Third, the radar system must be able to be deployed and survive for long periods on infrastructure that may be in harsh environmental conditions. Fourth, the radar may operate in a situation where the distance between the radar and the target is much larger than the target distance in medical applications. CW radar is promising for measuring structural displacement; however, SHM applications impose stricter requirements for the radar system design than for other applications. Thus the original radar hardware and software design used for medical applications must be advanced to address new challenges associated with SHM applications.

Even with changes in design of the original CW radar, adequate signal strength for SHM applications is challenging to achieve. To address this limitation, an active transponder was designed to work with the CW radar to amplify its signal and improve measurement accuracy.

10.3.1 CW radar system

The CW radar system consists of a CW radar sensor and a base station to conduct structural displacement measurement. There have been two generations of the CW radar sensors designed for wireless SHM applications: AC-coupled (the first generation) and DC-coupled radars (the second generation). Both the AC- and DC-coupled radars were designed and fabricated by Dr. Changzhi Li's group at Texas Tech University. The CW radar has been integrated with commercial integrated circuits (also referred to as ICs or chips). The radars are fabricated in a compact size, which is convenient for transportation and installation.

The hardware design of the CW radar follows the standard framework of a wireless sensor system: sensing component, microprocessor, power supply, and wireless communication device. Correspondingly, the CW radar system has four major components: (1) radio frequency (RF) board and a pair of patch antenna (sensing component), (2) baseband board with MSP430 family microprocessor (microprocessor), (3) battery board with DC battery pack (power supply), and (4) XBee radio pair (wireless communication device).

In brief, the CW radar hardware system operates as follows: the RF board generates the continuous microwave. The patch antennae are connected to the RF board via subminiature version A (SMA) connectors or cables. One of the patch antennae transmits the continuous microwave and the other captures the signal

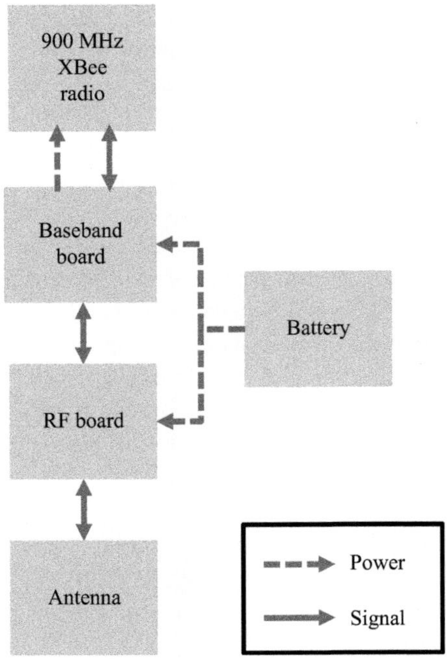

Figure 10.3 Block diagram of CW radar

reflected from the target. The baseband board is stacked on the top of the RF board using the socket connectors. The baseband board is responsible for controlling all the radar's operations by using the MSP430 family microprocessor. The base station is controlling all the radars operations. The XBee radio receives operation requests from the base station and transmits all the sensing data back to the base station wirelessly. The battery board with a battery pack assembled is used to supply a DC power for the radar system. The block diagram of the CW radar operating is shown in Figure 10.3.

The CW radar system base station is a USB XBee explorer with an XBee Pro radio. The XBee radio is a bidirectional wireless communication device that works as the transceiver. The major role of the base station is controlling all the radars' operation wirelessly, such as sending the requests to the CW radar and requiring the data from the radar through the XBee protocol. The base station is connected to a PC via the USB port. The board is embedded with an FT231X USB-to-Serial converter which translates data between the PC and the XBee. The PC stores the data received by the base station. The assembled base station is shown in Figure 10.4.

10.3.1.1 Antenna

In the design of the radar system, one of the critical factors is the carrier frequency of the microwave signal. Some commonly adopted CW microwave basebands are 900 MHz, 2.4 GHz, 5.8 GHz, and even higher frequency bands.

Figure 10.4 XBee base station for CW radar

In general, the higher baseband frequency results in the higher energy of the microwave and better displacement measurement accuracy. However, the fabrication of hardware for generating high-frequency baseband signals is costly and consumes more power. After balancing the system power consumption, the microwave signal strength, and fabrication complexities, 2.4-GHz continuous microwave was selected. The patch antennae each consist of a rectangular patch mounted on a larger rectangular substrate layer between the patch and the ground plane. The dimensions of the patch are a function of the operating frequency and the properties of the substrate; for the same substrate, a lower transmission frequency results in a larger antenna [74]. While a higher frequency radar would allow smaller antennae, the increase in power consumption, and potentially cost, must be considered.

Maximum radiation occurs normal to the patch surface while backward radiation is limited by the ground plane. The larger the ground plane is relative to the patch, the more gain and directivity is achieved by the antenna [74]. Based on the design parameters of the antennae for the radar sensors, the 3-dB beamwidth is conservatively estimated at 45°.

10.3.1.2 RF board

On the RF board, the 2.4-GHz microwave is generated by the voltage-controlled oscillator (VCO). The microwave is transmitted continuously by one front-end antenna which is connected to the VCO on the RF board via the SMA connectors or cables. Then the continuously transmitted microwave will reach the target and be reflected by the target surface. After the reflected signal is received by the second antenna, the RF board mixes the received signal with the transmitted microwave in the quadrature mixer and then downconverts the mixed signal to the baseband analog *I*/*Q* signals. Then the analog baseband *I*/*Q* signals are extracted by two external pins on the RF board to the top baseband board. The schematics of the RF board is shown in Figure 10.5.

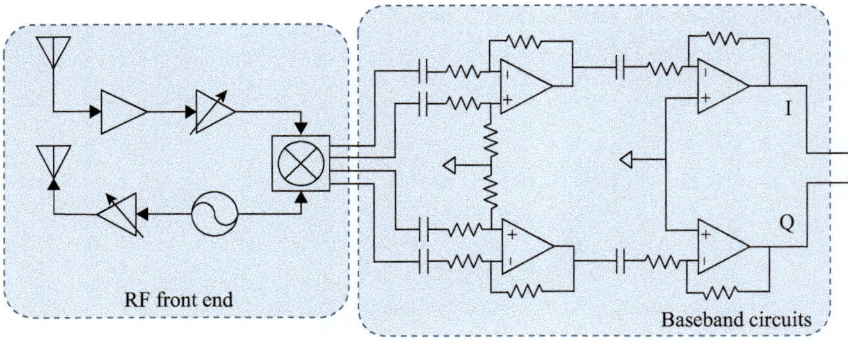

Figure 10.5 Block diagram of CW radar

10.3.1.3 Baseband board

After the baseband *I/Q* signals are transferred from the RF board to the baseband board, they are amplified by the operational amplifier (Op Amp). The signals are then sent to the TI MSP430F2619 microprocessor, which controls all the operations of the radar system. The baseband signals are digitalized by the built-in 10-bit analog-to-digital convertor (ADC) on the MSP430 microprocessor. The maximum clock speed of the MSP430 is 32,768 Hz. The sampling rate of the ADC is reconfigurable up to 100 Hz to satisfy different SHM displacement measurement requirements. The microprocessor can be embedded with different algorithms to process the signals and output the final displacement information.

On the baseband board, the two output signal pins from the microprocessor are connected to the XBee radio, which can be used to export the data to other external wired or wireless transmission devices. With the help of the external data interface board, the radar system can be integrated with other wireless smart sensor platforms to conduct multiple measurements (temperature, acceleration, etc.), analyze a large amount of data, and adopt other wireless transmission protocols (Wi-Fi, cellular, etc.). The block diagram of baseband board is shown in Figure 10.5.

10.3.1.4 Wireless communication device

The XBee radio on the baseband board works as the wireless communication channel between the base station and the radar system. There are two types of XBee radios available, based on the carrier frequency: 900 MHz and 2.4 GHz. The 900-MHz XBee radio is selected because of its longer transmission distance than the 2.4-GHz radio. The 900-MHz XBee radio also avoids any potential interference with the 2.4-GHz CW radar signal. The 900-MHz XBee Pro radio of the radar system is plugged on the top of the baseband board. Another 900-MHz XBee Pro radio with the USB XBee explorer serves as a base station and receives the transmitted baseband signals.

10.3.1.5 Power system

To prolong the radar system operation with limited power, the power system has been carefully designed in both hardware and software. The hardware components of the radar's circuits have been selected to minimize the power consumption while still maintaining good system performance. The hardware power supply components for the radar sensor system consist of a DC power supply and a DC voltage regulator. The radar sensor system requires approximately 5.0 V DC to operate. A battery board with four AAA batteries is used to supply the power for the radar system and keeps the system compact. The DC regulator controls the output voltage from the batteries to constant 5.0 V, to avoid overloading the system and also ensuring adequate supply voltage as the batteries deplete.

The baseband signals of CW radar are affected by DC offsets which may cause the raw output signals to fall outside of the input voltage range of the ADC on the microcontroller. To avoid this signal saturation, the DC offset must be corrected before obtaining accurate displacement results. Two common sources of the DC offset are the transmitted signal from VCO leakage into the received signal, and direct coupling from transmitter to receiver. In addition, the radar receives signals not only from the moving target but also from signals reflected by the stationary objects in the environment, which adds extra DC offset to the original baseband signals. This part of DC offset value may vary with different environments, making it more challenging to cancel the DC offset due to its inconstant value. The AC- and DC-coupled CW radars were designed to minimize the DC offset problem; however, the AC- and DC-coupled radars adopted two different approaches to cancel the DC offset.

10.3.2 AC-coupled radar

The AC-coupled radar was the first CW radar developed to measure structural displacement wirelessly. The AC-coupled radar adopts the general framework of the CW radar. The configurations of the antenna, XBee radio, and the power system of the AC-coupled radar are the same as the general CW radar previously described. The AC-coupled radar fits in a compact size (7.8 cm × 4.8 cm × 2.5 cm). The assembled AC-coupled radar is shown in Figure 10.6.

10.3.2.1 AC coupling design

An AC coupling approach is adopted to minimize the DC offset problem. A coupling capacitor is used in the analog circuit of the RF board to block the DC component of the signal. For the AC-coupled radar, the cutoff frequency is 1.75 Hz, meaning signals with less than 1.75 Hz frequency are filtered out. The AC-coupled radar is a reasonable approach for many applications where low-frequency and static motion measurements are not required. The focus of our research, however, is on low-frequency motion, requiring the use of a DC-coupled radar.

The AC-coupled radar has good accuracy when the motion frequency is above 1.75 Hz since the radar board has a high-pass filter with 1.75-Hz cutoff frequency. However, for SHM applications, many cases require the measurement of low-frequency vibrations and even static deflections (DC signal). This design constraint led to the development of the DC-coupled radar presented in the next section.

10.3.3 DC-coupled radar

The goal of designing the second generation CW radar, i.e. the DC-coupled radar, was to provide broader SHM applications than the AC-coupled radar. The major

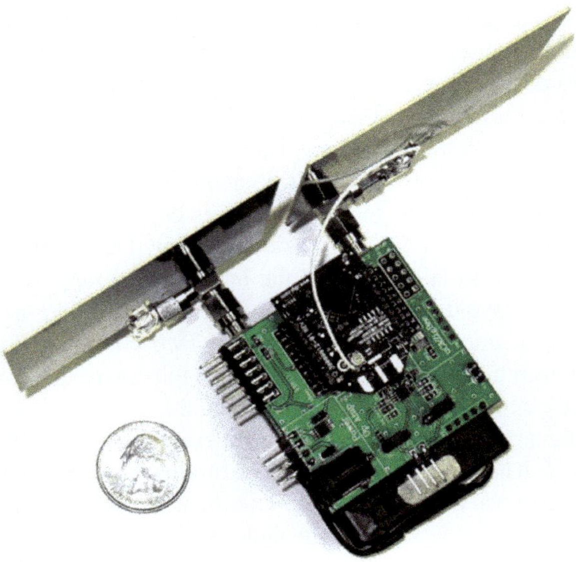

Figure 10.6 The assembled AC-coupled radar

advantage of DC-coupled radar over the AC-coupled radar is measuring low-frequency displacement and static deflections with less signal distortion. The DC-coupled radar adopts the same framework of the CW radar system described in earlier sections. The DC-coupled radar still applies the 2.4-GHz microwave to conduct displacement measurements. The antenna and the power system of the DC-coupled radar are the same as the AC-coupled radar. The DC-coupled radar has a compact size (5 cm × 5 cm × 2.5 cm) which is similar to the AC-coupled radar. The assembled DC-coupled radar is shown in Figure 10.7.

Different from the AC-coupled radar, the DC-coupled radar is used to measure low-frequency displacement and static deflection in addition to the higher frequency displacement. The DC-coupled radar needs to preserve low-frequency signals (including DC component) after canceling the DC offset. To achieve this goal, the original AC coupling approach was abandoned but the adaptive tuning architectures including RF coarse-tuning and baseband fine-tuning were added to DC-coupled radar hardware design. The adaptive tuning architectures help avoid signal saturation caused by the DC offsets. The DC offset due to the reflection from the stationary object in the environment can be removed by DC offset calibration process.

10.3.4 Active transponder

The passive backscattering mode provides the possibility of using AC- or DC-coupled radars for SHM applications; however, the limitation of the backscattering mode is that sometimes the signal strength is too low to provide accurate displacement measurement. To address this problem, the active transponder node was introduced and integrated with the existing CW radar system.

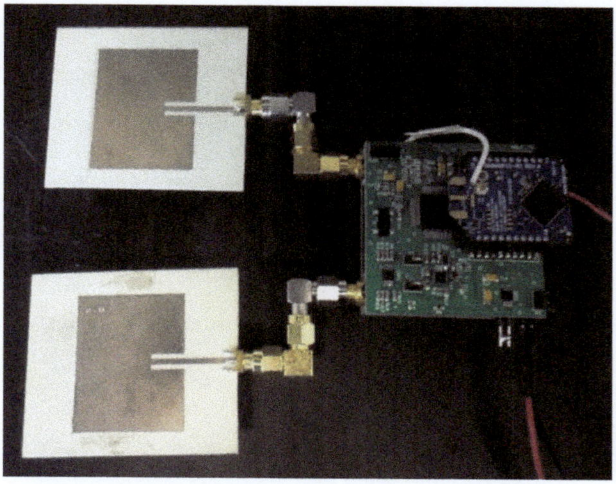

Figure 10.7 Fully assembled DC-coupled radar

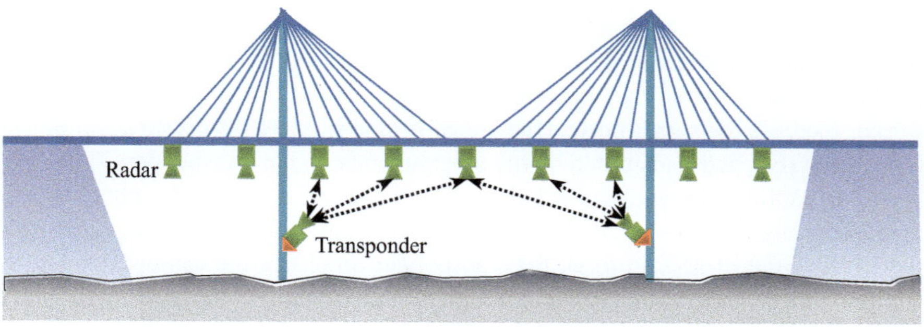

Figure 10.8 Diagram of active transponder strategy

The proposed active transponder strategy uses a radar and a transponder working together as a pair. The transponder acts as the target and operates by receiving and amplifying the radar's transmitted signal, then transmitting the amplified signal back to the radar's antenna. In contrast to the passive back-scattering mode, only the signal transmitted by the radar is amplified by the active transponder. This amplification results in a linear gain added to (2-2). An example of the active transponder configuration for bridge monitoring is shown in Figure 10.8.

The transponder node is composed of a printed circuit board, SMA connectors or SMA cables, a pair of 2.4-GHz patch antennae, and a DC power source (battery board with four AAA batteries). The fully assembled transponder is shown in Figure 10.9.

The transponder works as follows: first, the 2.4-GHz microwave signal is transmitted by one of the antennae of the DC-coupled radar, and then captured by

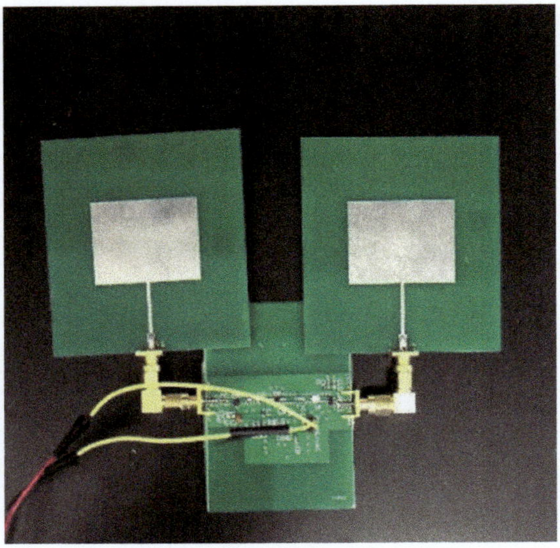

Figure 10.9 Active transponder node

the transponder's antenna. The signal is amplified 17 dB by a low-noise amplifier (LNA) and then filtered by a bandpass filter with −1-dB in-band insertion loss. To further increase the signal power, the signal is amplified 31 dB by two gain blocks (GB1 and GB2) and transmitted by the other antenna on the transponder. The return signal is captured by the DC-coupled radar's second antenna. To obtain better amplifications, the antenna on the transponder and the radar must be aligned within the beam angle of 45°. In total, the transponder amplifies the captured signal by 47 dB before transmitting back to the radar. The supply voltage range for the transponder is regulated to 4.8 V to stabilize the power for the transponder.

As the DC-coupled radar's signal is amplified by the transponder node, the radar can work at a larger target distance compared with using the passive backscattering mode. More importantly with the help of the transponder, the passive reflective surface is not necessary. The active transponder mode could be applied for the cases such as when there is traffic or water flow under the bridge. Even though the radar's signal is amplified by introducing the active transponder mode, there are still some cases that the radar's signal strength is not adequate to obtain the displacement. To ultimately solve the problem, the hardware of the CW radar needs to be redesigned and the carrier frequency of the radar signal should be increased.

10.4 Continuous radar sensor software

Built on the hardware layer of the continuous radar system, an advanced software system has been developed to control all sensors' operations and perform automatic data processing. There are multiple types of topologies to create a wireless sensor network: centralized data collection, independent data processing, and coordinated computing strategy [75]. The current Continuous Radar system adopts the centralized data acquisition system. For the centralized data collection system, the continuous radar sensors only conduct the data acquisition. Then all the data are sent back to the base station to process. This centralized data acquisition system is convenient to control multiple sensors simultaneously with a single base station. The software system has been developed to establish a centralized data acquisition architecture.

10.4.1 Signal-processing algorithms

Outputting processed displacement measurements is the core function of the continuous radar. Currently, the framework of data flow is as follows: first, each sensor outputs the digital *I*/*Q* signals and transmits them back to the base station. On the base station, the signal processing is conducted and the displacement information is saved. In the future, all the signal-processing code of converting the raw *I*/*Q* signals to the displacement will be embedded on the sensor nodes with the higher clock speed microprocessor. The signal-processing algorithm consists of the automated DC tuning algorithm and the automated displacement output algorithm. The idea of designing the algorithms is to maximize the performance with minimum power consumption.

10.4.1.1 Automated DC tuning process

The DC offset of baseband signals is a limitation in the application of the continuous radar. To ensure baseband signals are not saturated of the microprocessor ADC and have the largest dynamic range, the DC tuning process must be conducted correctly. The original DC tuning process was conducted through LabVIEW software manually as introduced in Chapter 2. To improve this process, an automated tuning algorithm was designed for the continuous radar system. With the WiseMote integrated, the automated DC tuning algorithm can be conducted with the help of the microprocessor on the continuous radar.

The general idea of the auto-tuning algorithm is first converting the baseband analog signal to digital signal via the ADC on the microprocessor and then evaluating the signal level between the upper and lower ADC limits. Since a 16-bit ADC is used to convert the baseband signals, the upper and lower limits of the ADC are 65,535 and 0, respectively. Second, the difference is taken between the signal level and the middle of the range (32,768) to determine whether to increase or decrease the signal level to move it closer toward the middle of the range. Third, the digitized *I*/*Q* signal is converted to analog signal via the DAC on the microprocessor and the signal level is shifted by adjusting the reference voltage of the ADC and then redigitizing the signal. The updated signal is compared with the signal from previous cycle, to ensure that it is moving closer to the middle of the ADC range. The steps are repeated until the signal reaches the middle of the ADC range which will provide the largest dynamic range and flexibility for using the radar. The flowchart of the auto-tuning algorithm is shown in Figure 10.10.

The DC tuning process takes less than a few seconds to complete, based on previous experiments.

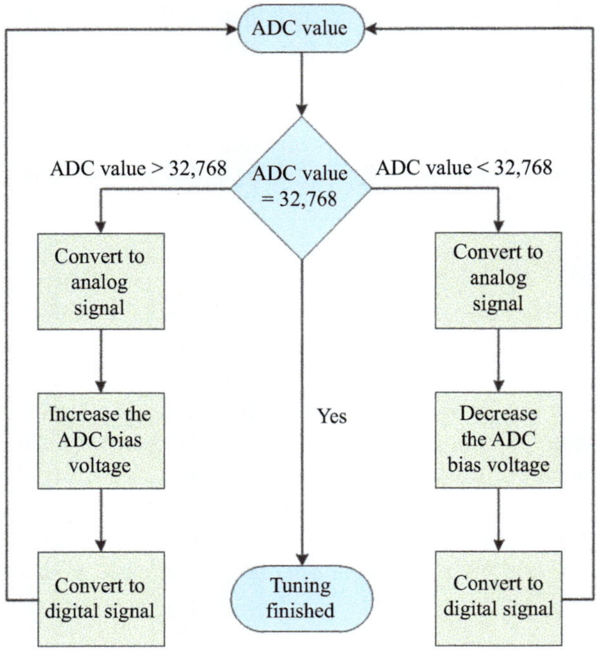

Figure 10.10 Automated DC tuning algorithm

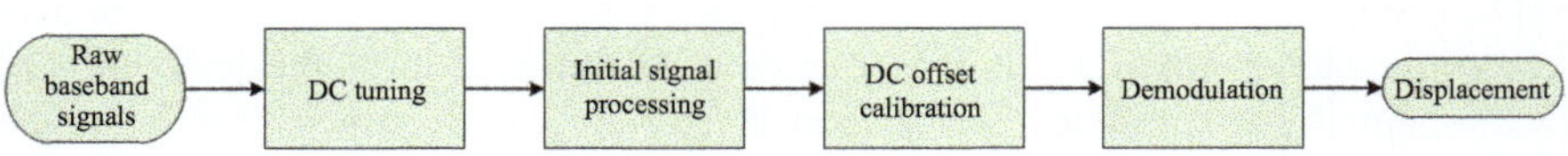

Figure 10.11 Flowchart of the signal-processing steps

10.4.1.2 Automated displacement-processing algorithm

After the DC tuning is conducted, the baseband *I*/*Q* signal is ready for processing to obtain the displacement. After conducting a series of laboratory experiments, some key factors, such as the noise level, signal power, and the method to conduct the DC offset calibrations, are found to be critical to the accuracy of final displacement results. The signal processing that must be conducted to convert raw baseband signals to displacement consists of several steps shown in Figure 10.11. To evaluate the effects of each factor listed above, each of the signal-processing steps has been studied individually. The goal of the signal processing is to ensure data accuracy with reasonable computation effort.

Initial signal-processing step

After the DC tuning is conducted to adjust baseband signals to the desired level, the baseband signals are ready for post-processing. A high signal-to-noise ratio (SNR) of the baseband signals is critical to obtain good displacement results. One method to improve the SNR in the initial processing step is the implementation of a

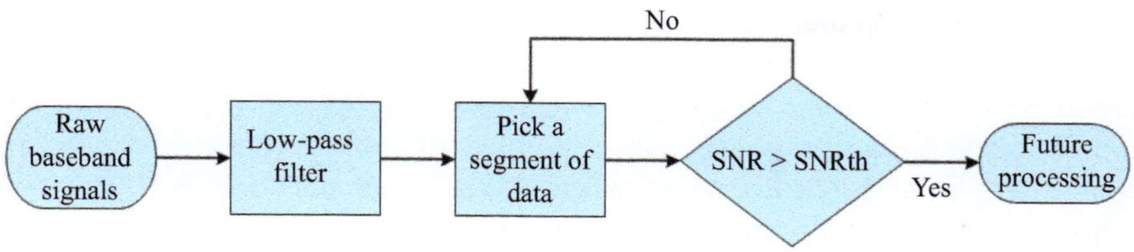

Figure 10.12 Flowchart of the initial signal processing

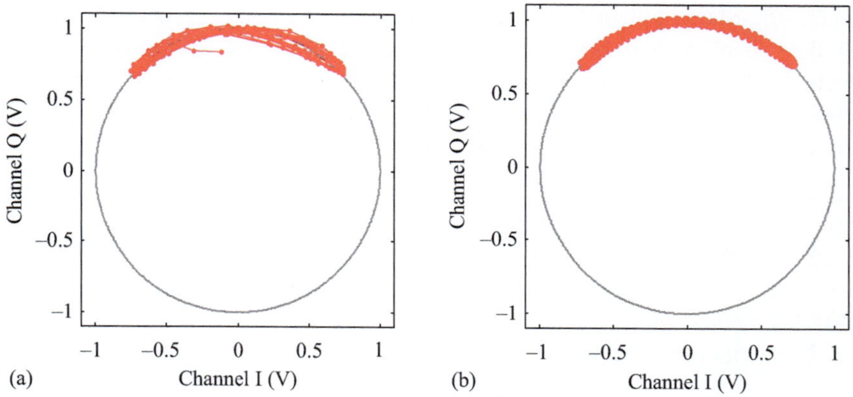

Figure 10.13 Before and after the baseband signals are lined up

low-pass filter to block high-frequency noise. Then SNR of baseband signals is calculated. A threshold value of SNR to ensure the accuracy has been set. Only a segment of data which has an SNR higher than the threshold value is selected to conduct the next step. The flowchart of the initial signal processing is shown in Figure 10.12.

One phenomenon that was observed in some of the experimental data was a "bow-tie phenomenon" when baseband signals are plotted in the constellation graph. This phenomenon created challenges in fitting the data to the unit circle, and thus DC offset calibration, resulting in erroneous displacement estimates. The source of this phenomenon was found to be a phase delay between the baseband signals. To eliminate this problem and ensure that the baseband signals are synchronized with one another, they are resampled and aligned following data segmentation. The data before and after the bow-tie correction are shown in Figure 10.13.

DC offset calibration

After the preliminary signal-processing step, the critical step of DC offset calibration should be conducted carefully. From (10.3) and (10.4), the corrected baseband signals are expected to have amplitudes, A_I and A_Q, where the ratio between A_I and A_Q is 1.0 and DC offset values, DC_I and DC_Q, of zero. To understand how important DC offset calibration is to achieve accurate displacement results some simulation work has been completed. The simulation model has demonstrated that slight changes of the DC offset calibration values will result in considerable errors in the final displacement measurement [76].

It is clear that a manual approach to calibrate DC offset is time-consuming and impractical for embedded wireless sensing [77]. To automate the DC offset calibration process (i.e. adjusting the baseband signals until their trajectory lies on the unit circle), a number of methods have been proposed: (1) A method for linear demodulation proposed by Massagram *et al.* [78]—in this proposal, this method is called the linear demodulation method; (2) the least squares method; (3) the Levenberg–Marquardt method [79]; (4) a compressed sensing method [80]; and (5) the Yuen method [81]. The Yuen method does not demonstrate accurate results [79]; therefore in this dissertation, the first four methods are discussed in detail. We have developed an automated DC offset calibration strategy based on the above methods to achieve the optimal solution.

Phase demodulation

After calibrating the DC offsets of baseband signals, the phase information is determined through demodulation. There are two approaches to demodulate the phase information: arctangent demodulation and differentiate and cross multiply (DACM) method [82].

The arctangent demodulation is the most commonly used, shown as

$$\psi(t) = \arctan\left[B_{Q,corr}(t)/B_{I,corr}(t)\right] = \theta + 4\pi x(t)/\lambda + \Delta\varphi \tag{10.7}$$

$$\varphi(t) = 4\pi x(t)/\lambda \tag{10.8}$$

where $\varphi(t)$ is proportional to the relative displacement between the target and the radar. The relative displacement can be obtained as $x(t) = x_0 + \varphi(t)\lambda/4\pi$, where x_0 is the initial distance to the target.

DACM demodulation calculates the forward differential values of each channel to calculate the displacement. The discrete signal expression of DACM is as follows:

$$\varphi[n] = \sum_{k=2}^{n} \frac{I[k]\{Q[k] - Q[k-1]\} - \{I[k] - I[k-1]\}Q[k]}{I[k]^2 + Q[k]^2} \tag{10.9}$$

The arctangent demodulation is straightforward to apply. However, from (10.7) when the trajectory crosses the boundary between the first and the fourth quadrants in the constellation plot, the phase angle of the trajectory is out of the range $[-\pi, \pi]$, which is the range for applying arctangent demodulation. An unwrapping process must be conducted before arctangent demodulation to eliminate discontinuities in the displacement results. According to the characteristics of 2.4-GHz radar sensors, the wavelength is approximately 12.25 cm corresponding to 2π range. When the displacement is larger than 12.25 cm, the unwrapping process must be implemented. For many practical applications, the structural displacement is very small and likely less than the wavelength of the 2.4-GHz CW radar (12.5 cm). The arctangent demodulation is selected as the default approach to demodulate the radar signal.

From (10.9), we can deduce that the DACM method does not have the discontinuity issue. It is a more accurate method than the arctangent method to deal with the case when the displacement is larger than 12.5 cm. However, the DACM method requires a high level of computation due to its order of N^2 complexity which is not appropriate to be used as the priority option for the continuous radar system. For the displacement demodulation process of the continuous radar system, the arctangent demodulation is used for most cases as the default, but the DACM method is still reserved as an option to deal with the unwrapping problem when the displacement is larger than 12.5 cm.

The entire signal-processing approach, with automated steps to DC tuning that preprocess, calibrate, and demodulate the raw baseband signals into displacement measurements, has been implemented on the microprocessor of the continuous radar.

10.5 Continuous radar sensor measurement characterization

The continuous radar system can measure displacement with the RF board and acceleration with the WiseMote board. Since the displacement measurement is of interest and the acceleration measurement has been evaluated in previous research [83], only the displacement measurement accuracy of the CW radar (DC-coupled) is investigated in this chapter. The continuous radar sensor can measure low-frequency displacement and static deflections. The signal characteristics and accuracy of the displacement measurement are evaluated by these two types of motions. The accuracy of continuous radar is related to several factors such as motion amplitude, motion frequency, and distance between the target and radar. Laboratory experiments have been conducted to quantify the accuracy based on these factors.

10.5.1 Dynamic displacement experiments

A series of laboratory experiments have been conducted to characterize radar's accuracy related to factors including the motion amplitude, the distance between the radar and the target, and the target surface material. For each test, one factor was changed while the others were kept constant. The radar's sampling rate can be adjusted easily; however, since we focus on the new low-frequency (less than 2 Hz) measurement capability of the DC-coupled radar, the sampling rate was set at 20 Hz for all tests. The following sections describe these tests for both dynamic displacement and static deflection.

A vertical shaker was utilized to generate low-frequency sinusoidal motion. A rigid aluminum beam was clamped to the top of the shaker and the DC-coupled radar was attached to the other end of the beam to measure the shaker's motion. The antennae of the radar were pointed toward the ground, which acted as the stationary target, at a distance of 50 cm. An linear variable differential transducer (LVDT) was attached to the beam to provide a reference measurement of the shaker's motion. The motion was generated by the shaker, and LVDT and radar

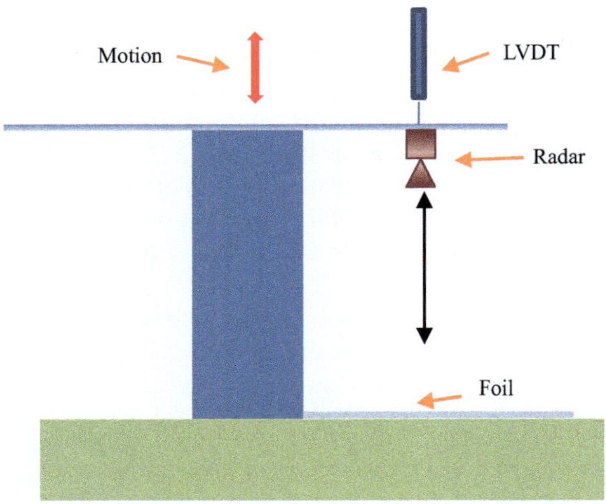

Figure 10.14 Experimental setup of dynamic displacement experiment

measured the displacement simultaneously. The experimental setup is shown in Figure 10.14.

A series of motions were generated by a shaker with different measurement conditions. A 1.25-Hz sinusoidal motion was generated by the shaker and the amplitude was varied from 1 to 10 mm with 1-mm increments, while aluminum foil acted as the target surface. The frequency of the motions in the second set of tests was varied from 0.25 to 2 Hz with 0.25-Hz increments while the amplitude was kept constant at 6.5 mm.

Results from both the radar and LVDT measuring a representative data set of 1.25-Hz sinusoidal motion are shown in Figure 10.15.

For each measurement condition, three trials were conducted and measurements from the radar and LVDT are compared in the time domain. From the time history record in Figure 10.15, the frequency of the radar matches well with the LVDT while some differences in the measured amplitude are observed. To compare the amplitude difference in the time domain, the mean and standard deviation of measurement errors between the radar and the LVDT are calculated and shown in Figure 10.16(a) and (b). As clear from Figure 10.16(a), the mean and standard deviation errors of radar's measurements decrease while the motion amplitude increases because the higher amplitude results in higher SNR value. In Figure 10.16(b), it is observed that the mean measurement errors are less than 1 mm for all motion frequencies and there is no trend in measurement error related to the motion frequency.

To study the radar's accuracy related to its distance from the target, the radar was attached to a horizontal shaker to measure motion of the shaker. A constant 1-Hz sinusoidal motion with a 10-mm amplitude was generated by the horizontal shaker. A 2 m × 1.5 m metallic board served as a reflection surface. The distance between the radar's antenna and the target surface was varied from 1 to 7 m in 1-m increments. The experimental setup is shown in Figure 10.17.

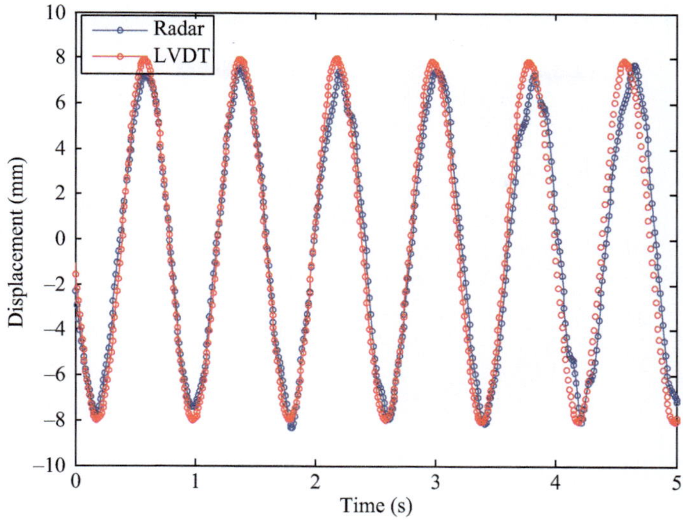

Figure 10.15 A representative measurement results of 1.25-Hz sinusoidal motion by radar and LVDT

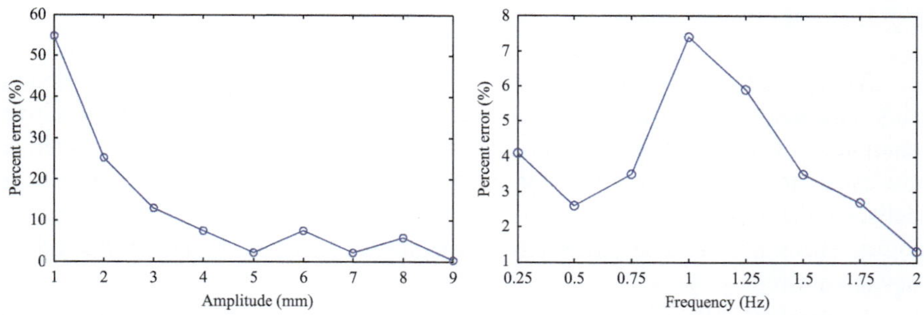

Figure 10.16 Measurement errors between radar and LVDT vs. different amplitudes and frequencies

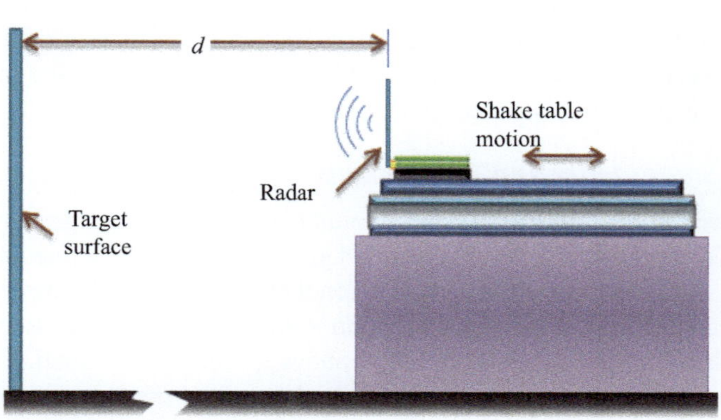

Figure 10.17 Experimental setup of radar's measurement performance vs. target distance

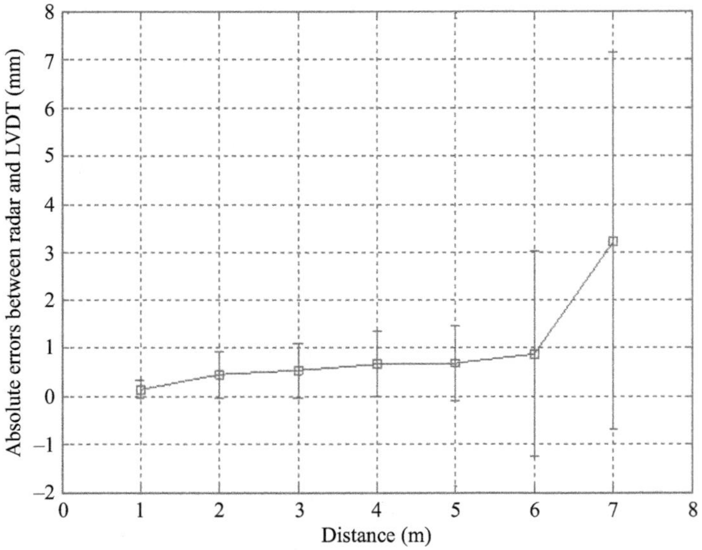

Figure 10.18 Absolute errors between radar and VLDT vs. target distance, with one standard deviation indicated

To quantify the radar's errors, an LVDT measured the shaker's motion simultaneously. Results from both the radar and LVDT measuring a representative data set of 1.0-Hz sinusoidal motion with 1-m distance between the radar and target are calculated. The mean and standard deviation errors of the motion amplitude between the radar and the LVDT are shown in Figure 10.18. The points in Figure 10.18 are the mean value and the magnitude of the error bar is the standard deviation errors.

From Figure 10.18, the measurement from the radar has a sub-millimeter level accuracy when the distance is less than 6 m. As the distance increases, both the mean and standard deviation of absolute errors increase due to attenuated signal power and increased signal noise, demonstrating that the radar has less measurement accuracy and consistency at larger target distances. For practical SHM applications, if the distance between the radar and target is less than 6 m, sub-millimeter accuracy may be achieved. Approaches for increasing the radar's measurement distances are discussed in the conclusions.

10.5.2 Static deflection experiments

Static deflection measurement is an important metric for bridge design and maintenance. To demonstrate the DC-coupled radar's potential for measuring static deflection, a series of static tests were conducted on a scaled simply supported beam. An aluminum beam of about 180 cm × 30 cm × 0.6 cm was used as the test bed, with the ability for the beam height above ground to be varied. Two masses (each about 2.25 kg) were loaded at the third points of the beam to generate the beam deflections. The LVDT was attached to the top of the middle span. For the radar measurements, only the antennae were attached to the bottom of the middle span and connected to the remainder of the radar mounted on top of the beam by cables. The ground below the beam acted as the target to reflect the radar's signal, with both bare floor tiles and aluminum foil serving as target surfaces for comparison. The radar and the LVDT detected the deflections of the mid-span simultaneously. The experimental setup is shown in Figure 10.19. One of the experimental results for a beam height of 1.25 m and a foil target surface is shown in Figure 10.20.

The accuracy of the static deflection measurements is related to a number of factors, including the magnitude of the deflection, the distance between the target

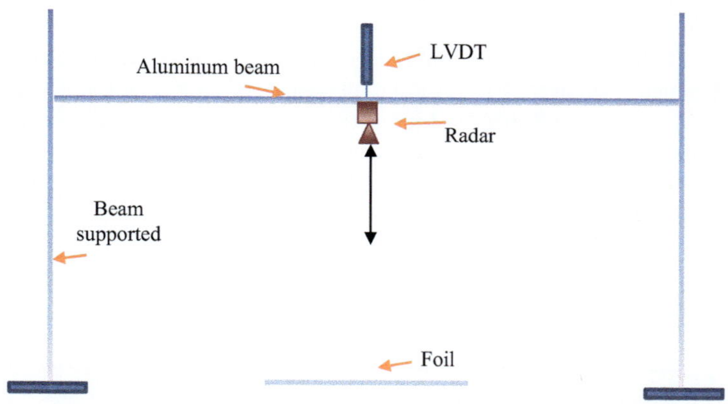

Figure 10.19 Experimental setup of static deflection measurements using radar and LVDT

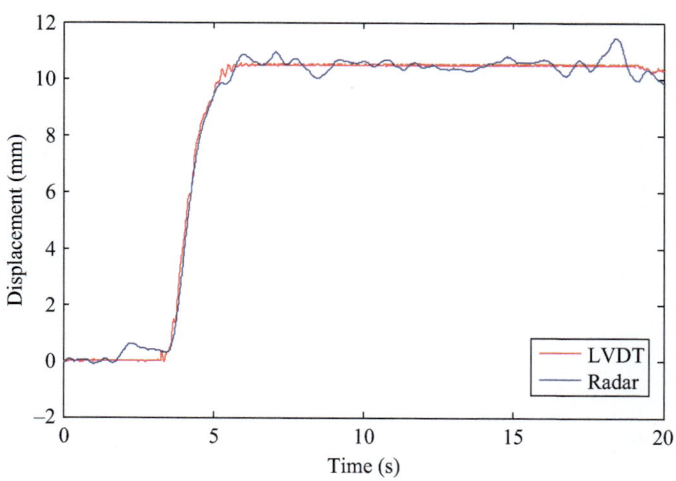

Figure 10.20 Deflection measured results from radar and LVDT

and the radar, and the reflection surface material. The amplitude of deflection affects the signal power and the distance between the target and the radar links to the noise level. The reflection surface is associated with how much signal power is reflected. To assess the effects of the target distance, the height of the beam was varied in the first set of tests. Three discrete heights were selected (0.5, 1, and 1.25 m), and the foil cover was used as the reflection surface. In the second series of tests, the test parameters remained the same while the deflections were varied. In the last series of tests, the effect of the target surface material was assessed with the height of the beam held constant at 1.25 m. For all tests, the LVDT and the radar measured the middle span deflections simultaneously. A summary of the measurement RMS errors between the radar and LVDT are given in Tables 10.1–10.3.

The experimental results indicate that the radar demonstrates good performance in measuring static deflections under different measurement conditions. As shown in Table 10.1, the radar exhibits reasonable measurement accuracy as the target distance increases, even in the presence of increasing noise levels. As expected, the results in Table 10.2 indicate better measurement accuracy when the displacement is larger. The results presented in Table 10.3 demonstrate the importance of the target surface material in acquiring accurate results; in this case, foil is preferable to the floor tiles.

Table 10.1 RMS error vs. target distance

Height between beam and ground (m)	**RMS error between radar and LVDT (mm)**
0.5	0.47
1.0	1.13
1.25	1.24

Table 10.2 RMS error vs. deflection amplitude

Deflection amplitude (mm)	**RMS error between radar and LVDT (mm)**
5	1.54
10	0.47

Table 10.3 RMS error vs. different reflection surfaces

Reflection surface material	**RMS error between radar and LVDT (mm)**
Ground	1.24
Foil	0.29

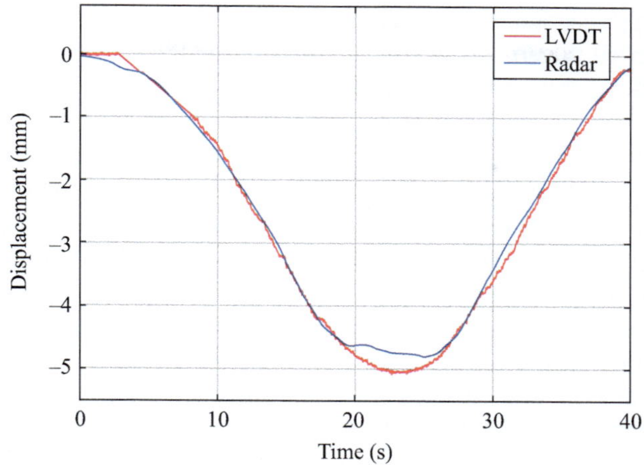

Figure 10.21 Measurement results of moving load test

10.5.3 Moving load experiment

To simulate the process of measuring deflections of a moving vehicle traveling across a bridge, a moving load laboratory experiment was conducted. The aluminum beam height was set at 1.25 m. A small moving car was used to create deflections of the aluminum beam. A light rope in tension was used to guide the loaded car along the bridge at a slow speed. The total weight of the car is about 0.7 kg. A DC-coupled radar was attached to the mid-span of the beam with the antennae pointing toward the floor below, and an LVDT provided the reference measurement. The car moved from one end of the beam to the other while the sensors simultaneously measured the mid-span deflection of the beam. The experimental setup is shown in Figure 10.19. The measurements of both the radar and the LVDT at mid-span are shown in Figure 10.21. From Figure 10.21, the radar captures the whole process of the car moving across the bridge, the shape of the deflection measured by radar and LVDT are similar while the maximum difference of the amplitude is about 0.3 mm.

10.5.4 Oblique angle tests

In addition to providing signal amplification over the passive backscattering approach, the use of the transponder enables a range of system configurations, providing flexibility in test design. The ideal configuration is for the radar to transmit signals toward a target along a path that is perpendicular to the motion it is measuring. In some cases, a perfectly perpendicular alignment is not possible, such as when there is water or traffic directly below the bridge. In such cases, the transponder may be placed in a more practical location, such as on a pier, median, or abutment below the bridge deck (as illustrated in Figure 10.8). With the use of free angle SMA connectors, the angle of the radar and transponder's antennae may be adjusted based on practical application conditions. Using a simple transformation equation, the measurement results from an oblique configuration can be converted to vertical deck displacement. The transformation equation used to calculate the vertical structural displacement from the radar's measurement is

$$x_{\text{vertical}}(t) = x_{\text{radar}}(t)/\cos(\theta) \tag{10.10}$$

where x_{vertical} is the structural displacement, x_{radar} is the relative displacement between the radar and transponder, and θ is the angle between radar's measurement and structural motion direction.

To verify the accuracy of the radar–transponder vertical displacement measurement when configured with an angle between the antenna and the structural motion direction, an oblique angle test was conducted on an aluminum beam in the laboratory. The radar was attached at the mid-span of the beam with its antennae at a 45° angle from vertical direction. The transponder was attached to the beam support column with its antennae at a 45° angle from the vertical direction. The result is that the antennae of the radar and the transponder were directly facing one another at a distance of 2.5 m, with the alignment ensured by a laser. An impact hammer was used to generate vertical motion in the beam and an LVDT was used as a reference sensor to compare the vertical displacement measurements with the radar. The experimental setup is shown in Figure 10.22.

With $\theta = 45°$, the vertical displacement measurement results from both the radar and LVDT are shown in Figure 10.23.

From Figure 10.23, the radar–transponder pair successfully measured the vertical motion of the bridge. With the motion decaying, the accuracy of the radar measurement decreases. Similar to the measurement results from the full-scale bridge test, the radar has measurement accuracy (less than 1 mm errors) compared

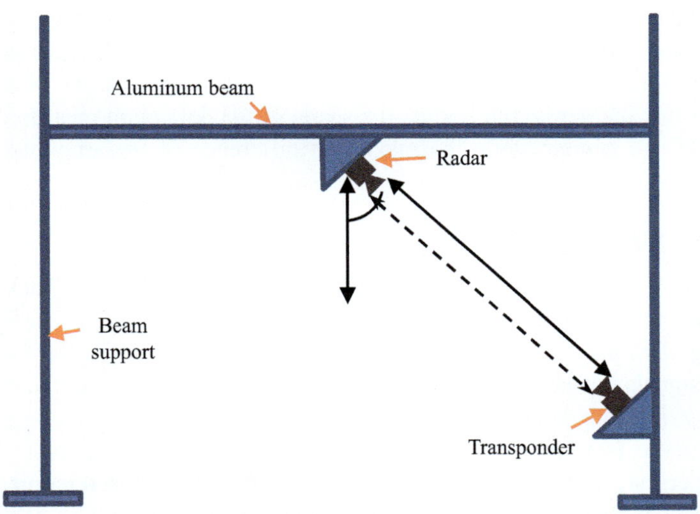

Figure 10.22 Experimental setup of the oblique angle test with the transponder strategy

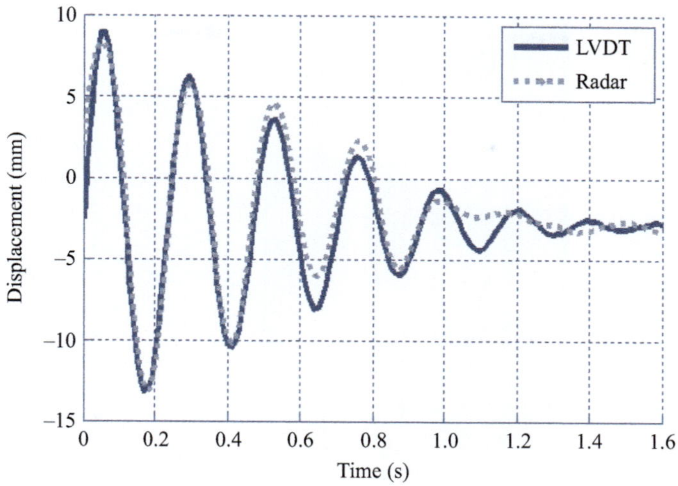

Figure 10.23 Vertical displacement measurement results from the sensors

to the measurement results from the LVDT when the relative motion amplitude is larger than about 2 mm, which corresponds to vertical motion of approximately 2.8 mm when θ is 45°. The larger the angle between the signal path and the vertical motion, the more error is introduced into the signal for small motion amplitudes.

10.6 Continuous radar full-scale structural experiments validation

The CW radar sensor and continuous radar are novel sensors for displacement measurements, especially for SHM applications. The sensors' performance has been characterized by a series of laboratory experiments described in Chapter 5. The laboratory experiments often provide a controllable setup, while the full-scale structure measurements are more challenging. The environment of the full-scale experiment is closer to the scenario of a practical application using the sensor. To apply the radar sensor or continuous radar broadly, the sensors' performance in full-scale structure experiments must be verified. First, a series of full-scale pedestrian bridge experiments were conducted to demonstrate the potential usage of the CW radar to measure structural displacement. More recently, a series of full-scale bridge experiment was conducted to verify the continuous radar sensor network's performance. These two sets of experiments are introduced in this chapter.

10.6.1 Sweetwater Park Bridge experiment

To verify the performance of the continuous radar system, a series of full-scale bridge tests have been conducted on a pedestrian bridge at Sweetwater Park, in Gainesville, FL. The goal of this experiment was not only to verify the performance

Figure 10.24 Sweetwater Park Bridge

of the single continuous radar sensor but also to assess the performance of multiple sensors working simultaneously. The Sweetwater Park pedestrian bridge was selected since it exhibits displacements values that are more realistic, in contrast to the very flexible O'Leno State Park Bridge. In addition, it is easier to predict the bridge response since it can be modeled as a simply supported structure.

10.6.1.1 Sweetwater Park Bridge

The bridge is a steel girder bridge with wood deck. The bridge was designed by CONTECTH and completed in 2015. The total length of the bridge span is approximately 31.6 m. Both ends of the bridge are bolted in the concrete block supports. The bridge is about 2 m above the water surface. There is a concrete weir under the bridge to block the water flow. The bridge structure is shown in Figure 10.24.

10.6.1.2 Instrumentation

For this experiment, the continuous radar sensors were fully assembled and mounted inside the waterproof enclosure. The enclosure was attached to the bottom of the bridge with magnets shown in Figure 10.25. The continuous radar's antenna was facing down to the ground. To measure the maximum displacement of the bridge, one continuous radar was installed close to the mid-span of the bridge. To verify the continuous radar sensor network's performance, the other two sensors were installed close to the quarter span points of the bridge. The base station was set close to the end of the bridge, approximately 15 m away from the closest sensor.

(a) (b)

Figure 10.25 Experimental setup at Sweetwater Park Bridge

All three sensors were installed at the center of the beams at mid-span and quarter-span points. The three sensors were installed along the longitudinal direction of the bridge shown in Figure 10.25. To amplify the continuous radar's signal, three transponder nodes were set up for corresponding continuous radar sensors below the bridge on the weir. The distance between the continuous radar's antenna and the transponder node was about 170 cm. To compare the measurement results from the center continuous radar, a string potentiometer was installed at the mid-span. One end of the string potentiometer was mounted on the concrete weir. The other end was connected to the continuous radar enclosure. Due to the height between the weir and the enclosure, an extension cable was required. The experimental setup on the concrete weir is shown in Figure 10.25. All the sensors measured the vertical displacement of the bridge simultaneously.

10.6.1.3 Description of load conditions

To verify the performance of the continuous radar's measurement accuracy, two types of motions were generated and measured: dynamic displacement and static displacement. Similar to the excitation approach of the O'Leno State Park Bridge, the bridge was excited by jumping around the mid-span of the bridge with the frequency close to 1.5 Hz. Since the bridge at the Sweetwater Park is more rigid, smaller displacement was expected. The near static deflection measurement caused by the vehicle load was also studied. A truck load worked as the live load on the bridge. A truck was driven across the bridge with a constant 10 mph (4.47 m/s) speed to create deflection. Different from the dynamic displacement setup, only two continuous radars (one at mid-span and one at quarter-span) and the string potentiometer were deployed to monitor the process of the vehicle driving on the bridge.

10.6.1.4 Measurement results

For the dynamic experiments, the displacement results from the continuous radar and the string potentiometer are shown in Figure 10.26. From the time history

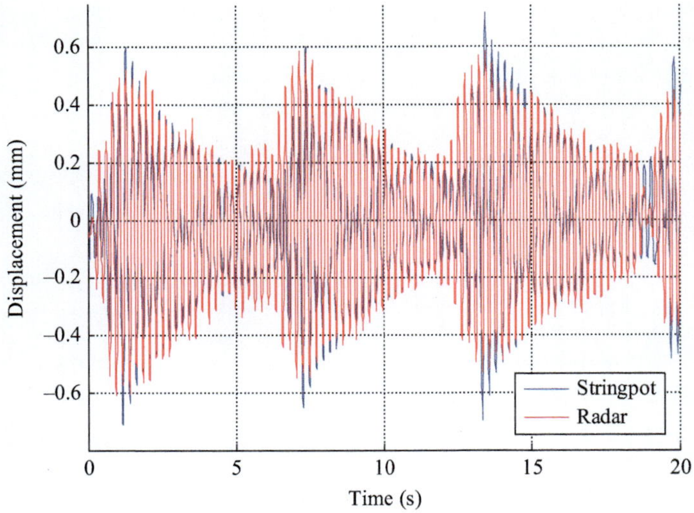

Figure 10.26 Displacement results from stringpot and LVDT

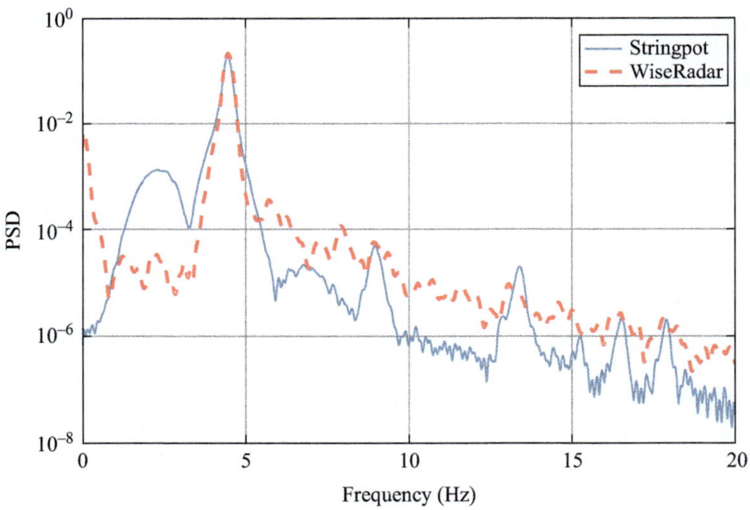

Figure 10.27 PSD of displacement measurement results

measurement results, the continuous radar has a good match with the string potentiometer measurement. They both successfully detected the jumping motions. For most of the measurements, the displacement errors were less than 0.1 mm. When the motion amplitude was less than 0.2 mm, the errors increased. From the frequency domain results shown in Figure 10.27, both sensors successfully measured the first natural frequency of the bridge, which was close to 4.4 Hz.

Compared with the stringpot, the continuous radar did not detect any higher frequency modes due to its high noise level.

The measurement results from three continuous radar sensors were also compared. The measurement results are shown in Figure 10.28. From Figure 10.28, all three radar sensors successfully measure the motion of the bridge. The continuous radar (WR1) at the mid-span measured the largest displacement. The continuous radars (WR2 and WR3) at the quarter-span had similar displacement results since they were symmetrical to the mid-span. From the frequency result comparisons shown in Figure 10.29, all three continuous radars successfully detected the first natural frequency of the bridge, but none of the sensors detected any higher mode of the structure. WR2 and WR3 had similar noise level while the WR1's noise level was slightly higher.

In addition to comparison with the reference sensor (stringpot), the practical measurement results from continuous radars are also characterized with the theoretical calculation results from the finite element model (FEM) analysis. The bridge structure is created as a three-dimensional (3D) bridge model in Abaqus. The model is discretized with linear (two-node) beam elements. Based on the structural design information, two nodes at the bottom of each end of the bridge provide the constraints for the bridge. All four nodes constrain motion in the vertical direction. One of the four nodes has additional longitudinal and lateral constraints. The other node at the same end of the bridge provides another longitudinal constraint along the bridge. All four nodes allow for rotational movement. The railing and the bridge girder are modeled directly. The wood deck is modeled as a distributed mass on the bridge structure without adding any stiffness. The bridge model is shown in Figure 10.30.

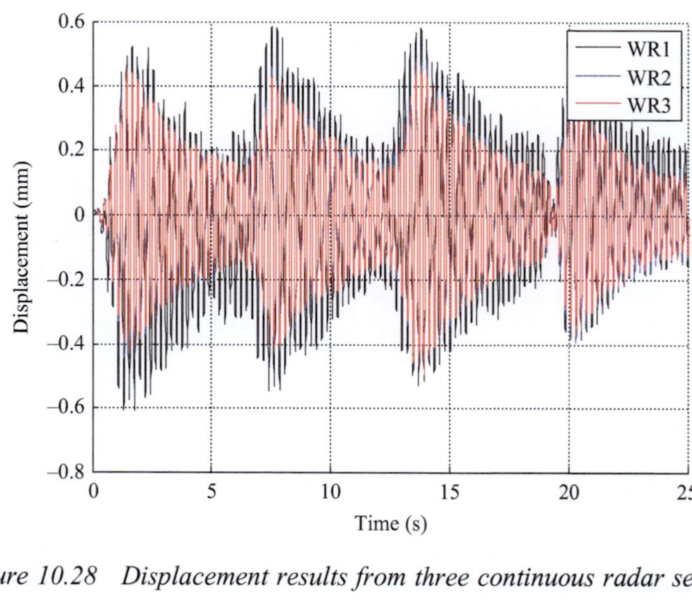

Figure 10.28 Displacement results from three continuous radar sensors

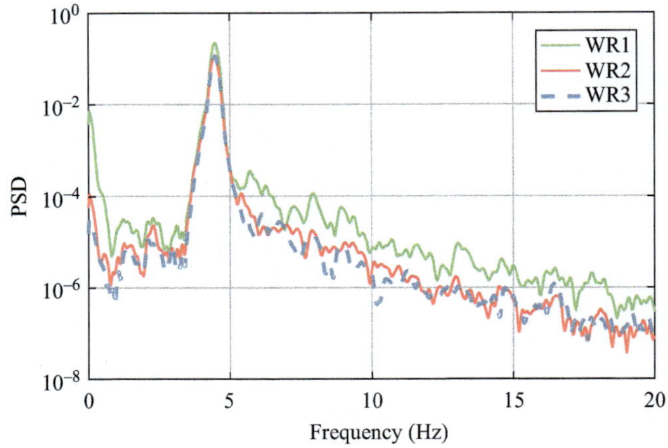

Figure 10.29 PSD of displacement results from three continuous radar sensors

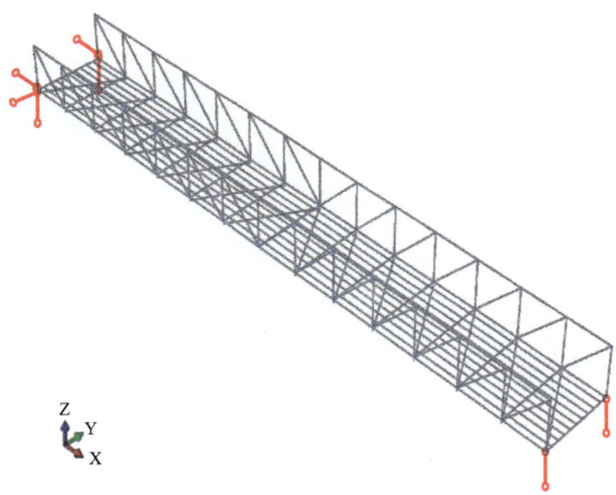

Figure 10.30 3D FEM bridge model

After creating the bridge model, dynamic analysis was conducted to calculate the natural frequencies and mode shapes. The first natural frequency derived from the FEM is 4.2 Hz, which is slightly lower than 4.4 Hz of the measurement results from both continuous radar sensors and the stringpot. This lower frequency is likely because the wood deck in the model only contributes to the structural mass but not the stiffness. As a result, there may be a reduction in the actual stiffness of the bridge in the model, thereby lowering the natural frequency results.

In addition to the natural frequencies, the first mode shape of the bridge is also calculated from the FEM model and compared with the results from the continuous radar sensors (Figure 10.31). From Figure 10.31, at the three locations where the

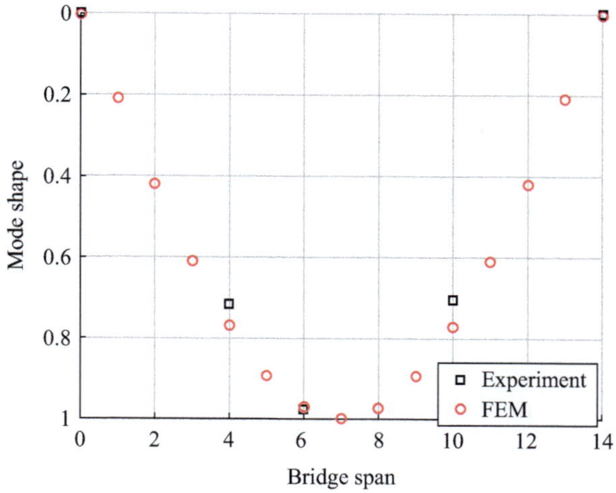

Figure 10.31 Mode shape comparison between the FEM and experimental results

sensors were installed, the mode shape calculated from the measurement results are close to the mode shape calculated from the FEM analysis. To obtain better mode shape estimates, more sensors must be installed.

In addition to the displacement measurement performance, the synchronization accuracy of the continuous radar sensor network has also been characterized. The synchronization accuracy is evaluated by measuring the time delay between different sensors' measurements. To quantify the time delay, the cross-correlation values of the displacement measurements between the continuous radars 2 and 3 with different sample delays were calculated and shown in Figure 10.32. From Figure 10.32, the cross-correlation values varied greatly with the sample delay changing. As the two continuous radars were supposed to measure the similar motion, when the sample delay became the real-time delay value, the correlation value would reach the maximum. From the calculation results, there was only one data sample delay for the two sensors when the correlation value reached the maximum corresponding to 5.9 ms for time delay. This synchronization error is much smaller than the original CW radar sensor network with more than a hundred milliseconds error.

10.6.2 Vehicle load experiment

The displacement measurement results of two continuous radars and one stringpot from the vehicle load experiment are shown in Figure 10.33. For this experiment, to ensure the entire vehicle crossing captured by the sensors, sensing was initiated before the vehicle started crossing the bridge and ended after the vehicle completely left the bridge. From Figure 10.33, all the sensors successfully measured the bridge deflections. The measurement results from the continuous radar and the stringpot at the mid-span were very close with less than 0.3-mm errors.

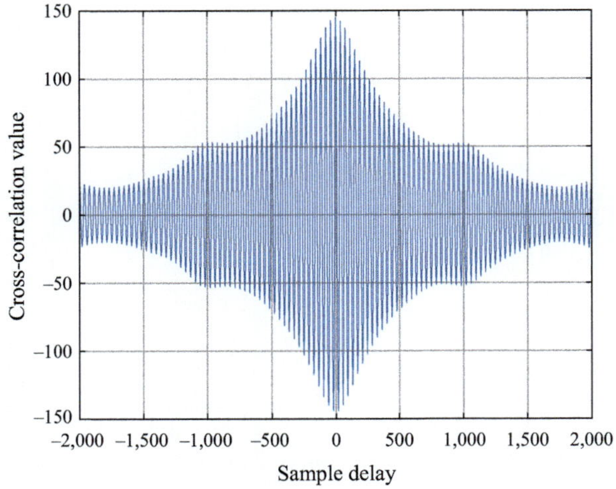

Figure 10.32 Cross-correlation value between continuous radars 2 and 3

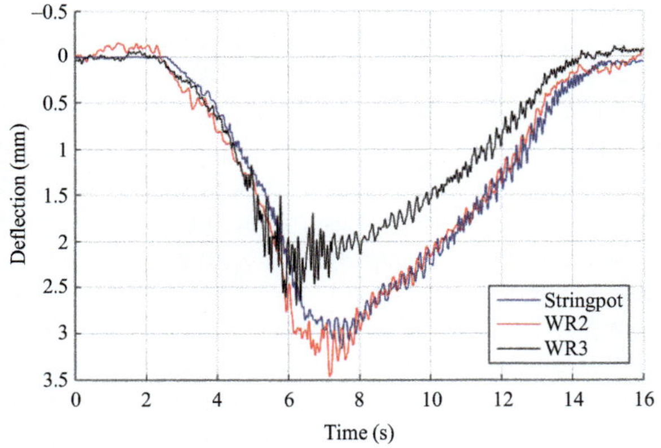

Figure 10.33 Static deflection measurement results from all the sensors

The measurement results from both the continuous radars reached the maximum amplitude at different times. The maximum deflection from the continuous radar at quarter-span was smaller than the maximum deflection from the continuous radar at mid-span, as expected.

Focusing on the measurement results from both the continuous radar sensors, a low-pass filter was applied on both results to smooth the displacement measurements. The results, after using the low-pass filter, are shown in Figure 10.34. With the distance between the sensors known (4.52 m) and the traveling time between the sensors being 1.01 s measured from Figure 10.34, the estimated vehicle speed was calculated at approximate 9.9 mph (4.43 m/s). This result matched closely with the approximate speed of the vehicle (10 mph).

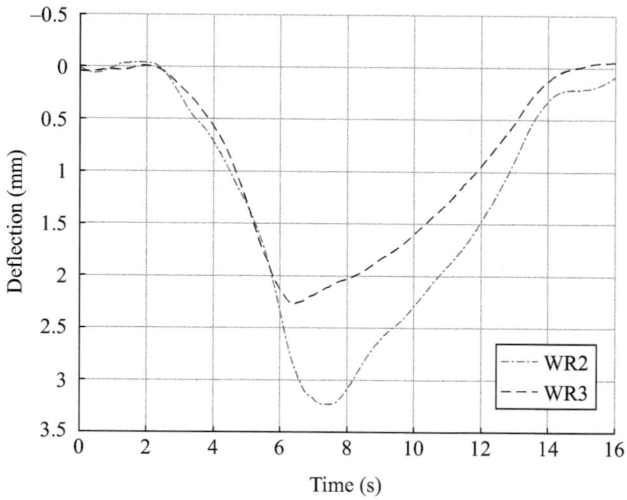

Figure 10.34 Deflection results from two continuous radar sensors with low-pass filter applied

10.7 Conclusions

Structural displacement measurements (both dynamic and static displacement) are of interests to civil engineers. The ability to measure structural displacement, especially static displacement, with high accuracy at a low cost is a challenge. Addressing this challenge was the initiative of the continuous radar sensor network. A wireless, smart, CW radar sensor was developed and verified to measure structural displacement. In general, the continuous radar sensor system demonstrates some advantages including low-cost, wireless operation, reasonable accuracy, and compact size over current displacement measurement approaches.

Two types of 2.4-GHz CW radars, including the AC- and DC-coupled radars, for measuring structural displacement were designed and evaluated. The AC-coupled radar was targeted for measuring nonstatic displacement. Due to its AC coupling design, the AC-coupled radar has better accuracy of measuring motion above 1.75 Hz. The DC-coupled radar was developed to measure low frequency and static displacement, in addition to higher frequency displacement. To apply the CW radar for SHM, especially bridge monitoring, the passive backscattering strategy and active transponder strategy were developed. The passive backscattering strategy is a straightforward approach using a stationary target as the reflection surface but results in a low signal strength that limited its application for many full-scale structure experiments. The active transponder strategy was developed by introducing the transponder node to improve the radar's signal strength. The active transponder also provides more flexibility for using the radar system when there is traffic or water flow under the bridge.

In addition to the hardware and software development of continuous radar system, its routine performance, including the power management, wireless transmission, and operating time, has been investigated by a series of laboratory experiments. The operating time of the continuous radar was related to factors such as the number of sensors, the number of sample points, and the number of wireless communication retransmissions. The operations have been optimized to avoid unnecessary operating time. The wireless transmission performance was assessed by examining the transmission range of the selected radios. The system demonstrated a reliable communication performance for distances within 120 m in the absence of any obstacles. The power management of the continuous radar has also been evaluated. The approach of using sleep cycling demonstrated a large reduction in the system power consumption (more than 50%). The battery life of the system was also investigated with a duty cycle scenario. The results demonstrate that the D-cell batteries are reasonable power source for medium-term usage.

In addition to the operations, the continuous radar displacement measurement accuracy has been evaluated by a series of dynamic and static displacement experiments. The continuous radar using passive backscattering strategy has demonstrated a sub-millimeter accuracy when the motion amplitude is large (more than 2 mm) and the target distance is small (less than 7 m). With the help of the active transponder strategy, the target distance can be extended to approximate 15 m with similar accuracy. With the active transponder strategy, the continuous radar can be more flexible for measuring the displacement even with an oblique angle from the motion direction. The AC-coupled radar can only measure the motion with frequency above 1.75 Hz. The DC couple radar can measure the low-frequency motion and even static deflection with a sub-millimeter accuracy. For both AC- and DC-coupled radar sensors, the frequency information of the measurement is accurate with a very small error (less than 0.1 Hz). The radar's signal demonstrates the presence of multipath effects caused by the reflection from the objects in the environment. This phenomenon indicates that the distance between the radar and the target should be selected carefully to obtain the optimal radar signal.

In addition to laboratory experiments, a series of full-scale bridge experiments have been conducted with the continuous radar system to verify its performance for measuring slow moving and dynamic displacement. This is the first time a CW radar sensor has been used to measure structural displacement on an in-service structure. The measurement results demonstrate that the CW radar is a promising technology for measuring dynamic and static displacement. The continuous radar sensor network can detect motion with sub-millimeter accuracy when the motion amplitude is larger than 2 mm. The frequency information obtained from the continuous radar is close to the natural frequency of the bridge and verified by the reference sensor. In addition to the displacement measurement from the single radar, the displacement information obtained from multiple continuous radars can be further used for dynamic analysis such as mode shape calculation. The mode shape calculation results from continuous radars measurements have been validated for reasonable accuracy compared with FEM analysis. The static deflection measurement results illustrate the continuous radar's good performance for moving load tests and tracking vehicle speeds.

References

[1] Johnson, C. (2001). *I-35W and Mississippi River Bridge anti-icing project operational evaluation report* (No. 2001-22).

[2] ASCE. From https://www.infrastructurereportcard.org/

[3] Youssef, N. F., Bonowitz, D., and Gross, J. L. (1995). *A survey of steel moment-resisting frame buildings affected by the 1994 Northridge earthquake*: US National Institute of Standards and Technology.

[4] Tremblay, R., Filiatrault, A., Bruneau, M., Nakashima, M., Prion, H. G., and DeVall, R. (1996). Seismic design of steel buildings: lessons from the 1995 Hyogo-ken Nanbu earthquake. *Canadian Journal of Civil Engineering, 23*(3), 727–756.

[5] Ko, J. M., and Ni, Y. Q. (2005). Technology developments in structural health monitoring of large-scale bridges. *Engineering Structures, 27*(12), 1715–1725.

[6] Chang, P. C., Flatau, A., and Liu, S. (2003). Review paper: health monitoring of civil infrastructure. *Structural Health Monitoring, 2*(3), 257–267.

[7] Doebling, S. W., Farrar, C. R., Prime, M. B., and Shevitz, D. W. (1996). Damage identification and health monitoring of structural and mechanical systems from changes in their vibration characteristics: a literature review (No. LA-13070-MS). Los Alamos National Lab., NM (United States).

[8] Giurgiutiu, V., Zagrai, A., and Jing Bao, J. (2002). Piezoelectric wafer embedded active sensors for aging aircraft structural health monitoring. *Structural Health Monitoring, 1*(1), 41–61.

[9] Wolowicz, C. H. (1966). *Analysis of an emergency deceleration and descent of the XB-70-1 airplane due to engine damage resulting from structural failure.*

[10] Wykes, J., and Mori, A. (1965). *Techniques and results of an analytical investigation into controlling the structural modes of flexible aircraft.* Paper presented at the Symposium on Structural Dynamics and Aeroelasticity.

[11] Breuer, P., Chmielewski, T., Górski, P., Konopka, E., and Tarczyński, L. (2008). The Stuttgart TV Tower—displacement of the top caused by the effects of sun and wind. *Engineering Structures, 30*(10), 2771–2781.

[12] Brownjohn, J., Dumanoglu, A., Severn, R., and Taylor, C. (1987). Ambient vibration measurements of the Humber Suspension Bridge and comparison with calculated characteristics. *Proc. of the Institution of Civil Engineers (London), 83*, 561–600.

[13] Adeli, H., and Jiang, X. (2006). Dynamic fuzzy wavelet neural network model for structural system identification. *Journal of Structural Engineering, 132*(1), 102–111.

[14] Calebi, M. (2002). *Seismic instrumentation of buildings (with emphasis on federal buildings)*. United States Geological Survey. Technical report No. 0-7460-68170.

[15] Soh, C., Tseng, K. K., Bhalla, S., and Gupta, A. (2000). Performance of smart piezoceramic patches in health monitoring of a RC bridge. *Smart Materials and Structures, 9*(4), 533.

[16] Brownjohn, J. M. (2005). Lateral loading and response for a tall building in the non-seismic doldrums. *Engineering Structures, 27*(12), 1801–1812.

[17] Nayeri, R. D., Masri, S. F., Ghanem, R. G., and Nigbor, R. L. (2008). A novel approach for the structural identification and monitoring of a full-scale 17-story building based on ambient vibration measurements. *Smart Materials and Structures*, *17*(2), 025006.

[18] Darbre, G. R., and Proulx, J. (2002). Continuous ambient-vibration monitoring of the arch dam of Mauvoisin. *Earthquake Engineering & Structural Dynamics*, *31*(2), 475–480.

[19] De Sortis, A., and Paoliani, P. (2007). Statistical analysis and structural identification in concrete dam monitoring. *Engineering Structures*, *29*(1), 110–120.

[20] Jawhar, I., Mohamed, N., Mohamed, M. M., and Aziz, J. (2008). *A routing protocol and addressing scheme for oil, gas, and water pipeline monitoring using wireless sensor networks.* Paper presented at the Wireless and Optical Communications Networks, 2008. WOCN'08. 5th IFIP International Conference on.

[21] Stoianov, I., Nachman, L., Madden, S., Tokmouline, T., and Csail, M. (2007). *PIPENET: A wireless sensor network for pipeline monitoring.* Paper presented at the Information Processing in Sensor Networks, 2007. IPSN 2007. 6th International Symposium on.

[22] Fujihashi, K., Kurihara, K., Hirayama, K., and Toyoda, S. (2005). Monitoring system based on optical fiber sensing technology for tunnel structures and other infrastructure. *Sensing Issues in Civil Structural Health Monitoring* (pp. 185–195): Dordrecht, Netherlands, Springer.

[23] Sharma, J., Hefny, A., Zhao, J., and Chan, C. (2001). Effect of large excavation on deformation of adjacent MRT tunnels. *Tunneling and Underground Space Technology*, *16*(2), 93–98.

[24] Carstensen, J., Henriksen, O., and Teilmann, J. (2006). Impacts of offshore wind farm construction on harbour porpoises: acoustic monitoring of echolocation activity using porpoise detectors (T-PODs). *Marine Ecology Progress Series*, *321*, 295–308.

[25] Ciang, C. C., Lee, J.-R., and Bang, H.-J. (2008). Structural health monitoring for a wind turbine system: a review of damage detection methods. *Measurement Science and Technology*, *19*(12), 122001.

[26] Liu, T., Chiang, W., Chen, C., Hsu, W., Lu, L., and Chu, T. (2011). RETRACTED: Identification and monitoring of bridge health from ambient vibration data. *Journal of Vibration and Control*, *17*(4), 589–603.

[27] Fan, W., and Qiao, P. (2011). Vibration-based damage identification methods: a review and comparative study. *Structural Health Monitoring*, *10*(1), 83–111.

[28] Allison, D. K. (1981). *New eye for the Navy: The origin of radar at the Naval Research Laboratory* (No. NRL-8466). Washington DC: Naval Research Lab. Available at: https://apps.dtic.mil/docs/citations/ADA110586

[29] Gong, C., and Chan, W. (2002). *Using flight manual data to derive aero-propulsive models for predicting aircraft trajectories.* Paper presented at the AIAA's Aircraft Technology, Integration, and Operations (ATIO) 2002 Technical Forum.

[30] Elsherbini, A., and Sarabandi, K. (2012). Compact directive ultra-wideband rectangular waveguide based antenna for radar and communication applications. *IEEE Transactions on Antennas and Propagation*, *60*(5), 2203–2209.

[31] Giurgiutiu, V., and Bao, J. (2004). Embedded-ultrasonics structural radar for in situ structural health monitoring of thin-wall structures. *Structural Health Monitoring*, *3*(2), 121–140.

[32] Hughes, J. E., Kim, Y., El-Korchi, T., and Cyganski, D. (2017). Radar-based impact load prediction for damage mitigation of infrastructure. *Journal of Vibration and Control*, *23*(12), 1908–1924.

[33] Yu, T.-Y., and Büyüköztürk, O. (2008). A far-field airborne radar NDT technique for detecting debonding in GFRP–retrofitted concrete structures. *NDT & E International*, *41*(1), 10–24.

[34] Rhim, H. C. (2001). Condition monitoring of deteriorating concrete dams using radar. *Cement and Concrete Research*, *31*(3), 363–373.

[35] Maierhofer, C. (2003). Nondestructive evaluation of concrete infrastructure with ground penetrating radar. *Journal of Materials in Civil Engineering*, *15*(3), 287–297.

[36] Maser, K. R. (1996). Condition assessment of transportation infrastructure using ground-penetrating radar. *Journal of infrastructure systems*, *2*(2), 94–101.

[37] Alani, A. M., Aboutalebi, M., and Kilic, G. (2013). Applications of ground penetrating radar (GPR) in bridge deck monitoring and assessment. *Journal of Applied Geophysics*, *97*, 45–54.

[38] Tapete, D., Fanti, R., Cecchi, R., Petrangeli, P., and Casagli, N. (2012). Satellite radar interferometry for monitoring and early-stage warning of structural instability in archaeological sites. *Journal of Geophysics and Engineering*, *9*(4), S10–S25.

[39] Tarchi, D., Ohlmer, E., and Sieber, A. (1997). Monitoring of structural changes by radar interferometry. *Research in Nondestructive Evaluation*, *9*(4), 213–225.

[40] Motagh, M., Beavan, J., Fielding, E. J., and Haghshenas, M. (2014). Post-seismic ground deformation following the September 2010 Darfield, New Zealand, earthquake from TerraSAR-X, COSMO-SkyMed, and ALOS InSAR. *IEEE Geoscience and Remote Sensing Letters*, *11*(1), 186–190.

[41] Yun, S.-H., Hudnut, K., Owen, S., *et al.* (2015). Rapid damage mapping for the 2015 Mw 7.8 Gorkha earthquake using synthetic aperture radar data from COSMO–SkyMed and ALOS-2 satellites. *Seismological Research Letters*, *86*(6), 1549–1556.

[42] Atkins, N. T., Butler, K. M., Flynn, K. R., and Wakimoto, R. M. (2014). An integrated damage, visual, and radar analysis of the 2013 Moore, Oklahoma, EF5 tornado. *Bulletin of the American Meteorological Society*, *95*(10), 1549–1561.

[43] Palojarvi, P., Maatta, K., and Kostamovaara, J. (1997). Integrated time-of-flight laser radar. *IEEE Transactions on Instrumentation and Measurement*, *46*(4), 996–999.

[44] Makynen, A. J., Kostamovaara, J. T., and Myllyla, R. A. (1994). Tracking laser radar for 3-D shape measurements of large industrial objects based on time-of-flight laser rangefinding and position-sensitive detection techniques. *IEEE transactions on instrumentation and measurement*, *43*(1), 40–49.

[45] Shariff, K., and Wray, A. (2002). Analysis of the radar reflectivity of aircraft vortex wakes. *Journal of Fluid Mechanics*, *463*, 121–161.

[46] Lund, B., Graber, H. C., and Romeiser, R. (2012). Wind retrieval from shipborne nautical X-band radar data. *IEEE Transactions on Geoscience and Remote Sensing*, *50*(10), 3800–3811.

[47] Wu, M., Dai, X., Zhang, Y. D., Davidson, B., Amin, M. G., and Zhang, J. (2013). *Fall detection based on sequential modeling of radar signal time-frequency features.* Paper presented at the Healthcare Informatics (ICHI), 2013 IEEE International Conference on.

[48] Brooker, G. M. (2005). *Understanding millimetre wave FMCW radars.* Paper presented at the 1st International Conference on Sensing Technology.

[49] Jaeschke, T., Bredendiek, C., Küppers, S., and Pohl, N. (2014). High-precision D-band FMCW-radar sensor based on a wideband SiGe-transceiver MMIC. *IEEE Transactions on Microwave Theory and Techniques*, *62*(12), 3582–3597.

[50] Roehr, S., Gulden, P., and Vossiek, M. (2008). Precise distance and velocity measurement for real time locating in multipath environments using a frequency-modulated continuous-wave secondary radar approach. *IEEE Transactions on Microwave Theory and Techniques*, *56*(10), 2329–2339.

[51] Pieraccini, M., Fratini, M., Parrini, F., Atzeni, C., and Bartoli, G. (2008). Interferometric radar vs. accelerometer for dynamic monitoring of large structures: An experimental comparison. *NDT & E International*, *41*(4), 258–264.

[52] Gentile, C. (2010). Deflection measurement on vibrating stay cables by non-contact microwave interferometer. *NDT & E International*, *43*(3), 231–240.

[53] Dei, D., Pieraccini, M., Fratini, M., Atzeni, C., and Bartoli, G. (2009). Detection of vertical bending and torsional movements of a bridge using a coherent radar. *NDT & E International*, *42*(8), 741–747.

[54] Gentile, C., and Bernardini, G. (2008). Output-only modal identification of a reinforced concrete bridge from radar-based measurements. *NDT & E International*, *41*(7), 544–553.

[55] Gentile, C., and Bernardini, G. (2009). An interferometric radar for non-contact measurement of deflections on civil engineering structures: laboratory and full-scale tests. *Structure and Infrastructure Engineering*, *6*(5), 521–534.

[56] Pieraccini, M., Parrini, F., Fratini, M., Atzeni, C., Spinelli, P., and Micheloni, M. (2007). Static and dynamic testing of bridges through microwave interferometry. *NDT & E International*, *40*(3), 208–214.

[57] Pieraccini, M., Tarchi, D., Rudolf, H., *et al.* (2000). Structural static testing by interferometric synthetic radar. *NDT & E International*, *33*(8), 565–570.

[58] Torfs, T., Sterken, T., Brebels, S., *et al.* (2013). Low power wireless sensor network for building monitoring. *IEEE Sensors Journal*, *13*(3), 909–915.

[59] Atzeni, C., Barla, M., Pieraccini, M., and Antolini, F. (2015). Early warning monitoring of natural and engineered slopes with ground-based synthetic-aperture radar. *Rock Mechanics and Rock Engineering*, *48*(1), 235–246.
[60] Wang, H., Ren, L., Mao, E., and Fathy, A. E. (2016). *Phase based motion characteristics measurement for fall detection by using stepped-frequency continuous wave radar.* Paper presented at the Biomedical Wireless Technologies, Networks, and Sensing Systems (BioWireleSS), 2016 IEEE Topical Conference on.
[61] Rice, J. A., Li, C., Gu, C., and Hernandez, J. C. (2011). *A wireless multi-functional radar-based displacement sensor for structural health monitoring.* Paper presented at the SPIE Smart Structures and Materials + Nondestructive Evaluation and Health Monitoring.
[62] Droitcour, A. D., Boric-Lubecke, O., Lubecke, V. M., Lin, J., and Kovacs, G. T. (2004). Range correlation and I/Q performance benefits in single-chip silicon Doppler radars for noncontact cardiopulmonary monitoring. *IEEE Transactions on Microwave Theory and Techniques*, *52*(3), 838–848.
[63] Li, C., and Lin, J. (2008). Random body movement cancellation in Doppler radar vital sign detection. *IEEE Transactions on Microwave Theory and Techniques*, *56*(12), 3143–3152.
[64] Li, C., Lubecke, V. M., Boric-Lubecke, O., and Lin, J. (2013). A review on recent advances in Doppler radar sensors for noncontact healthcare monitoring. *IEEE Transactions on Microwave Theory and Techniques*, *61*(5), 2046–2060.
[65] Li, C., Xiao, Y., and Lin, J. (2006). Experiment and spectral analysis of a low-power Ka-band heartbeat detector measuring from four sides of a human body. *IEEE Transactions on Microwave Theory and Techniques*, *54*(12), 4464–4471.
[66] Gu, C., and Li, C. (2014). From tumor targeting to speech monitoring: accurate respiratory monitoring using medical continuous-wave radar sensors. *IEEE Microwave Magazine*, *15*(4), 66–76.
[67] Gu, C., Li, R., Zhang, H., *et al.* (2012). Accurate respiration measurement using DC-coupled continuous-wave radar sensor for motion-adaptive cancer radiotherapy. *IEEE Transactions on Biomedical Engineering*, *59*(11), 3117–3123.
[68] Lu, L., Li, C., and Rice, J. A. (2011). *A software-defined multifunctional radar sensor for linear and reciprocal displacement measurement.* Paper presented at the Wireless Sensors and Sensor Networks (WiSNet), 2011 IEEE Topical Conference on.
[69] Moll, J., Bechtel, K., Hils, B., and Krozer, V. (2014). *Mechanical vibration sensing for structural health monitoring using a millimeter-wave Doppler radar sensor.* Paper presented at the EWSHM-7th European Workshop on Structural Health Monitoring.
[70] MOLL, J., and KROZER, V. (2016). *Radar-based mechanical vibration sensing for structural health monitoring applications: A comparison of radar transceiver measurements at 24 GHz and 100 GHz.* Paper presented at the 8th European Workshop on Structural Health Monitoring.

[71] Rice, J. A., Guan, S., Li, C., and Gu, C. (2012). *A MIMO radar sensor network for distributed bridge monitoring*. Paper presented at the Proc. ANCRiSST Workshop.

[72] Rappaport, T. S. (1996). *Wireless Communications: Principles and Practice* (*Vol. 2*): Prentice Hall PTR, New Jersey.

[73] Changzhi, L., Xiaogang, Y., Chien-Ming, L., Dong, L., Ran, L., and Jenshan, L. (2010). High-sensitivity software-configurable 5.8-GHz radar sensor receiver chip in 0.13-m CMOS for noncontact vital sign detection. *Microwave Theory and Techniques, IEEE Transactions on*, *58*(5), 1410–1419.

[74] Breed, G. (2009). The fundamentals of patch antenna design and performance. *High Frequency Electronics*, *3*(12), 49–51.

[75] Nagayama, T., Sim, S.-H., Miyamori, Y., and Spencer Jr., B. (2007). Issues in structural health monitoring employing smart sensors. *Smart Structures and Systems*, *3*(3), 299–320.

[76] Rice, J. A., Gu, C., Li, C., and Guan, S. (2012, April). A radar-based sensor network for bridge displacement measurements. In *Sensors and Smart Structures Technologies for Civil, Mechanical, and Aerospace Systems 2012* (Vol. 8345, p. 83450I). International Society for Optics and Photonics.

[77] Guan, S., Rice, J. A., Li, C., and Gu, C. (2014). Automated DC offset calibration strategy for structural health monitoring based on portable CW radar sensor. *IEEE transactions on instrumentation and measurement*, *63*(12), 3111–3118.

[78] Massagram, W., Lubecke, V. M., Host-Madsen, A., and Boric-Lubecke, O. (2009). Assessment of heart rate variability and respiratory sinus arrhythmia via Doppler radar. *Microwave Theory and Techniques, IEEE Transactions on*, *57*(10), 2542–2549.

[79] Zakrzewski, M., Raittinen, H., and Vanhala, J. (2012). Comparison of center estimation algorithms for heart and respiration monitoring with microwave Doppler radar. *Sensors Journal, IEEE*, *12*(3), 627–634.

[80] Xu, W., Gu, C., Li, C., and Sarrafzadeh, M. (2012). Robust Doppler radar demodulation via compressed sensing. *Electronics Letters*, *48*(22), 1428–1430.

[81] Yuen, P. C., and Feng, G.-C. (1996). A novel method for parameter estimation of digital arc. *Pattern Recognition Letters*, *17*(9), 929–938.

[82] Wang, J., Wang, X., Chen, L., Huangfu, J., Li, C., and Ran, L. (2014). Noncontact distance and amplitude-independent vibration measurement based on an extended DACM algorithm. *IEEE Transactions on Instrumentation and Measurement*, *63*(1), 145–153.

[83] Hoover, D. P., Bilbao, A., and Rice, J. A. (2012). WiSeMote: A novel high fidelity wireless sensor network for structural health monitoring. *Smart Structures and Systems*, *10*(3), 271–298.

Chapter 11
Short-distance radar sensing application

Chen Song[1], Zhengxiong Li[1], and Wenyao Xu[1]

11.1 Introduction

Human-related sensing plays a more and more important role because of the emerging applications that include search and rescue, security and surveillance, as well as other important operations. Radar offers a number of unique advantages compared with the other technologies (e.g., infrared cameras [1] and lidar systems [2]). For instance, the Doppler radar system is cost-effective and can sense the target unobtrusively by penetrating obstacles with lower propagation loss. Due to these characteristics, researchers usually employ the Doppler radar in the long-distance object detection (e.g., pedestrian detection [3,4] and human localization [5]). But recently, researchers start to find out that the use of the Doppler radar is also a natural choice for human micro-motion sensing. Because Doppler information represents the time rate of change of a range, the Doppler radar can suppress stationary clutters to highlight human-related micro-scale movements. The employment of the Doppler radar in short-range micro-motion sensing widely broadens the domain of human sensing network. In the following, we highlight the emerging radar sensing applications in the area of home-based smart-health and human biometrics.

11.1.1 Smart healthcare

In the healthcare domain, the quality of sleep has gained increasing attention because there is a growing recognition of the adverse effects from poor sleep quality and sleep disorders. Patients with sleep disorders are prone to suffer from chronic diseases such as obesity, diabetes, and hypertension. Vorona *et al.* [6] demonstrated the relationship between obesity and sleep time. Spiegel *et al.* [7] showed that sleep loss increases the risk for diabetes and obesity. Unfortunately, people are usually not aware of sleep disorders because they happen during sleep. Therefore, it has become a chronic, underexplored but critical health challenge in modern life.

[1]University at Buffalo, SUNY, USA

To date, there are several methods to perform sleep monitoring based on polysomnography devices or real-time infrared camera. However, the obtrusiveness [8,9] and lack of privacy [10,11] of these methods prevent people from using current sleep monitoring systems in daily life. Therefore, it is interesting to explore how to accurately detect the very small magnitude of human sleep activities without interrupting the user's sleep at night.

One research group successfully addressed this challenge by developing a Doppler radar-based sleep monitoring system, called SleepSense, which is non-contact and cost-effective [12,13]. Sleep is a period of inactivity and rejuvenation. During sleep, the breathing pattern and movement distribution pattern are particularly being interested, as they characterize the different sleep states and are closely associated the sleep quality. Based on the movement distribution pattern and breathing pattern, they defined three sleep events: on-bed movement event, bed-exit event, and breathing event. The on-bed movement event contains the movements such as turnover and arm trebling. In the breathing event, the subject keeps still on the bed without body movements. The bed-exit event refers to the event with bed-exit movement, which indicates the interruption of sleeping state. Successful recognition of these three interested sleep events is the basis for obtaining the breathing pattern and movement distribution pattern.

To perform sleep monitoring in a remote and cost-effective manner, they instrumented an electromagnetic probe as shown in Figure 11.1. In their case, the electromagnetic probe was developed based on the commercial of-the-shelf (COTS) components. Figure 11.2 demonstrates a practical using scenario of the implemented sleep monitoring system. Specifically, the electromagnetic probe generates a single-tone carrier signal which is transmitted to the subject. When the microwave hits the subject, the body displacement (caused by the movement or the respiration) of the subject enables the microwave to generate a phase shift. This phase shift is proportional to the corresponding body displacement. By demodulating

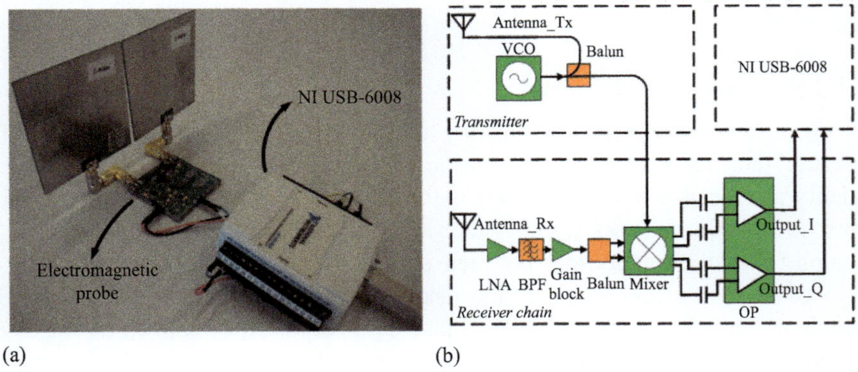

Figure 11.1 (a) The electromagnetic probe and the DAQ NI-USB6008 device. (b) The block diagram of electromagnetic probe. The electromagnetic probe captures the sleep-related signal and outputs the baseband signal, which is digitized by a DAQ device [12]

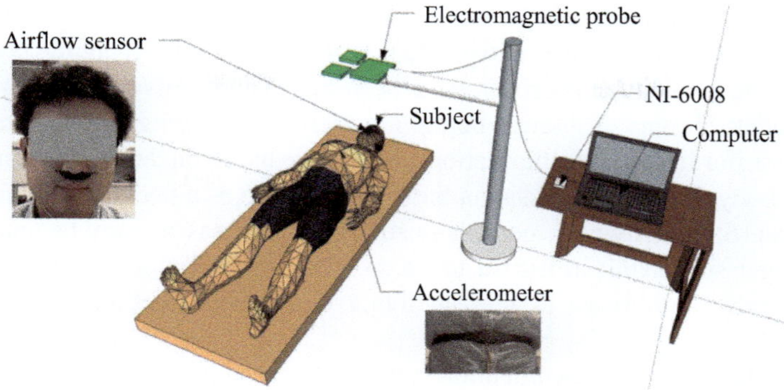

Figure 11.2 The experimental setup: the subject lies his back on a mattress. The electromagnetic probe locates on the top of the chest with the distance of one meter. In the meanwhile, an airflow sensor and an accelerometer sensor are attached to serve as the ground truth measurements [12]

this phase information properly, they were able to obtain the displacement information. The electromagnetic probe adopted direct-conversion radar architecture to capture the subject movement and breathing signal. In the circuit implementation, the voltage-controlled oscillator (VCO) in the transmitter generates a carrier signal at 2.4-GHz space. The VCO also provides local oscillator to the mixer in the receiver chain. The output power of this transmitter is around 0 dBm. The system consists of a low-noise amplifier (LNA), a band-pass filter (BPF), a gain block, a balun, a mixer, and two baseband operational amplifiers (OPs) form the receiver chain. The LNA amplifies the received signal at 2.4 GHz. The interferences with frequencies outside the 2.4-GHz band are removed by the BPF. A gain block is adopted to further amplify the received signal. Two OPs with the same gain of 40 dB are used to amplify the downconverted $I(t)$ and $Q(t)$ baseband signals. Lastly, an national instruments (NI) data acquisition (DAQ) device, NI USB-6008, digitizes the baseband $I(t)$ and $Q(t)$ signal.

Overall, their system can monitor and classify the sleep-related events by detecting the on-bed movement activities during sleep based on the radar signal. The Doppler effect [14] has been employed in various motion detection applications, such as gait assessment, vital signal detection, and hand gesture recognition. In SleepSense, a specialized Doppler radar was developed to measure subject's chest displacement remotely. The sampled baseband radar signal went through a demodulation module that uses the extended differentiate and cross multiple (DACM) algorithm to obtain the body movement information from sleep status. Afterward, the sleep status recognition framework processed the displacement signal and accurately recognized three sleep stages, i.e., on-bed movement, bed exit, and breathing section. By deploying such a sleep monitoring system at home, SleepSense can help people to assess sleep quality, even diagnose the sleep disorders at the earliest stages.

Another work also proposed a routine long-term home monitoring solution based on radar sensing technology [15]. With the growth of aging population, innovative technology approaches have been increasingly investigated for the last two decades aiming at human-being long-term monitoring. However, current solutions suffer from critical limitations as they are based on devices attached to the patient's body, involving pressing a button, e.g., worn as a necklace, in emergency situations [16,17]. However, persons in such situation may already be unconscious or no longer sufficiently reflexive to do so. The ideal solution is therefore a contactless health monitoring system, avoiding the need for actions by the elderly person.

In the last two decades, attention has been focused mainly on contactless vital sign monitoring for remote fall which is also in connection with health monitoring in home environments. Many relevant academic developments are based on radar techniques implemented as a single device sensor, e.g., continuous-wave (CW) Doppler radar or ultra-wideband impulse-radio (UWB-IR) radar in the controlled condition. Other investigations are systems based on video cameras [18], floor vibration [19], and acoustic sensors [20]. Besides privacy concerns, the use of video camera systems is still troubled by issues related to low light environments, field of view, and image processing, resulting in a success rate of 90% using two cameras [18]. Successes in floor vibration and acoustic sensors are similarly limited due to environmental interference and background noise. Moreover, they are also less effective in detecting cases of soft human falls, defined as a fall after the individual collides with an object (e.g., table, chair, or carpet).

As a result, the authors in [15] proposed a full system which enables indoor, noninvasive fall detection, and tagless localization. The system combined radar, wireless communications, and data-processing techniques. Moreover, it has been designed to satisfy the European and Federal Communications Commission (FCC) UWB spectrum masks, and it can also be potentially connected to medical monitoring personnel to provide a prompt alert in the event of emergencies. Figure 11.3 shows a simplified block diagram of the proposed health monitoring system. It consists of a sensor, combining both radar and wireless communications features, and a base station for data processing. A radar waveform is generated and sent to the target, and then its reflected echo, containing speed and absolute distance information, is collected by the receiver. The resulting baseband signals are digitized and transmitted wirelessly to a base station that consists of a Zigbee module, a laptop, and a microcontroller. The latter collects and transfers the data received from the Zigbee module to the laptop to determine remotely the target's absolute distance and to distinguish a fall event from normal movements (e.g., walking or sitting down).

The data processing is not performed by the sensor in order to avoid complex processor on board, reducing costs, size, and energy consumption. Moreover, this represents a flexible solution if multiple sensors will be used in the future. In fact, the base station should combine and process multiple pieces of information simultaneously. The radar waveform was designed based on a previous hybrid approach [21]. It consisted of a single tone, at 5.8 GHz in the industrial, scientific

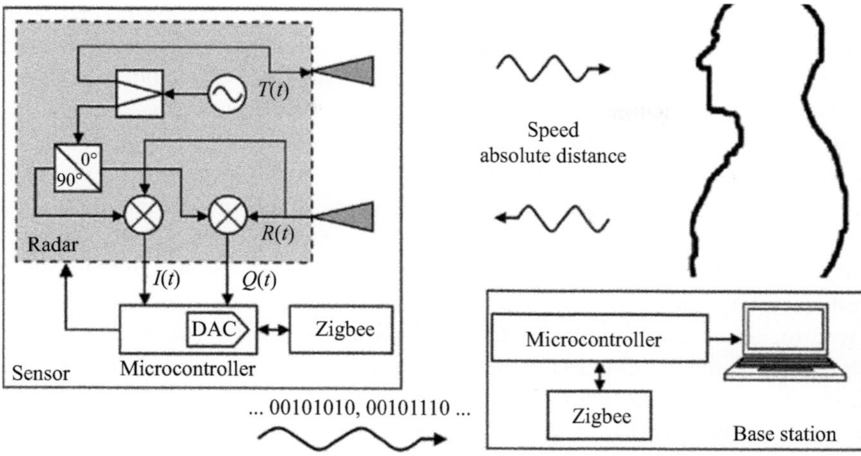

Figure 11.3 The simplified block diagram of the proposed health monitoring system [15]

and medical radio band (ISM band), alternated with a stepped frequency CW (SFCW) waveform working in the UWB band. Each tone lasts 1 s and was used to continuously detect the speed of a person using the Doppler concept. The SFCW waveform was used to detect the target's absolute distance. It consisted of $N = 40$ coherent CW pulses (called burst), the frequencies of which are increased from pulse to pulse by a fixed increment $\Delta f = 25$ MHz. Each pulse was $T = 50$ μs long, resulting in a burst duration $N \cdot T$ of 2 ms, while its total band $N \cdot \Delta f$ is 1 GHz positioned between 6 and 7 GHz, enabling a smallest resolution of 15 cm. The full waveform is 1.002 s. Afterward, they carried out the waveform spectral analysis and utilized specific considerations to justify the choice of the waveform parameters and to explain the operations of compensation and calibration in data processing.

In the experiment evaluation, they recruited real human volunteers who were allowed to move freely about the whole room. The sensor was fixed to the wall at a height of 1.5 m. Furniture and metallic shelves were deliberately included to enable the existence of clutter and reflections, mimicking a typical room setting. Falls were mimicked with two different human volunteers, with a similar 1.75 m height but different weights to enable the evaluation of different fall speeds. The first and second subjects' weights were 90 and 75 kg, respectively. However, only frontal falls have been evaluated at different locations at radial distances in the whole room, using an inflated mattress to avoid injuries. Also, only one person was present in the room at a time. The tagless localization evaluation showed that after applying the compensation and the calibration process, their system can perfectly distinguish the target's signal peak, which indicates the target's absolute distance, and will not be overwhelmed by the undesired reflections originating from the clutter and the antenna's cross coupling. Then, they validated whether their system can detect the fall event. Specifically, a data set was built containing 70 activities measured from two persons. In particular, 20 walking signals have been acquired for each person, who was allowed free movement in the whole room, and 30 fall signals have been acquired with each subject located at known distances from the antennas. The least-squares support-vector machine (LS-SVM) model was trained using the data of a single person (Target 1) and then validated using the data from the other person (Target 2). The above process was repeated two times since data from two persons were available. The results demonstrated that the LS-SVM model with the Global Alignment kernel, which incorporates time-dependent information, presented a success rate in distinguishing fall events from normal movements of 94.3% outperforming the linear and radial basis function (RBF) kernels. This is consistent with the expectation since falls exhibit a time-dependent structure as the speed increases continuously until the sudden moment when it stops abruptly. An alternative method called dynamic time warping (DTW) combined with a Euclidean distance measure is frequently used to classify sequences of vectors. In order to compare the LS-SVM with GA solution to this standard method, they ran an additional experiment and the results showed that the GA kernel slightly outperforms the DTW alternative on their data set. They also pointed out that by further reducing the cross talk, it will be possible to increase the total receiver gain and, consequently, to improve the accuracy of velocity detection, besides the detection range extension. Moreover, the system's sampling rate influenced the accuracy of the speed detection, i.e., the higher it is, the more accurate is its detection. However, this will significantly complicate the system since the duration of the SFCW pulses must be shortened, and the analog-to-digital convertor (ADC) sample rate increased. This will involve higher power consumption and a larger number of transmitted frames to the base station. However, it is important to note that in their fall detection application, the main goal is to detect the changes in speed and not to determine how accurate is the value of the instantaneous speed.

To sum up, they proposed a new radar-based system for contactless fall detection and tagless localization in an indoor environment. This research work is in line with the growing need for novel healthcare solutions. The full system combined radar, wireless communications, and data-processing techniques to satisfy the European and FCC UWB spectrum masks. They proposed unique hardware structures and algorithms to address the practical problems, such as backscattering and cross talk. Experimental evaluations with real human subjects demonstrated accurate detection of the target's absolute distance and fall events.

11.1.2 Biometric authentication

Human authentication has been unprecedentedly important in the era of the Internet of Things (IoT) as IoT devices are becoming increasingly attractive targets for cybercriminals. Taking smartphone as an example, hundreds of millions of Americans have had their private or financial information compromised and more than 12 new people are hacked each second [22,23]. Meanwhile, it is proven that Internet-connected cars can be compromised, and hackers can carry out any number of malicious activities, including taking control of the vehicle system and unlocking the doors [24]. Wearables can also become a source of threat since hackers can use the motion sensors embedded in smartwatches to steal information you're typing, or gather health data from health tracker devices you might be using. Therefore, how to securely protect IoT devices without interrupting user activities has achieved great attention.

The current widely applied security defense is the so-called human authentication, which is a mechanism to secure the access of an IoT device by determining whether a person is who he or she claims to be. Among different kinds, biometric authentication is believed to be more secure because it is a "what you are" factor based on unique individual characteristics. Two types of biometric properties have been proliferating in customer devices. Specifically, physical biometrics emphasizes the unique physiological attributes in human body, which include fingerprints [25], facial recognitions [26], and eye scans (iris [27], retina [28]). Behavioral biometrics, on the other hand, represents the way people do things and the typical examples are keystroke rhythms [29], gait patterns [30] and handwritten signatures [31].

Because of the unobtrusive and sensitive nature of radar sensing, researches begin to explore the usage of radar in human identification for better usability and performance. For instance, one nascent work instrumented a system that can identify a subject accurately by using the multiple-input and multiple-output (MIMO) array sensor for the safety monitoring [32]. The authors set an 8 × 8 MIMO radar array around the target, and observed waves reflected and scattered from the target. As depicted in Figure 11.4, a single-pole 64 throw (SP64T) switch was used at the transmitting side. A CW signal at 2.47125 GHz was used, and transmitted power at the antennas was set to −12 dBm. The CW signal was later split to the receiver side since accurate synchronization between transmitting and receiving sides is required. At the receiver side, the received signals were input to a downconverter unit by way of an LNA. The downconverted baseband signals (I_1, Q_1 ~ I_8, Q_8) were digitized by a DAQ unit with sampling frequency of 20 kHz. The snapshot rate for the MIMO channel is determined by the switching speed of the SP64T. In the experiment, the

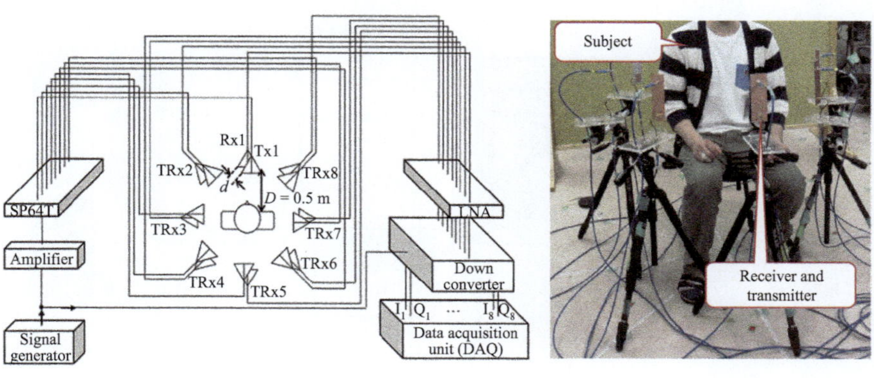

Figure 11.4 The system setup of the proposed radar MIMO array in [32]. (a) The system diagram. (b) The experiment setup

rate for taking a snapshot of the MIMO channel was set to 100 Hz. In their experiment, they found that the target generates a time-variant channel due to respiration and heartbeat. Based on the spatial and temporal characteristics of the measured channel, the system was able to identify different subjects. Their experiments were carried out in an indoor environment to evaluate the recognition performance of the system. The results show that the proposed method can identify eight individuals and reject four impostors with 0% equal error rate. This work has proved the feasibility of detecting human biometrics through the reflected radar signals.

One research team also presented an interesting work [33] and we hereafter introduce their work in detail. They pointed out that existing biometric authentications often follow the principle of one-step validation. After the initial login, there is no further identification procedure as long as the device remains active. This is mainly because the existing biometrics (e.g., fingerprint reading, iris scanning, or face recognition) are often intrusive and disruptive so that users are only willing to perform the authentication once at the beginning of the usage. Consequently, an impostor can gain physical access to the device if the legitimate user leaves it unlocked and unattended. For this reason, this is critical to propose a continuous authentication scheme.

After the wide exploration, the team placed their attention on the cardiac motion, which is a periodical 3D automatic heart deformation caused by the unique self-excitement of the cardiac muscle in individuals [34] (see Figure 11.5). The cardiac motion is unique (i.e., distinguishable across subjects), non-volitional (i.e., unknown to the user), secure (i.e., difficult to counterfeit), and present in all living individuals (i.e., intrinsic liveness). However, they identify three main challenges before further utilizing it as a secure human biometric: (1) How to obtain the high-resolution cardiac motion information unobtrusively? (2) How to extract invariant geometric-based features for each heart with regard to the cardiac motion

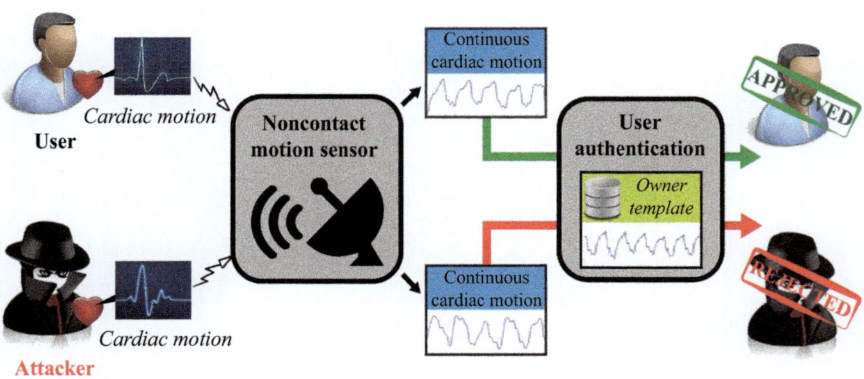

Figure 11.5 Cardiac Scan directly and continuously detects the heart motion via the Doppler radar signal. The system utilizes the micro-motion signal as the unique human biometric to secure the account security through the entire login session [33]

mechanism? (3) How to examine the usability and security of the continuous authentication scheme?

Noncontact monitoring of human body motion, such as respiration and heartbeat rates using a Doppler radar motion sensor, has gone through a few decades of scientific study [35–39]. Efforts have been devoted to the development of radar front-end hardware, signal processing algorithms, and system on-chip/on-board integration. Compared with other techniques such as noncontact laser vibrometer [40] and infrared imager [41] that can only detect motion at body surface, it has been shown that the Doppler radar sensor can directly measure the motion of internal organs [42] and heart [43,44]. However, research results in those works are incomprehensive for a real authentication system; e.g., the impact of random body movement is not considered. Although random body movement and clutter noise still require significant efforts to resolve, some progress has been achieved [45,46] and preliminary clinical studies have been reported [47]. However, existing cancellation approaches either compromise the quality of the baseband signals [45] or require sweeping the carrier frequency and adjusting the target position [46], which is not applicable to capture the high-fidelity cardiac motion in a real-world setup.

Instead, this team proposed that DC-coupled interferometry radar and Doppler radar with digital intermediate frequency (digital-IF) architecture can avoid frequency-selective signal distortion and thus make it possible to recover accurate motion patterns using CW Doppler radar sensors. In detail, to monitor cardiac motion pattern, a smart DC-coupled CW radar sensor was employed by taking advantage of real-time signal processing and mixed-signal design in modern devices. For cardiac motion sensing, the DC offset due to reflection from other parts of the body not related to cardiopulmonary activities may easily saturate the receiver and create frequency-dependent distortion, and is an important factor for the central intelligence unit to handle. As shown in Figure 11.6, the DC-coupled adaptive tuning architecture includes radio frequency (RF) coarse-tuning and baseband fine-tuning. For RF tuning, the electronically controlled phase shifter and attenuator add a portion of the transmitted signal to the receiver signal to cancel most of the DC offset caused by clutter reflections. However, due to quadrature imbalance, the phase variation of

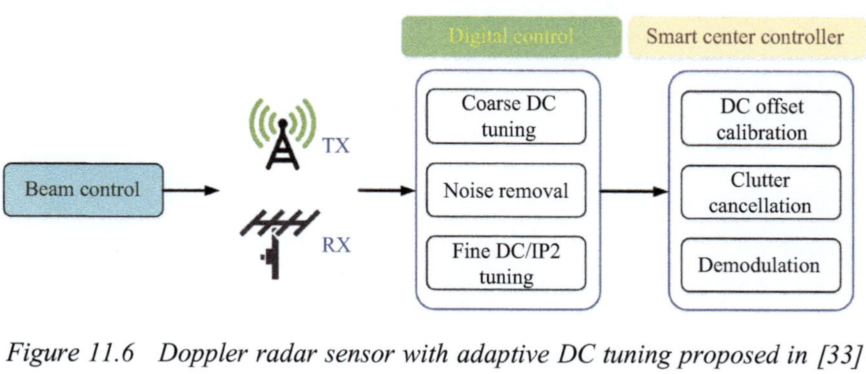

Figure 11.6 Doppler radar sensor with adaptive DC tuning proposed in [33]

the received signals, and the limited resolution of the phase shifter and the attenuator, the RF tuning cannot completely remove all the DC offsets. To further eliminate the remaining DC offsets, a baseband fine-tuning block was implemented to dynamically adjust the amplifier bias to the desired level that allows the maximum dynamic range. With the above DC tuning realized by a smart center in real time, the radar will precisely measure cardiac motion pattern. The integration of the DC-tuning technique into portable devices will be addressed with the help of logic control circuits coordinated by the I2C bus and complementary metal–oxide–semiconductor (CMOS)-integrated calibration Digital-to-analog convertor (DAC).

Besides manipulating the penetration depth, radar carrier frequency also determines the modulation sensitivity. Therefore, they carried out experiments to compare the performance of carrier frequencies ranging from 2.4 to 40 GHz. It should be noted that increasing the carrier frequency beyond 40 GHz may not help because as the wavelength approaches physiological motion amplitude, strong nonlinear phase modulation will generate harmonic interference [48]. Moreover, cardiac sensing should be realized from different angles to obtain sufficient information for biometrics applications. Also, multiple radars around a subject may "probe" cardiac signals simultaneously. To achieve this, they believed that a radar can configure the radiation beam to precisely point at the location of interest. As shown in Figure 11.6, digital beam control was implemented on the radar front end. Conventional beamforming systems directly adjust the phase and amplitude of the signal of each element antenna. They demonstrated that it is much more convenient to simultaneously adjust the phase and amplitude in the complex domain than to adjust them separately. For a complex signal $x = \exp(-j2\pi ft)$ sent into each element antenna (where f is the signal frequency), a vector multiplier was used to realize phase and amplitude modulation by first splitting the signal into in- and out-of-phase components and then by multiplying each one using a variable gain amplifier. Finally, by adding the amplified in- and out-of-phase components together, complex modulation to the original signal can be achieved thus effectively realizing radar beam control. In their work, they used a laser pointer to indicate the beam direction by aligning the radar beam with the user.

Afterward, they elaborated on the radio signal processing schemes and correspondingly investigated user authentication methods to achieve secure and usable authentication results. As depicted in Figure 11.7, their proposed approach was

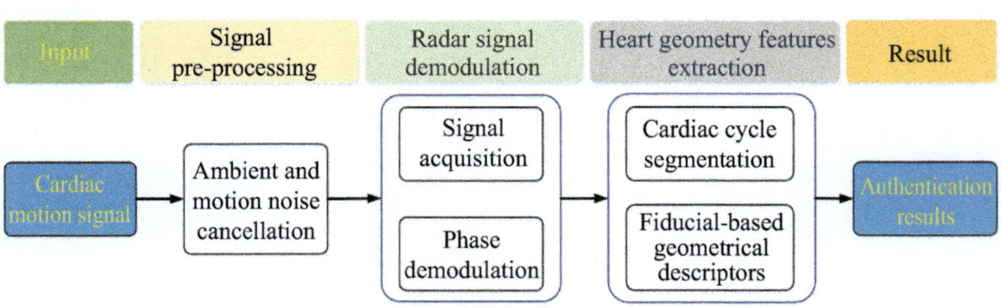

Figure 11.7 The flowchart of Cardiac Scan, a heart-biometric-based continuous user authentication system [33]

mainly comprised of three modules. First, the original sequential signal was preprocessed for noise reduction. The noise includes low-band components (e.g., baseline wander), high-band components (e.g., power-line interference), and unpredictable-band components (e.g., arbitrary motion in the scene). Considering diverse and known frequency bands of the noise spectrum, we have addressed the noise-level reduction in two areas: one-pass noise-reduction techniques (e.g., a Butterworth BPF) and adaptive noise-canceling techniques. Second, they performed de-noising-aware radar signal demodulation by investigating an extended DACM algorithm to avoid the phase unwrapping problem. Their proposed algorithm computes a derivative to the arctangent-demodulated phase information and reconstructs the desired phase information, which represents cardiac motion. Third, they extracted fiducial-based descriptors from the periodical signal segments. The fiducial-based method extracts intrinsic geometrical descriptors (e.g., temporal, amplitude, area, or angle) from fiducial points in the cardiac motion signal. Specifically, fiducial points are the biomarkers with physical meanings in clinics during the cardiac motion cycle. Fiducial points contain the biological information that is unique and nonvolatile for individuals, and are also independent of the sensor location or state of the individual such as anxiety, nervousness, or excitement. Lastly, they obtained authentication results. Note that the existing heart-based biometrics, such as electrocardiography (ECG), is recording the electrical activity of the heart, whose descriptors are extracted on the basis of the QRS complex [49]. As a new biometric modality, their noncontact cardiac motion is substantially different from the typical ECG signal in that it is a direct heart motion activity measured by an RF sensor.

To elaborate the system usage, they devised three scenarios in particular for Cardiac Scan enabled continuous authentication, including *Authentic user is present*, *Authentic user leaves*, and *Adversary is present*.

- *Authentic user is present:* When an authentic user has logged into the system and is present within the range of the radar sensor, Cardiac Scan is able to detect sensed cardiac motions are from the same person who was initially authorized. Thus, the permission of using the system for the user can be continuously granted without any interruption, unless the user logs off intentionally or leaves, as shown in Figure 11.8(a). By designing the false negative tolerance, Cardiac Scan allows one single "classified as adversary" event given that the classification results just before and after this event are both positive as "classified as authentic user." In case two or more than two consecutive "classified as adversary" events occur, though it has a low probability, Cardiac Scan will log out the initial user. Under such circumstance, the user is asked to confirm its identity again using other complementary existing biometrics system, such as PIN, fingerprint, or face.
- *Authentic user leaves:* When the authentic user is away from the system and the radar sensor has detected the user's absence, as shown in Figure 11.8(b), Cardiac Scan will first check whether the user has logged off and the system has been locked up. If so, Cardiac Scan will classify the user's absence as a legitimate action and no further action needs to be taken. Otherwise, the

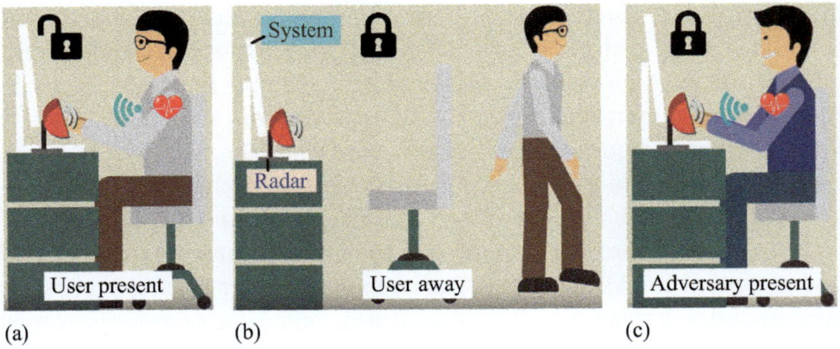

Figure 11.8 *Continuous authentication scenarios: (a) Authentic user is present and the system remains unlocked. (b) Authentic user leaves and the system locks up. (c) Adversary is present and the system locks up. Green screen means the system is unlocked and red screen means the system is locked [33]*

system is at risk of unauthorized access, hence necessary actions such as locking the session, logging out the original user, or notifying the administrator [50], which depend on the system policy, have to be considered to address the security risks.

- *Adversary is present:* In this scenario, an unauthorized adversary (the dark one in Figure 11.8(c)) is present and close to the system, and the system has been logged in initially by an authentic user. This can happen when the authentic use is forced to be present or the adversary takes over the system before the system automatically locks up when the authentic user leaves. Therefore, immediate action is demanded to keep the adversary outside the system and prevent the leakage of sensitive information. In this case, Cardiac Scan will immediately log out the initial user and lock up the system once the false negative tolerance threshold Th_{nt} is exceeded.

In the experiment, they conducted a pilot study to prove identifiability in cardiac motion. A Doppler radar vital sign detection system was developed for the study. As shown in Figure 11.9, a subject sat in a chair in a relaxed condition. The customized Doppler radar sensor was placed in front of the subject with a distance of 1 m and the sampling frequency was 20 Hz. An iPhone was placed close to a radar to record the subject identity and label the ground truth. Motion compensation was carried out for the baseband complex signal obtained from subjects who breathed normally but randomly moved their body. A pulsed sensor (UFI 1010 pulse transducer) was attached to the subject's finger to provide a heartbeat reference, and a chest belt (UFI 1132 piezo-electric respiration transducer) was used to provide a respiration reference. Seventy-eight healthy subjects (46 males and 32 females) without heart disease participated in the study. Each subject has 20 trials, and each trial lasts 8 s including 8 to 10 cardiac cycles. In total, there are 14,886 cardiac cycle samples in the evaluation.

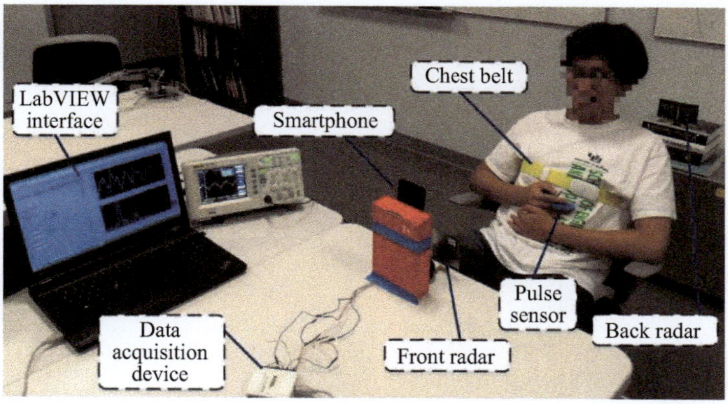

Figure 11.9 Experimental setup for cardiac motion sensing. A subject is sitting one meter away from both radar sensors, a chest belt and a pulsed sensor is attached to the subject [33]

They first studied the effectiveness of random body movement compensation because the subject will perform free movements in the practical continuous monitoring system. Compared with cardiac motion, body movement may result in a large perturbation to the output DC offset, and thus confuse the radar demodulation algorithm or even saturate the baseband circuit. In the experiment, the time-domain signal had fluctuations due to the random body motion as shown in Figure 11.10(a). This phenomenon was first observed in [51] and the authors observed the strong near-DC spectral components and found that the heartbeat was invisible in the spectrum as shown in Figure 11.10(b). Simply reducing the front-end gain, as adopted in some communication systems, does not work because the radar will lose the sensitivity to the weak cardiac motion signal.

Because biomedical radar can detect cardiac motion from four sides of a human body, multiple radars can be installed at different locations around the human body to cancel out random body motion based on the different patterns of body motion and cardiac motion [45]. In the view of the two radars, the heartbeat-and-respiration-caused body movements are in phase, while the random body movements are out of phase. For example, in case of two radars detecting from the front and the back of the body, when the body is drifting toward one radar, it is moving away from the other; whereas heartbeat presents similar expansion/contraction patterns to the two radars [60]. Therefore, random body motion creates an opposite Doppler frequency shift to the signals of the radars, while cardiac motion leads to the same polarity. By properly combining the low-speed baseband signals from the radars, one type of motion can be canceled and the other type will be enhanced. Figure 11.10(c) shows the results after random body motion cancellation, where not only are the near-DC interferences suppressed but also the heartbeat signal is clearly visible based on the proposed solution. Moreover, respiration signal can also be identified within a range of frequencies which is the most prominent peak than the other peaks in the spectrum.

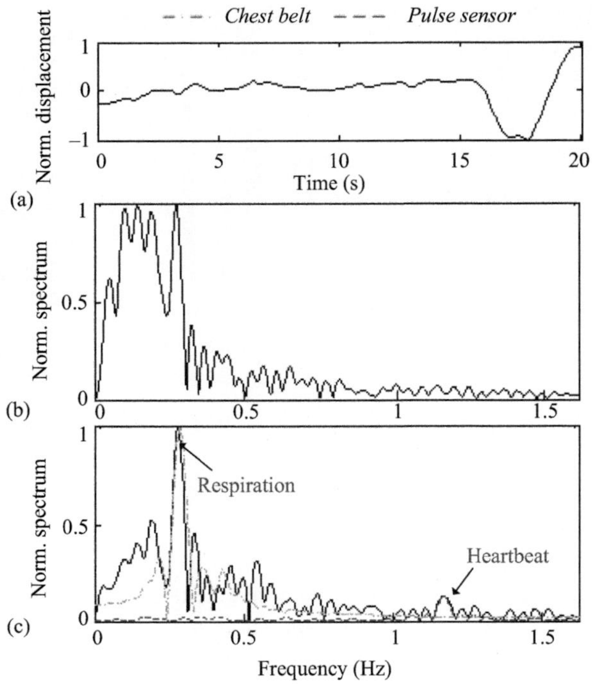

Figure 11.10 (a) Time-domain signal of 20 s under random body motion; (b) Spectrum without random body motion cancellation; (c) Spectrum with the random body motion cancellation [51]

They also emphasized that the baseband output of each radar has very low frequencies because human physiological motion is very slow (no more than a few hertzs). In the meantime, the "coherence detection" feature of the radar is maintained at the front end of each individual radar to achieve a sub-millimeter-scale detection sensitivity, which is not affected by the synchronization among different radars. Therefore, for practical noncontact cardiac password applications, either wired or wireless secure data communication can transfer the data to a central computing unit signal processing.

Afterward, they verified the validity of the collected data from their system. When the radar sensor detects the cardiac motion, the fingertip sensor simultaneously collects a signal as the ground truth signal. Both the radar sensor and fingertip pulse sensor were sampled at 20 Hz. They observed that the cardiac motion circles are in similar shape with the corresponding fingertip signal which is served as the ground truth. The detected cardiac motion patterns appeared periodically, each of which precisely matches the peaks in the fingertip signal. To this end, they verified that Cardiac Scan could accurately detect the cardiac motion signal in a noncontact way. Then, they evaluated the performance, usability, and vulnerabilities in practice of Cardiac Scan. Here, we'd like to introduce the vulnerability study to demonstrate the robustness of their radar sensing system for authentication.

Investigating the vulnerability of Cardiac Scan is crucial. Although cardiac motion is invisible and might possess better safety and security than other authentication approaches (e.g., PIN and fingerprint), it could become fallible under direct or spoofing attacks [52]. One immediate attack approach is the presentation of human characteristics to the acquisition device, including different living traits (i.e., zero-effort impostor attempts that try to take advantage of the false acceptance rate of biometric systems) [53]. Specifically, they discussed one advanced spoofing approach, i.e., replay attack.

Replay attack

One major risk of using biometrics is the danger that the biometric token can be intercepted and replayed by an unauthorized party. Compared to visual-based still biometrics (face/fingerprint/iris), the cardiac motion is more complex and dynamic to fake or replicate. However, there is still a chance to compromise Cardiac Scan under some extreme scenarios. For example, attackers might hack into the database and obtain cardiac motion patterns or engineer the same cardiac motion sensing device to extract a user's cardiac motion. The authors conducted this experiment to prove the possibility of a replay attack on Cardiac Scan if a legitimate user's cardiac signals are obtained by attackers. The team investigated the method of synthesizing cardiac motion and developed a programmable actuator to imitate the cardiac motion. As shown in Figure 11.11, a linear actuator (ZABER TNA08A50) and a linear translational stage (ZABER TSB28-1) were placed 30 cm from the cardiac motion-sensing device. The actuator was programmed to perform a harmonic back-and-forth motion toward the fixed position radar for mimicking cardiac motion patterns.

Anti-spoofing (liveness artifacts)

Given the above attack, the team also investigated a set of anti-spoofing approaches against a replay attack. The general idea of anti-spoofing is liveness detection [54,55]. Liveness detection has been applied to existing biometrics systems by using living traits of humans. Pan *et al.* proposed the method to extract liveness information through eye blinks in face recognition [56]. Wei *et al.* detected

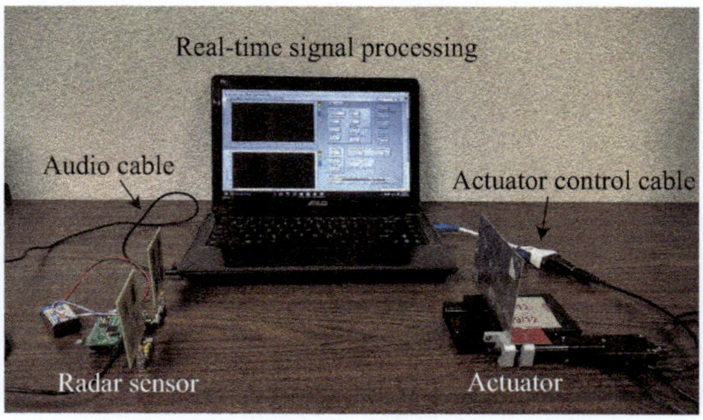

Figure 11.11 A linear actuator imitates cardiac motion as a replay attack [33]

counterfeit iris through texture analysis [57]. In their work, they exploited the uniqueness of living traits in human cardiac motion to defend the above adversarial model. Specifically, they tackled the challenge from two dimensions: hardware-based and software-based approaches. First, they integrated assisted sensors in Cardiac Scan, so that they can leverage additional information from these sensors to examine the legitimacy of subjects and capture the characteristics of multi-dimensional cardiac motions for liveness simultaneously. Specifically, the system employed multichannel radars for noise reduction. Since the linear actuator only moves in rectilinear directions, the direction of arrival (DoA) [58] measurements with the linear actuator on these radars are different from DoA measured with real cardiac motion. Second, we have investigated software-based approaches. Based on the experiment, the sensor data from a live subject inevitably include vital sign (e.g., respiration) and other motion artifacts (e.g., body sway). Utilizing these vital sign detection and motion artifacts, liveness detection is conducted against the replay attack [59]. They programmed the actuator working with different moving amplitudes and frequencies to imitate cardiac motions of 12 subjects. All replay attacks were successfully rejected by the proposed liveness detection method.

Overall, their group prototyped a novel biomedical radar system, Cardiac Scan, featured with DC-coupled interferometry radar and Doppler radar with digital-IF architecture. Such a design can avoid frequency-select signal distortion and thus accurately recover the internal cardiac motion via remote sensing. Cardiac Scan represents a new security mechanism which significantly enhances the security level of the system by continuously monitoring the user's trait during the entire login session without compromising the usability. It holds the potential to transform existing authentication systems into a more undeceivable, disclosure-resistant, and user-friendly solution. Moreover, Cardiac Scan can be conveniently integrated with existing one-pass user verification techniques (e.g., personal identification number, fingerprint, iris scan, and face recognition) to enhance the continuous authentication capability of existing systems.

References

[1] P. Rudol and P. Doherty, "Human body detection and geolocalization for UAV search and rescue missions using color and thermal imagery," in *Aerospace Conference, 2008 IEEE*. IEEE, 2008, pp. 1–8.

[2] C. Premebida, O. Ludwig, and U. Nunes, "Lidar and vision-based pedestrian detection system," *Journal of Field Robotics*, vol. 26, no. 9, pp. 696–711, 2009.

[3] D. Geronimo, A. M. Lopez, A. D. Sappa, and T. Graf, "Survey of Pedestrian Detection for Advanced Driver Assistance Systems," in *IEEE Transactions on Pattern Analysis and Machine Intelligence*, vol. 32, no. 7, pp. 1239–1258, 2010.

[4] H. Rohling, S. Heuel and H. Ritter, "Pedestrian detection procedure integrated into an 24 GHz automotive radar," in *2010 IEEE Radar Conference*, IEEE, 2010, pp. 1229–1232.

[5] R. Zetik, S. Crabbe, J. Krajnak, P. Peyerl, J. Sachs, and R. Thomä, "Detection and localization of persons behind obstacles using m-sequence through-the-wall radar," in *Sensors, and Command, Control, Communications, and Intelligence (C3I) Technologies for Homeland Security and Homeland Defense V*, vol. 6201. International Society for Optics and Photonics, 2006, p. 62010I.

[6] R. D. Vorona, M. P. Winn, T. W. Babineau, B. P. Eng, H. R. Feldman, and J. C. Ware, "Overweight and obese patients in a primary care population report less sleep than patients with a normal body mass index," *Archives of internal medicine*, vol. 165, no. 1, pp. 25–30, 2005.

[7] K. Spiegel, K. Knutson, R. Leproult, E. Tasali, and E. V. Cauter, "Sleep loss: a novel risk factor for insulin resistance and type 2 diabetes," *Journal of applied physiology*, vol. 99, no. 5, pp. 2008–2019, 2005.

[8] M. J. Sateia, K. Doghramji, P. J. Hauri, and C. M. Morin, "Evaluation of chronic insomnia. An American Academy of Sleep Medicine review," *Sleep*, vol. 23, no. 2, pp. 243–308, 2000.

[9] M. Rofouei, M. Sinclair, R. Bittner, *et al.*, "A non-invasive wearable neck-cuff system for real-time sleep monitoring," in *Body Sensor Networks (BSN), 2011 International Conference on*. IEEE, 2011, pp. 156–161.

[10] C.-W. Wang, A. Hunter, N. Gravill, and S. Matusiewicz, "Unconstrained video monitoring of breathing behavior and application to diagnosis of sleep apnea," *IEEE Transactions on Biomedical Engineering*, vol. 61, no. 2, pp. 396–404, 2014.

[11] T. Hao, G. Xing, and G. Zhou, "iSleep: unobtrusive sleep quality monitoring using smartphones," in *Proceedings of the 11th ACM Conference on Embedded Networked Sensor Systems*. ACM, 2013, p. 4.

[12] Y. Zhuang, C. Song, A. Wang, *et al.*, "SleepSense: non-invasive sleep event recognition using an electromagnetic probe," in *Wearable and Implantable Body Sensor Networks (BSN), 2015 IEEE 12th International Conference on*. IEEE, 2015, pp. 1–6.

[13] F. Lin, Y. Zhuang, C. Song, *et al.*, "SleepSense: a noncontact and cost-effective sleep monitoring system," *IEEE Transactions on Biomedical Circuits and Systems*, vol. 11, no. 1, pp. 189–202, 2017.

[14] M. I. Skolnik, *Radar Handbook*, New York, McGraw-Hill Book Co., 1970.

[15] M. Mercuri, P. J. Soh, G. Pandey, *et al.*, "Analysis of an indoor biomedical radar-based system for health monitoring," *IEEE Transactions on Microwave Theory and Techniques*, vol. 61, no. 5, pp. 2061–2068, 2013.

[16] E. Kańtoch, J. Jaworek, and P. Augustyniak, "Design of a wearable sensor network for home monitoring system," in *Computer Science and Information Systems (FedCSIS), 2011 Federated Conference on*. IEEE, 2011, pp. 401–403.

[17] V. M. Lubecke and O. Boric-Lubecke, "Wireless technologies in sleep monitoring," in *Radio and Wireless Symposium, 2009. RWS'09. IEEE*. IEEE, 2009, pp. 135–138.

[18] M. Yu, S. M. Naqvi, and J. Chambers, "A robust fall detection system for the elderly in a smart room," in *Acoustics Speech and Signal Processing (ICASSP), 2010 IEEE International Conference on*. IEEE, 2010, pp. 1666–1669.

[19] Y. Zigel, D. Litvak, and I. Gannot, "A method for automatic fall detection of elderly people using floor vibrations and sound—proof of concept on human mimicking doll falls," *IEEE Transactions on Biomedical Engineering*, vol. 56, no. 12, pp. 2858–2867, 2009.

[20] Y. Li, Z. Zeng, M. Popescu, and K. Ho, "Acoustic fall detection using a circular microphone array," in *Engineering in Medicine and Biology Society (EMBC), 2010 Annual International Conference of the IEEE*. IEEE, 2010, pp. 2242–2245.

[21] M. Mercuri, D. Schreurs, and P. Leroux, "Optimised waveform design for radar sensor aimed at contactless health monitoring," *Electronics letters*, vol. 48, no. 20, pp. 1255–1257, 2012.

[22] B. Hill, "Warning: 180 million smartphone users just hacked," 2017. [Online]. Available: https://banyanhill.com/warning-smartphone-users-hacked/

[23] R. Legal, "Top data breaches of 2018: hackers find new methods," 2018. [Online]. Available: https://revisionlegal.com/data-breach/2018-statistics/

[24] F. Paul, "Connected cars have an 'indefensible' security vulnerability," 2017. [Online]. Available: https://www.networkworld.com/article/3218712/internet-of-things/connected-cars-have-an-indefensible-security-vulnerability.html

[25] M. Jakobsson, E. Shi, P. Golle, and R. Chow, "Implicit authentication for mobile devices," in *Proceedings of the 4th USENIX conference on Hot topics in security*, 2009, pp. 9–9.

[26] P. Bigioi, E. Steinberg, and P. Corcoran, "Face recognition training method and apparatus," Jan. 29 2013, US Patent 8,363,952.

[27] S. Sanderson and J. Erbetta, "Authentication for secure environments based on iris scanning technology," *IEE Colloquium on Visual Biometrics*, 2000, p. 8.

[28] A. C. Weaver, "Biometric authentication," *IEEE Computer*, vol. 39, no. 2, pp. 96–97, 2006.

[29] S. Zahid, M. Shahzad, S. A. Khayam, and M. Farooq, "Keystroke-based user identification on smart phones," in *International Workshop on Recent Advances in Intrusion Detection*. Springer, Berlin, Heidelberg, 2009, pp. 224–243.

[30] M. O. Derawi, C. Nickel, P. Bours, and C. Busch, "Unobtrusive user-authentication on mobile phones using biometric gait recognition," in *Intelligent Information Hiding and Multimedia Signal Processing (IIH-MSP), 2010 Sixth International Conference on*. IEEE, 2010, pp. 306–311.

[31] G. Bailador, C. Sanchez-Avila, J. Guerra-Casanova, and A. de Santos Sierra, "Analysis of pattern recognition techniques for in-air signature biometrics," *Pattern Recognition*, vol. 44, no. 10–11, pp. 2468–2478, 2011.

[32] D. Sasakawa, N. Honma, T. Nakayama, and S. Iizuka, "Human identification using MIMO array," *IEEE Sensors Journal*, vol. 18, no. 8, pp. 3183–3189, 2018.

[33] F. Lin, C. Song, Y. Zhuang, W. Xu, C. Li, and K. Ren, "Cardiac scan: a non-contact and continuous heart-based user authentication system," in

Proceedings of the 23rd Annual International Conference on Mobile Computing and Networking. ACM, 2017, pp. 315–328.

[34] J. E. Hall, *Guyton and Hall Textbook of Medical Physiology* e-book: with Student Consult Online Access. Elsevier Health Sciences; 2010 Jul 19.

[35] J. C. Lin, "Microwave sensing of physiological movement and volume change: a review," *Bioelectromagnetics*, vol. 13, no. 6, pp. 557–565, 1992.

[36] K. Chen, Y. Huang, J. Zhang, and A. Norman, "Microwave life-detection systems for searching human subjects under earthquake rubble or behind barrier," *Biomedical Engineering, IEEE Transactions on*, vol. 47, no. 1, pp. 105–114, 2000.

[37] C. Li and J. Lin, *Microwave Noncontact Motion Sensing and Analysis*. New Jersey: John Wiley & Sons, 2013.

[38] B. Lohman, O. Boric-Lubecke, V. Lubecke, P. Ong, and M. Sondhi, "A digital signal processor for Doppler radar sensing of vital signs," *Engineering in Medicine and Biology Magazine, IEEE*, vol. 21, no. 5, pp. 161–164, 2002.

[39] A. D. Droitcour, O. Boric-Lubecke, V. M. Lubecke, J. Lin, and G. T. Kovacs, "Range correlation and I/Q performance benefits in single-chip silicon Doppler radars for noncontact cardiopulmonary monitoring," *Microwave Theory and Techniques, IEEE Transactions on*, vol. 52, no. 3, pp. 838–848, 2004.

[40] L. Scalise, I. Ercoli, P. Marchionni, and E. P. Tomasini, "Measurement of respiration rate in preterm infants by laser Doppler vibrometry," in *Medical Measurements and Applications Proceedings (MeMeA), 2011 IEEE International Workshop on*. IEEE, 2011, pp. 657–661.

[41] M. Martinez and R. Stiefelhagen, "Breath rate monitoring during sleep using near-IR imagery and PCA," in *Pattern Recognition (ICPR), 2012 21st International Conference on*. IEEE, 2012, pp. 3472–3475.

[42] F. Pfanner, J. Maier, T. Allmendinger, T. Flohr, and M. Kachelrieß, "Monitoring internal organ motion with continuous wave radar in CT," *Medical physics*, vol. 40, no. 9, p. 091915, 2013.

[43] D. Rissacher and D. Galy, "Cardiac radar for biometric identification using nearest neighbour of continuous wavelet transform peaks," in *Identity, Security and Behavior Analysis (ISBA), 2015 IEEE International Conference on*. IEEE, 2015, pp. 1–6.

[44] Y. Zhuang, C. Song, F. Lin, Y. Li, C. Li, and W. Xu, "On the feasibility of non-contact cardiac motion sensing for emerging heart-based biometrics," in *Radio and Wireless Symposium (RWS), 2016 IEEE*. IEEE, 2016, pp. 204–206.

[45] C. Li and J. Lin, "Random body movement cancellation in Doppler radar vital sign detection," *Microwave Theory and Techniques, IEEE Transactions on*, vol. 56, no. 12, pp. 3143–3152, 2008.

[46] F.-K. Wang, T.-S. Horng, K.-C. Peng, J.-K. Jau, J.-Y. Li, and C.-C. Chen, "Single-antenna Doppler radars using self and mutual injection locking for vital sign detection with random body movement cancellation," *Microwave Theory and Techniques, IEEE Transactions on*, vol. 59, no. 12, pp. 3577–3587, 2011.

[47] J. E. Kiriazi, O. Boric-Lubecke, and V. M. Lubecke, "Dual-frequency technique for assessment of cardiopulmonary effective RCS and displacement," *Sensors Journal, IEEE*, vol. 12, no. 3, pp. 574–582, 2012.

[48] C. Li and J. Lin, "Optimal carrier frequency of non-contact vital sign detectors," in *Radio and Wireless Symposium, 2007 IEEE*. IEEE, 2007, pp. 281–284.

[49] L. Biel, O. Pettersson, L. Philipson, and P. Wide, "ECG analysis: a new approach in human identification,"*Instrumentation and Measurement, IEEE Transactions on*, vol. 50, no. 3, pp. 808–812, 2001.

[50] K. B. Rasmussen, M. Roeschlin, I. Martinovic, and G. Tsudik, "Authentication using pulse-response biometrics," in *21st Annual Network and Distributed System Security Symposium, NDSS 2014, San Diego, California, USA, February 23–26, 2013*, 2014.

[51] C. Gu, G. Wang, Y. Li, T. Inoue, and C. Li, "A hybrid radar-camera sensing system with phase compensation for random body movement cancellation in Doppler vital sign detection," *IEEE Transactions on Microwave Theory and Techniques*, vol. 61, no. 12, pp. 4678–4688, 2013.

[52] S. Marcel, M. S. Nixon, and S. Z. Li, *Handbook of Biometric Anti-Spoofing*. Vol. 1. New York: Springer, 2014.

[53] P. Johnson, R. Lazarick, E. Marasco, E. Newton, A. Ross, and S. Schuckers, "Biometric liveness detection: framework and metrics," in *International Biometric Performance Conference*, vol. 1, 2012.

[54] X. Tan, Y. Li, J. Liu, and L. Jiang, "Face liveness detection from a single image with sparse low rank bilinear discriminative model," in *European Conference on Computer Vision*. Springer, Berlin, Heidelberg, 2010, pp. 504–517.

[55] G. L. Marcialis, A. Lewicke, B. Tan, *et al.*, "First international fingerprint liveness detection competition—LivDet 2009," in *International Conference on Image Analysis and Processing*. Springer, Berlin, Heidelberg, 2009, pp. 12–23.

[56] G. Pan, L. Sun, Z. Wu, and S. Lao, "Eyeblink-based anti-spoofing in face recognition from a generic webcamera," in *2007 IEEE 11th International Conference on Computer Vision*. IEEE, 2007, pp. 1–8.

[57] Z. Wei, X. Qiu, Z. Sun, and T. Tan, "Counterfeit iris detection based on texture analysis," in *Pattern Recognition, 2008. ICPR 2008. 19th International Conference on*. IEEE, 2008, pp. 1–4.

[58] A. Lin and H. Ling, "Doppler and direction-of-arrival (DDOA) radar for multiple-mover sensing," *IEEE Transactions on Aerospace and Electronic Systems*, vol. 43, no. 4, pp. 1496–1509, 2007.

[59] C. Gu and C. Li, "Assessment of human respiration patterns via noncontact sensing using Doppler multi-radar system," *Sensors*, vol. 15, no. 3, pp. 6383–6398, 2015.

[60] C. Li, and J. Lin, "Complex signal demodulation and random body movement cancellation techniques for non-contact vital sign detection.", *2008 IEEE MTT-S International Microwave Symposium Digest* (pp. 567–570). IEEE, 2008.

Chapter 12

Micro-Doppler signatures for sensing micro-motion

Victor C. Chen[1,2] *and William J. Miceli*[1,3]

Micro-Doppler signature is a distinctive characteristic of micro-motion represented with intricate time-varying Doppler frequency modulations. It can be used to extract motion kinematic features for identifying an object of interest.

In this chapter, we introduce the micro-Doppler signature in radar and its applications for sensing micro-motion.

12.1 An introduction to micro-motion and micro-Doppler effect

Recent psychophysical investigations of motion perception and activity recognition based on the human visual system motivate similar studies based on micro-Doppler kinematics in radar.

The approach of the former is to present the visual system with a dynamic spatial display consisting of a small number of point markers representing an object engaged in some activity. The markers are positioned at the major joints of the object being represented, and typical activities being depicted include walking, running, crawling, swimming, riding a bicycle, gesturing, etc.

When there is no motion, the human visual system registers a static configuration of point markers and cannot discern who or what is being represented nor the activity in progress. However, when a dynamic display of these point markers is viewed at translated and/or rotated positions, the brain is able to perceive who or what is being represented and usually can recognize what activity is being depicted.

The point markers are frequently implemented as point lights on a black (or inversely bright points on a white) background that masks the moving object. Experimental research employs this strategy: the observer sees only the moving point light markers, yet can interpret the underlying identity and can usually

[1]Ancortek Inc., USA
[2]Retiree, Radar Division, US Naval Research Laboratory, USA
[3]Retiree, US Office of Naval Research, USA

recognize the activity of the subject, e.g., running, walking, swimming, crawling, boxing, and juggling.

The connection between psychophysical vision studies and radar-based micro-Doppler is the following: the former is based on the human visual system that forms angle–angle imagery of a dynamic pattern created by point marker motion that the brain recognizes as some common activity. The latter is based on radar echoes (possibly with enhanced radar cross section) originating at the articulating joints of moving objects and presented as a range and micro-Doppler signature that upon viewing reveals the nature of the object and its associated activity.

Figure 12.1 illustrates the feasibility of recognizing that a human being is walking from a sequence of static images consisting of markers placed on the joints of a stick figure. Note that the "sticks" are shown to facilitate the perception of walking from static imagery; they are not necessary when viewing a dynamic display. Moreover, recent experimental psychophysical studies have determined that it is also feasible to determine gender and to recognize known individuals based on observations of bipedal motion, e.g., walking and running [1–8]. It was also determined that the recognition of motion is not based entirely on local motion, i.e., tracking the trajectories of markers placed on specific body joints, but also on the variation of aggregate shapes resulting from temporal changes in the spatial distribution of markers, i.e., the perception of motion is a spatiotemporal process. These studies also indicated the possibility of distinguishing different emotions based on extracted kinematic parameters, e.g., velocity, acceleration, and jerk.

The aim of this chapter is to establish that radar micro-Doppler signatures can provide similar information to the psychophysical vision if properly processed and interpreted.

The following sections will explain the phenomenology of radar micro-Doppler signatures, illustrate the equivalence of point scatterers (sometimes with enhanced radar cross section (RCS)) and point markers (lighted or not) used in psychophysical vision studies, and discuss the processing required to extract meaningful features from micro-Doppler signatures.

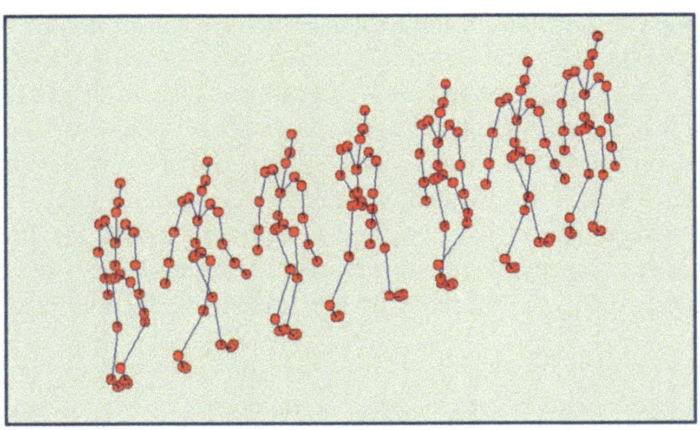

Figure 12.1 A walking person represented by a limited number of point displays

12.1.1 Micro-motion and micro-Doppler effect in radar

If an object or any structural component of the object has vibration, rotation, oscillation, or other small motions in addition to the bulk translation of the object, these small motions are called as the micro-motion. The source of micro-motion may be the swinging arms and legs of a walking human, gestures of human hand, flapping wings of a bird, vibrating surface of a vehicle, rotating rotor blades of a helicopter, or other causes. If an object or any structural component of the object has any micro-motion, the micro-motion can induce additional Doppler frequency modulation onto the Doppler frequency of the bulk translation. The additional Doppler frequency modulation is known as the micro-Doppler effect [9–12].

Due to the Doppler effect, the total Doppler frequency shift f_{D} of radar signal returned from a translational target is determined by the radar carrier frequency f_{c}, the speed of wave propagation c, and the radial velocity of the target $\boldsymbol{v}_{\mathbf{radial}}$:

$$f_D = -2\boldsymbol{v}_{\mathbf{radial}}\frac{f_c}{c} \tag{12.1}$$

If an object has a translational velocity $\boldsymbol{v}_{\mathbf{radial}}$ with respect to the radar and an angular rotation velocity ω, the Doppler frequency shift of a point scatterer of the object is approximately [12]:

$$f_D = \frac{2f_c}{c}[\boldsymbol{v}_{\mathbf{radial}} + \boldsymbol{\omega} \times \boldsymbol{r}] \cdot \boldsymbol{n} \tag{12.2}$$

where $\boldsymbol{n}$ is a unit vector of the radar line-of-sight (LOS) direction and $\boldsymbol{r}$ is the distance vector from the rotation center to the point scatterer of the object. The first term is the Doppler frequency shift due to the bulk translation:

$$f_{\mathrm{Trans}} = \frac{2f_c}{c}\boldsymbol{v}_{\mathbf{radial}} \cdot \boldsymbol{n} \tag{12.3}$$

and the second term is the micro-Doppler shift due to the rotation:

$$f_{mD} = \frac{2f_c}{c}[\boldsymbol{\omega} \times \boldsymbol{r}] \cdot \boldsymbol{n} \tag{12.4}$$

Usually, an object may consist of more than one point scatterers. Thus, the micro-Doppler term becomes the superposition of multiple micro-Doppler terms:

$$f_{mD} = \frac{2f_c}{c}\sum\nolimits_k [\boldsymbol{\omega}_k \times \boldsymbol{r}_k] \cdot \boldsymbol{n} \tag{12.5}$$

12.1.2 Micro-Doppler signatures

Micro-Doppler effect is characterized by its signature. Micro-Doppler signature is a distinctive characteristic of an object in any micro-motion and gives the moving object an identity [13,14]. The following examples are illustrations of micro-Doppler signatures generated by different objects in various micro-motions.

12.1.2.1 Micro-Doppler signatures of a walking human

An example of micro-Doppler signature of a walking human and its decomposed component signatures, called motion curves, are shown in Figure 12.2. The micro-Doppler signature is the superposition of multiple monocomponent signatures. Each of them is associated with an individual structural component of the walking person, such as the torso, the arm, the leg, the hand, and the foot.

For better performing target classification based on its micro-Doppler signature, the first step is to decompose the micro-Doppler signature into multiple monocomponent signatures. For example, for a micro-Doppler signature of a walking human, the decomposed monocomponent signatures must be associated with the torso, arms, legs, and feet of the human. Therefore, the classifier may be able to identify the human's actions, emotions, and even its gender information. Similarly, for a micro-Doppler signature of a hand gesture, the decomposed monocomponent signatures must be associated with the palm, thumb, index, middle, and other fingers.

Although a human observer can easily track a monocomponent signature from a multicomponent micro-Doppler signature, it is a challenge for a machine to track

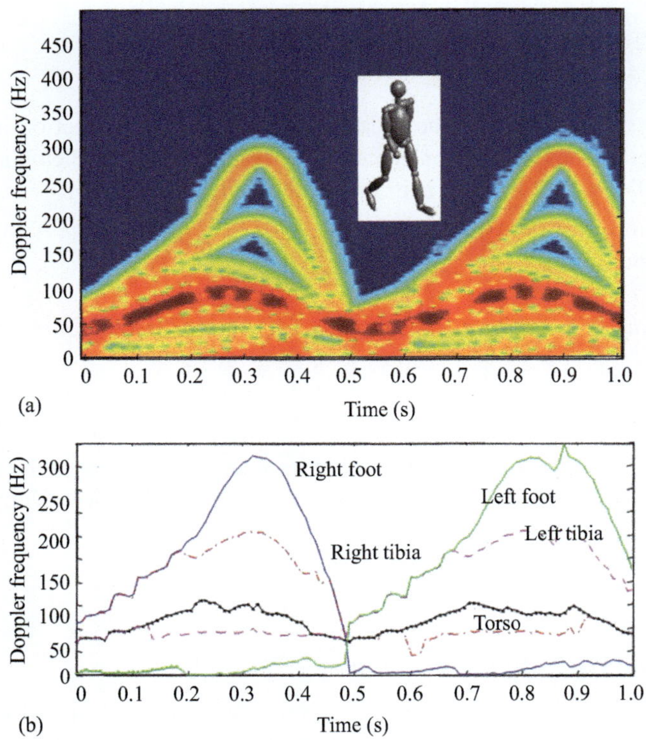

Figure 12.2 (a) Micro-Doppler signature of a walking person; (b) the decomposed motion curves from the micro-Doppler signature in (a)

and isolate any individual monocomponent signature from a multicomponent micro-Doppler signature. A recent method to track monocomponent is based on the Viterbi algorithm and time-frequency distribution. As a dynamic programming, the Viterbi algorithm can be used to estimate the motion curves in a joint time-frequency domain and has achieved some level of success [15–17].

12.1.2.2 Micro-Doppler signatures of swimmers

Compared to a human walking, the micro-motion of swimming has more choices. Basic types of swimming include backstroke, breaststroke, butterfly stroke, and freestyle. Human walk is a periodic motion with swinging arms and legs. Each cycle of walking consists of two phases: the stance phase and the swing phase. But each type of swimming may have more phases. For example, for the backstroke, each arm stroke cycle consists of entry/extension forward phase, downsweep phase, catch phase, upsweep phase, and recovery phase.

Figure 12.3(a), (b), and (c) shows micro-Doppler signatures of freestyle, butterfly stroke, and breaststroke swimmers, respectively. Different swimming styles have different micro-motions. The freestyle swim has downsweep, catch, insweep, and upsweep phases. The butterfly stroke has wave-like body movements and dolphin-kick-like legs' whipping movement. In the breaststroke swimming, the arm stroke phases are outsweep–catch–insweep.

12.1.2.3 Micro-Doppler signatures of hand gestures

The radial velocity is related to the Doppler frequency f_D by

$$\boldsymbol{v}_{\mathbf{radial}}(t) = -\frac{\lambda}{2} f_D(t) \tag{12.6}$$

where λ is the wavelength of the transmit signal. At each time instant, the micro-Doppler frequency shift is $f_D(t)$ and the radical velocity of the finger $\boldsymbol{v}_{\mathbf{radial}}(t)$ is related by a half wavelength $\lambda/2$. The sign of radial velocity indicates the moving direction, which relates to the type of gesture. The time interval of nonzero velocity represents the duration of the gesture. Based on the area covered by the radial velocity curve, the distance of the finger from the radar can also be estimated.

The micro-Doppler signature of the snap fingers in Figure 12.4 indicates that there is a higher negative Doppler peak and a lower positive Doppler peak located at the same time instant. The higher negative Doppler peak corresponds to flinging the middle finger onto the palm and the lower positive Doppler peak corresponds to flinging the thumb forward the radar. Then, the three fingers are resetting back to their initial positions. Thus, there is another low positive Doppler peak appeared as shown in the figure.

12.1.2.4 Micro-Doppler signatures of drones

Drones are usually small-sized and flying relatively slow and at a low altitude. The RCS of the main body of a drone is small and its rotor blade is even smaller (about 0.001 m^2). Birds are also flying relatively slow and at a low altitude that may cause false alarms when radar detecting drones.

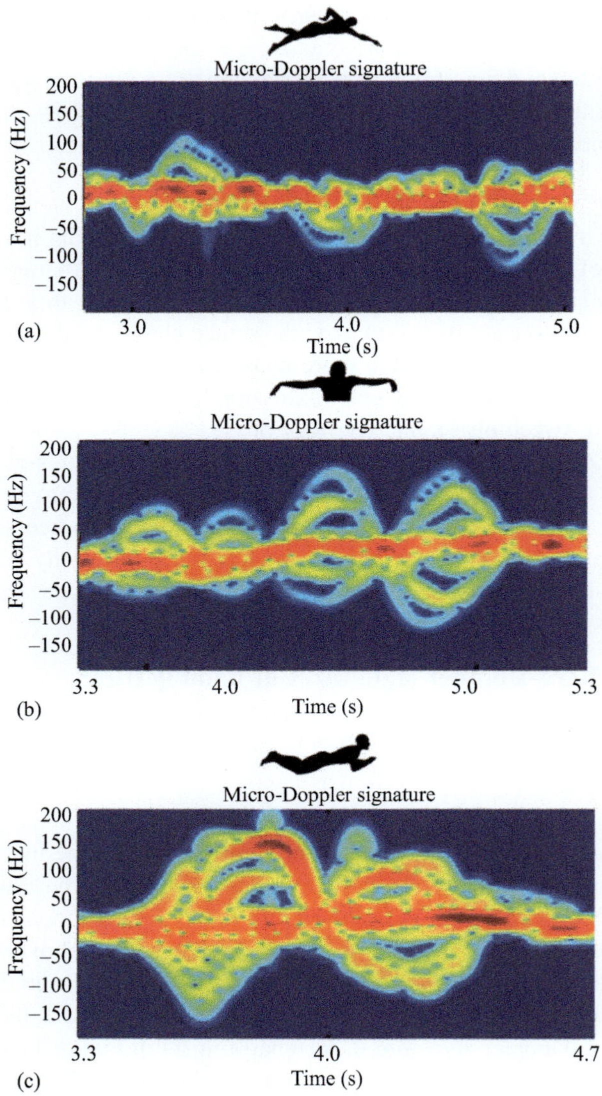

Figure 12.3 (a) Freestyle, (b) butterfly, and (c) breaststroke types of swimming and corresponding micro-Doppler signatures

The quadrotor can be modeled by a set of four rotors, where each rotor has N_B blades.

The radar returned signal from a quadcopter can be expressed by

$$s_{\sum}(t) = \sum_{j=1}^{N_R} \sum_{k=0}^{N_B-1} \sigma_{j,k}(t) exp\left\{ -j\frac{4\pi}{\lambda} R_j(t) \right\} exp\{-j\, \Phi_k(t)\} \tag{12.7}$$

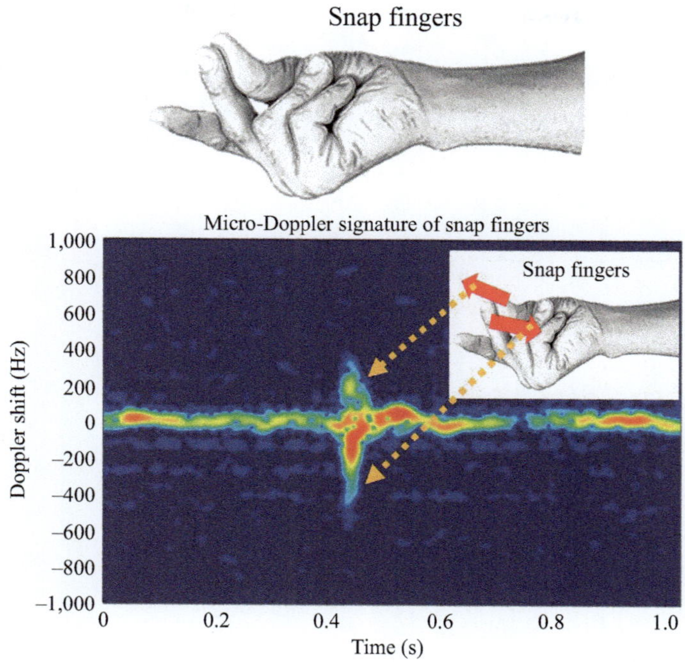

Figure 12.4 Interpretation of micro-Doppler signature of snap fingers

where the phase function $\Phi_k(t)$ is

$$\Phi_k(t) = \frac{4\pi}{\lambda} L\cos\beta\cos\left(\Omega_j t + \varphi_j + \frac{k2\pi}{N_B}\right), (j = 1, \ldots, N_R; k = 0, 1, \ldots, N_B - 1) \tag{12.8}$$

and $R_j(t)$ is the range from the radar to the center of the jth rotor, $\sigma_{j,k}(t)$ is the RCS of the kth blade on the jth rotor, L is the blade length, Ω_j is the rotation rate of the jth rotor, and φ_j is the initial rotating phase of the jth rotor.

When the drone is hovering, the rotation rate of rotors is about 100 rev/s. While at full power, it can be about 150 rev/s. Thus, for a C-band radar with a wavelength of 0.0517 m, if the rotation rate of the rotor is 150 rev/s, the length of each blade is 7.0 cm and its width is 2.5 cm, the tip velocity $V_{\text{tip}} = 2\pi L\Omega = 66$ m/s and the maximum Doppler shift $\{f_D\}_{\max}$ is 2.4 kHz. The micro-Doppler signature of the drone is shown in Figure 12.5.

From the micro-Doppler signature, a periodic pattern can be seen. The pattern is generated from the four rotating rotors with random initial rotation angles. Thus, the periodic pattern may vary, but the period is closely related to the rotation rate. The rotation rate of rotors can be estimated from the periodic pattern. As shown in the figure, the half period is 3.3 ms and, thus, the rotation rate is about 150 rev/s. Along with the maximum Doppler shift, these unique micro-Doppler features are useful for identifying a drone from flying birds.

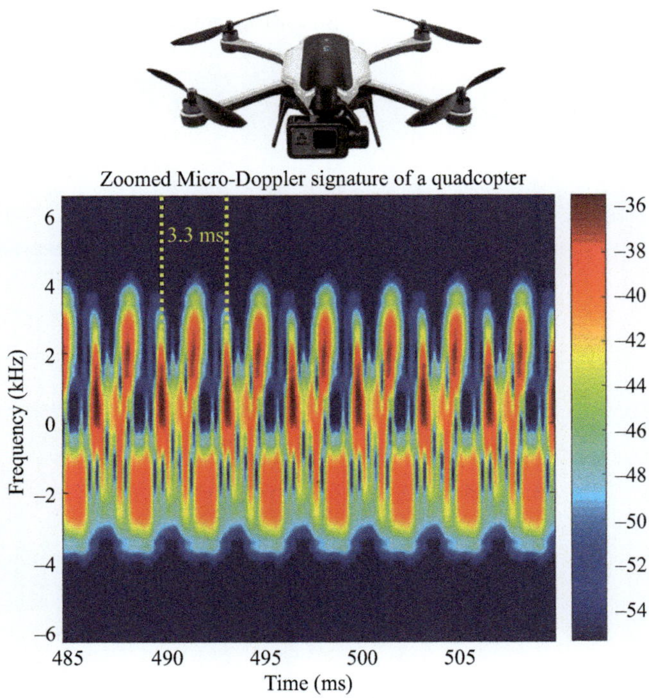

Figure 12.5 Micro-Doppler signature of a quadcopter

12.2 Angular velocity-induced micro-Doppler signatures

As well known, the micro-Doppler signature is induced by the radial velocity of moving objects. Without radial velocities, there are no Doppler and micro-Doppler shifts.

When an object is moving along a curved path as illustrated in Figure 12.6, the velocity of the object is determined by both the rate of change of the positional vector (i.e., the radial velocity) and the directional change (i.e., the angular velocity). For a curved path, the actual velocity of the object is the tangential velocity to the curve, known as the linear velocity. The linear velocity is determined by $v = r_c \Omega$, where r_c is the radius from the rotation center and Ω is the angular velocity of the object.

When a radar observes an object moving along a curved path, the actual linear velocity vector consists of two velocity vectors: the radial velocity vector v_r and the perpendicular velocity vector v_p. For a given linear velocity, while the radial velocity is decreased, the perpendicular velocity must be increased. The radial velocity is the radar LOS velocity, which can be easily measured by a Doppler radar. However, the perpendicular velocity v_p, as shown in Figure 12.6, is caused by the angular velocity, which cannot be directly measured by a monostatic Doppler radar.

A technique of measuring the angular velocity of an object was proposed in [18–20]. This technique uses an interferometric correlation receiver. The receiver

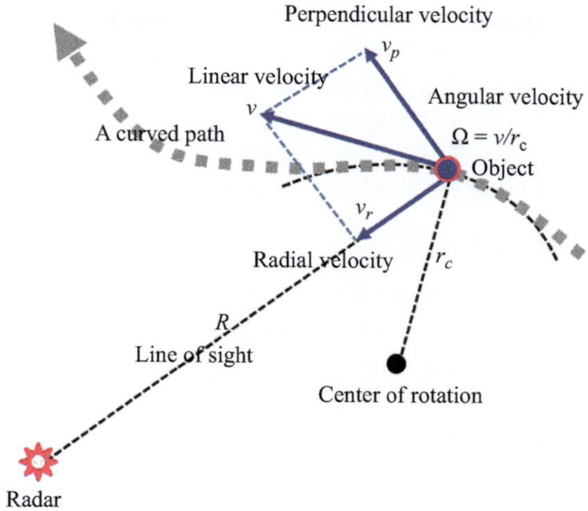

Figure 12.6 When an object is moving along a curved path, the object has both the radial velocity and the angular velocity

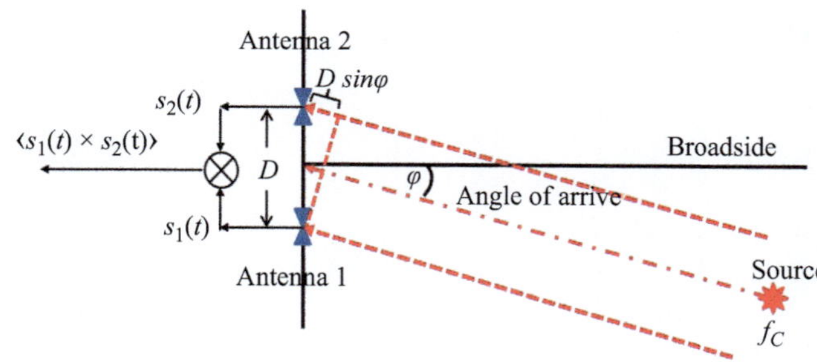

Figure 12.7 Interferometric radar receiver with two separated receiver channels

response frequency is proportional to the angular velocity. Actual measurement of the angular velocity is accomplished with an interferometric radar. The interferometric radar has two separated receiver channels with two antennas separated by a baseline D for observing a far-field source, as shown in Figure 12.7.

If the angle of an incidence wave from a far-field source is φ, the signal received at the first antenna is $s_1(t) = \exp\{j2\pi f_c t\}$ and that at the second antenna is $s_2(t) = \exp\{j2\pi f_c(t - \tau)\}$, where f_c is the source frequency and τ is the time delay compared to that at the first antenna determined by $\tau = D\sin\varphi/c$, where c is the wave propagation speed.

When the two received signals pass through a complex correlator, i.e., a complex multiplier and a low-pass filter, the output is the complex correlator response:

$$C(\varphi) = \langle s_1(t) \cdot s_2^*(t) \rangle = \exp\{j2\pi f_c \tau\} = \exp\{j2\pi f_c D \sin\varphi/c\} \tag{12.9}$$

The real part of the correlator output, $\Re e\{C(\varphi)\}$, forms a fringe pattern. It indicates that when the angle of the incidence wave φ increases from $-90°$ to $90°$, the correlator response is a sinusoidal type. The angular velocity of an object is defined by $\Omega = d\varphi(t)/dt$. Thus, the angle of the incidence wave can be replaced by Ωt.

The correlator output response can be rewritten by

$$C(\Omega t) = \exp\{j2\pi D \sin \Omega t/\lambda\} \tag{12.10}$$

When an object passes through the interferometer beam pattern, an oscillation occurs in the output response. The oscillation frequency is proportional to the angular velocity of the object.

The instantaneous frequency (IF) of the correlator output response is the angular velocity-induced frequency shift, which is called the interferometric frequency shift:

$$f_{IF} = \frac{1}{2\pi}\frac{d}{dt}\{2\pi D \sin \Omega t/\lambda\} = D\Omega \cos \Omega t/\lambda \tag{12.11}$$

For the near broadside of the antenna, cos Ωt or the near interferometric frequency shift becomes:

$$f_{IF} = D\Omega/\lambda \tag{12.12}$$

which has the similar form as the Doppler frequency shift:

$$f_D = 2v_r/\lambda \tag{12.13}$$

where v_r is the radial velocity.

The Doppler frequency shift f_D is proportional to the radial velocity. But the interferometric frequency shift f_{IF} is proportional to the angular velocity.

By combining the traditional Doppler measurement of radial velocity with the measurement of angular velocity using the interferometric receiver, the movement of objects can be measured directly, over a wide field of view. The radial Doppler shift can be measured by a single-channel receiver, and the angular Doppler shift uses the dual interferometric receivers. The interferometric radar provides a complementary way to Doppler radar in measurement of velocities.

Thus, the angular velocity-induced micro-Doppler modulation is a good complementary feature to the radial velocity-induced one. Especially for small drones, the angular micro-Doppler signature is always noticeable.

12.3 Feature extraction and motion decomposition from micro-Doppler signatures

To extract features from micro-Doppler signatures, an important procedure is to decompose or analyze micro-Doppler signatures. As mentioned in Section 12.1.2.1, a micro-Doppler signature is usually a multicomponent signature, which is a superposition of multiple monocomponent signatures. A monocomponent signature corresponds to the motion of a single component part of an object. Micro-Doppler signature of a monocomponent is a narrow-band time-varying frequency function. At each time, the function contains only a single-frequency component. Conversely, a multicomponent function has multiple isolated narrow frequency bands at the same time and is a superposition of multiple monocomponent functions. Thus, the purpose of decomposition is to decompose a multicomponent micro-Doppler signature into multiple addable monocomponent micro-Doppler signatures.

There are many decomposition methods, such as the kernel function method, the statistically independent decomposition method, and the empirical mode decomposition (EMD) method. The kernel function method decomposes a joint time-frequency distribution into independently controllable components. The statistically independent decomposition method, such as the principal component analysis (PCA), decomposes into uncorrelated components. The EMD method decomposes into the intrinsic mode functions (IMF). None of the abovementioned methods can decompose a multicomponent micro-Doppler signature into multiple monocomponent micro-Doppler signatures that are associated with physical component parts of an object, such as an arm, a foot, a leg, or a hand of a human.

12.3.1 Feature extraction from micro-Doppler signatures

Features that can be extracted from micro-Doppler signatures are all related to time and Doppler frequency parameters, such as time duration, Doppler frequency bandwidth, maximum Doppler frequency, and Doppler change rate.

The following example shows what exact features can be extracted from the micro-Doppler signature of *Rotating Rotor Blades*.

It is obvious that the Doppler patterns of the even number of blades and the odd number of blades are different. Even-number blades generate a symmetric Doppler pattern around the mean Doppler frequency, but odd-number blades generate an asymmetric pattern around it. From the micro-Doppler signature of rotating rotor blades, the number of blades, the length of blades, the rotation rate of the blades, and the speed of the tip can be estimated. These features are important for classification of an unknown helicopter.

The radar-returned signal from the rotor blades has short flashes when a blade has a specular reflection at the approaching (or advancing) points and the receding points. The time interval between two successive flashes is related to the rotation rate of the rotor. The width of the flash is determined by the blade length L, the wavelength λ, the elevation angle β, and the rotation rate Ω.

Because rotating rotor blades impart periodic modulations on the radar returned signal, the rotation-induced Doppler shifts occupy unique locations in the frequency domain relative to the Doppler shift of the fuselage. The Doppler shift of the fuselage and the hub should be around zero Doppler if the helicopter has no translational motion. The total width of the Doppler spectrum is two times of the maximum Doppler frequency, $2\{f_D\}_{max}$. The interval of frequency sampling is the number of blades times of the rotation rate, $N_B\,\Omega$.

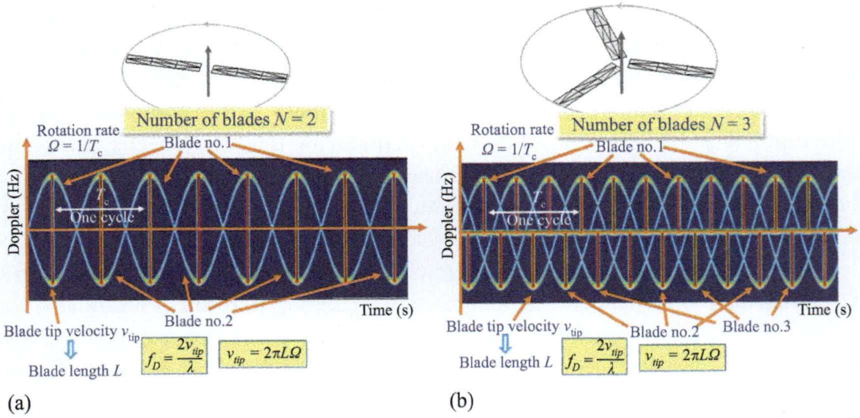

Figure 12.8 Micro-Doppler signatures of the rotating two-blade rotor (a) and three-blade rotor (b)

The rotation feature of rotor blades is considered an important feature for identifying helicopters of interest. The Doppler modulation induced by rotating rotor blades is regarded as such a unique signature of helicopters. Figure 12.8(a) is the micro-Doppler signature of a rotating two-blade rotor, where 8 flashes from the blade no.1 and other 8 flashes from the blade no.2 can be seen clearly. But there is no strong zero-Doppler point at the flashes due to 180° phase cancel-out in the two-blade case. For comparison, Figure 12.8(b) shows the micro-Doppler signature of a rotating three-blade rotor, where each blade has 8 flashes and the total number of fleshes is 24. A strong zero-Doppler zone can be seen for the three-blade case.

In summary, the features extracted from the micro-Doppler signature of rotating rotor blades may include the following:

1. The width of the Doppler spectrum, $2\{f_D\}_{max}$, is a function of rotation rate Ω, blade length L, elevation angle β, and the wavelength λ, i.e., $\{f_D\}_{max} = (4\pi L\Omega /\lambda) \cos \beta$.
2. There are always sinusoidal time-varying Doppler spectrum components: $\{f_D\}_{max} \sin(\Omega t + k2\pi/N)$, $k = 1, 2, \ldots, N$, associated with each of the blades.
3. Due to specular scattering when the blades are normal to the radar LOS, the flashes occur and show a train of strong scattered fields with maximum Doppler shifts. These flashed lines are located at the time instant $t = kT_{flash}/2$ where k is an odd integer ($k = 1, 3, 5, 7, 9, \ldots$) and $T_{flash} = \pi/\Omega$.
4. There is a strong Zero-Doppler zone, which is generated by the blade sections close to the center of rotation.

12.3.2 Motion decomposition from micro-Doppler signatures

A key issue in decomposing a multicomponent micro-Doppler signature into multiple monocomponent signatures is to estimate the true IFs and track the time-varying IF trajectory (i.e., motion curve) for finding a monocomponent signature.

As a dynamic programming, the Viterbi algorithm has been used to estimate the hidden state in a joint time-frequency domain and achieved some level of success [15–17,21,22]. Based on the Viterbi algorithm [23], a method on the IF estimation is to reduce the error caused by strong noise and decomposing in a joint time-frequency domain through detecting IFs at each time and finding the best paths to next IFs. Thus, the Viterbi algorithm becomes an effective tool for estimating IFs in the time-frequency domain. It can be expressed by

$$IF(n) = \mathrm{argmax}_k MD(n, k) \tag{12.14}$$

where $MD(n, k)$ is the micro-Doppler signature in the joint time and frequency domain, and n and k are the time index and the frequency index, respectively. By using this method, all IF points can be independently detected at corresponding time slices. However, in cases of strong noise environment, the estimated IF can be deviated from its true IF. Thus, an algorithm based on the Viterbi algorithm and time-frequency distribution was proposed to handle the strong noise [15,17,21].

Thus, from a decomposed micro-Doppler signature, the intensity value at each point in the two-dimensional (2-D) time and Doppler domain represents relative RCS of the related component part. The Doppler shift at each time instant is proportional to the radial velocity of the corresponding component part. Other parameters to be extracted from the micro-Doppler signature include the instant radial acceleration, the average radial velocity and acceleration, the oscillation cycle and amplitude, and others.

12.4 Micro-Doppler signature-based identification

Target classification is an extremely important research topic in radar. Classification means to resolve the class of a target, e.g., a ground vehicle vs. an aircraft. Recognition is to determine whether the ground vehicle is a truck or a tank. Identification is to identify the model of the target (T72 tank or M60 tank). Thus, the purpose of identifying a body movement is to determine the model of the movement or identify an individual movement different from others.

Commonly used statistical algorithms for target classification include linear discriminant, Naïve Bayes classifier, support vector machine (SVM), and kernel machine. Recent development on machine learning, artificial neural networks, and deep learning has been successfully applied to the classification, recognition, and identification.

12.4.1 Micro-Doppler signature-based classification

As a pattern in the statistical pattern recognition, micro-Doppler signatures have been used for classifying targets' movement [24–29]. To classify human activities, because different human body movement has different micro-Doppler signature, by exploiting these differences, a human activity classifier can be formulated. Features used in the human activity classifier may include the torso signature curve, the maximal Doppler shift of the signature, the offset of the signature, the maximum Doppler variation of the torso curve, the oscillation frequency or period of the human locomotion, the kinematic parameters of limbs, and other available features.

Despite of the successful tracking and interpretation of human movements reported in computer vision literature, the selection of suitable descriptive features for prediction of human actions and intention is still a challenge. Most of micro-Doppler signature-based classifiers do not utilize decomposed signatures associated with human body parts and limbs. According to biological motion perception study, if a kinematic information-based classifier is based on the features of human body parts and limbs extracted from micro-Doppler signatures, this will lead to a more impressive performance for identifying human actions, emotions, and even the gender information.

12.4.2 Motion identification from micro-Doppler signatures

The purpose of identifying a human body movement is to determine the model of the movement or identify an individual movement different from others, such as walking, running, or jumping, after classifying the movement has being made. Existing approaches to identify human body movements in the computer vision are divided into two different types: the exploitation of the structural information, and the use of motion information, such as the moments of human body parts, eigen-vectors, and the hidden Markov model [30–34].

12.4.3 Classification, recognition, and identification using deep learning neural networks

Humans and animals can learn to see, perceive, act, and communicate in an efficient way in the brain, and use a long (deep) chain of synaptic communications to make an action.

As described by A. Samuel, machine learning gives computers an ability to learn without being explicitly programmed [35]. Thus, machine learning becomes a subset of artificial intelligence. In the machine learning, deep learning architectures and unsupervised learning algorithms have been utilized.

Deep learning is a subset of machine learning and uses artificial neural networks to learning tasks. It attempts to give machine an artificial brain and nervous system.

One of the great advantages of the deep learning is the automatic feature learning. The deep learning neural networks can decide which features in the input data set (such as a 2-D image or micro-Doppler signature) can be used as indicators to label that data reliably. Therefore, the deep learning actually surpassed the conventional feature extraction algorithms.

Convolutional neural networks (CNNs) are deep, feed-forward artificial neural networks that have been successfully applied to analyzing visual imagery [36]. Based on micro-Doppler signatures, deep learning CNNs have been successfully applied to classifying human activities and fine-grained hand gestures [37,38].

As noticed, when using a micro-Doppler signature as an input data set to deep learning CNNs, micro-Doppler signatures are often formatted as a 2-D time-frequency image. Although this is a straightforward and natural way to transform a micro-Doppler signal function into a 2-D time-frequency image domain, it should be realized that any micro-Doppler signal function can also be transfer to other formats than a 2-D time-frequency image format, such as the Hilbert–Huang transform [39], the PCA, or even the independent component analysis [40].

References

[1] J. E. Cutting and L. T. Kozlowski, "Recognizing friends by their walk: Gait perception without familiarity cues," *Bulletin of the Psychonomic Society*, Vol. 9, pp. 353–356, 1977.

[2] L. T. Kozlowski and J. E. Cutting, "Recognizing the sex of a walker from a dynamic point-light display," *Perception & Psychophysics*, Vol. 21, pp. 575–580, 1977.

[3] F. Loula, S. Prasad, K. Harber and M. Shiffrar, "Recognizing people from their movement," *Journal of Experimental Psychology: Human Perception and Performance*, Vol. 31, pp. 210–220, 2005.

[4] G. Mather and L. Murdoch, "Gender discrimination in biological motion displays based on dynamic cues," *Proceedings of the Royal Society of London Series B*, Vol. 258, pp. 273–279, 1994.

[5] N. F. Torje, "Decomposing biological motion: A framework for analysis and synthesis of human gait patterns," *Journal of Vision*, Vol. 2, pp. 371–387, 2002.

[6] G. Johansson, "Spatio-temporal differentiation and integration in visual motion perception," *Psychological Research*, Vol. 38, pp. 379–393, 1976.

[7] S. V. Stevenage, M.S. Nixon, and K. Vince, "Visual analysis of gait as a cue to identity," *Applied Cognitive Psychology*, Vol. 13, pp. 513–526, 1999.

[8] F. E. Pollick, V. Lestou, J. Ryu and S.-B. Cho, "Estimating the efficiency of recognizing gender and affect from biological motion," *Vision Research*, Vol. 42, pp. 2345–2355, 2002.

[9] V. C. Chen, "Analysis of radar micro-Doppler signature with time-frequency transform," *Proc. of the IEEE Workshop on Statistical Signal and Array Processing* (SSAP), Pocono, PA, pp. 463–466, 2000.

[10] V. C. Chen and H. Ling, *Time-Frequency Transforms for Radar Imaging and Signal Analysis*, Artech House, Boston, MA, USA, 2002.

[11] V. C. Chen, F. Li, S.-S. Ho and H. Wechsler, "Micro-Doppler effect in radar – phenomenon, model, and simulation study," *IEEE Transactions on Aerospace and Electronics Systems*, Vol. 42, No. 1, pp. 2–21, 2006.

[12] V.C. Chen, *The Micro-Doppler Effect in Radar*, Artech House, Boston, MA, USA, 2011.

[13] V. C. Chen, F. Li, S.-S. Ho and H. Wechsler, "Analysis of micro-Doppler signatures," *IEE Proceedings – Radar, Sonar and Navigation*, Vol. 150, No. 4, pp. 271–276, 2003.

[14] V. C. Chen, D. Tahmoush, and W. J. Miceli (Editors), *Radar Micro-Doppler Signature – Processing and Applications*, IET, London, UK, 2014.

[15] L. Stankovic, M. Dakovic, T. Thayaparan and V. Popovic-Bugarin, "Signal decomposition of micro-Doppler signatures," in Chapter 10, *Radar Micro-Doppler Signature – Processing and Applications*, V. C. Chen, D. Tahmoush, and W. J. Miceli (eds.), Radar Series 34, IET, London, UK, pp. 273–328, 2014.

[16] P. Li, D. C. Wang, and L. Wang, "Separation of micro-Doppler signals based on time frequency filter and Viterbi algorithm," *Signal, Image and Video Processing*, Vol. 7, No. 3, pp. 593–605, 2013.

[17] P. Mazurek, "Estimation of micro-Doppler signals using Viterbi track-before-detect algorithm," *2017 22nd International Conference on Methods and Models in Automation and Robotics*, pp. 898–902, 2017.

[18] J. A. Nanzer, "Millimeter-wave interferometric angular velocity detection," *IEEE Transactions on Microwave Theory & Techniques*, Vol. 58, No. 12, pp. 4128–4136, 2010.

[19] J. A. Nanzer, "On the resolution of the interferometric measurement of the angular velocity of moving objects," *IEEE Transactions on Aerospace and Electronics Systems*, Vol. 60, No. 11, pp. 5356–5363, 2012.

[20] J. A. Nanzer and K. Zilevu, "Dual interferometric-Doppler measurements of the radial and angular velocity of humans," *IEEE Transactions on Antennas and Propagation*, Vol. 62, No. 3, pp. 1513–1517, 2014.

[21] I. Djurović, V. Popović-Bugarin, and M. Simeunović, "The STFT-based estimator of micro-Doppler parameters," *IEEE Transactions on Aerospace and Electronic Systems*, Vol. 53, No. 3, 2017, pp. 1273–1283.

[22] S. Abdulatif, F. Aziz, B. Kleiner and U. Schneider, "Real-time capable micro-Doppler signature decomposition of walking human limbs," *2017 IEEE Radar Conference*, pp. 1093–1098, 2017.

[23] G. D. Forney, "The Viterbi algorithm," *Proc. IEEE*, Vol. 61, No. 3, pp. 268–278, 1973.

[24] M. G. Anderson, *Design of multiple frequency continuous wave radar hardware and micro-Doppler based detection and classification algorithms*, PhD Dissertation, University of Texas at Austin, 2008.

[25] J. Zabalza and C. Clemente, "Robust PCA micro-Doppler classification using SVM on embedded systems," *IEEE Transactions on Aerospace and Electronic Systems*, Vol. 50, No. 3, pp. 2304–2310, 2014.

[26] S. Vishwakama and S. S. Ram, "Dictionary learning for classification of indoor micro-Doppler signatures across multiple carriers," *2017 IEEE Radar Conference*, pp. 0992–0997, 2017.

[27] S. Vishwakama and S. S. Ram, "Classification of multiple targets based on disaggregation of Micro-Doppler signatures," *2016 Asia-Pacific Microwave Conference*, pp. 1–4, 2016.

[28] S. Bjorklund, H. Petersson and G. Hendeby, "Features for micro-Doppler based activity classification," *IET Radar, Sonar and Navigation*, Vol. 9, No. 9, pp. 1181–1187, 2015.

[29] J. J. M. De Wit, R. Harmanny, and P. Molchanov, "Radar micro-Doppler feature extraction using the singular value decomposition," *Proceedings 2014 International Radar Conference*, Lille, France, pp. 1–6, 2014.

[30] D. Gavrila, "The visual analysis of human movement, a survey," *Computer Vision and Image Understanding*, Vol. 73, No. 1, pp. 82–98, 1999.

[31] J. Aggarwal and Q. Cai, "Human motion analysis, a review," *Computer Vision and Image Understanding*, Vol. 73, No. 3, pp. 428–440, 1999.

[32] A. Sunderesan, A. Chowdhury, and R. Chellappa, "A hidden Markov model-based framework for recognition of humans from gait sequences," *Proc. of 2003 IEEE Intl. Conf. on Image Processing*, Vol. 2, pp. 93–96, 2003.

[33] L. Lee and W. Grimson, "Gait analysis for recognition and classification," *Proc. Intl. Conf. on Automatic Face and Gesture Recognition*, Vol. 1, pp. 155–162, 2002.

[34] W. Li, B. Xiong, and G. Kuang, "Target classification and recognition based on micro-Doppler radar signatures," *2017 Progress in Electromagnetics Research Symposium*, 19–22 Nov., Singapore, 2017.

[35] A. Samuel, "Some studies in machine learning using the game of checkers," *IBM Journal of Research and Development*, Vol. 3, No. 3, pp. 210–229, 1959.

[36] Y. LeCun and Y. Bengio, "Convolutional networks for images, speech, and time-series," In M. A. Arbib (ed.), *The Handbook of Brain Theory and Neural Networks*. MIT Press, Cambridge, MA, USA, 1995.

[37] Y. Kim and T. Moon, "Human detection and activity classification based on micro-Doppler signatures using deep convolutional neural networks," *IEEE Geoscience and Remote Sensing Letters*, Vol. 13, No. 1, pp. 8–12, 2016.

[38] J. Lien, N. Gillian, M. E. Karagozler, *et al.*, "Soli: ubiquitous gesture sensing with millimeter wave radar," *ACM Transactions on Graphics*, Vol. 35, No. 4, Article pp. 142:1–142:19, 2016.

[39] N. E. Huang, Z. Shen, S. R. Long, *et al.*, "The empirical mode decomposition and the Hilbert spectrum for nonlinear and non-stationary time series analysis," *Proc. Roy. Soc. London*, Ser. A, Vol. 454, pp. 903–995, 1998.

[40] A. Hyvarinen and E. Oja, "Independent component analysis: Algorithms and applications," *Neural Networks*, Vol. 13, No. 4–5, pp. 411–430, 2000.

Chapter 13

Repurposing millimeter-wave communication devices for high-precision wireless sensing

Teng Wei[1] and Xinyu Zhang[2]

Radio-based passive-object sensing can enable many new applications in pervasive and ubiquitous computing. It is a promising solution especially when the power consumption, physical size and cost of active sensors become the dominant constraint.

In this chapter, we discuss techniques that enable tracking/locating of passive objects at super-resolution, using commodity 60-GHz millimeter-wave (mmWave) radios as the practical and fundamental carrier platform. Two example system designs are presented which achieve mmWave sensing at different scales. The first system enables a high-precision virtual trackpad on any desktop-size surface using highly directional 60-GHz mmWave radios. It can locate/track the passive object, e.g., a writing marker, with the 90-percentile error below 8 mm, enabling new applications such as wireless transcription. The second example system enables the room-level environment sensing. It reconstructs a coarse representation of dominant reflectors from the mmWave radios' eyes and predicts network performance of arbitrarily located links. The design optimizes the deployment of 60-GHz access points and thus enhances network capacity under user mobility and link blockage.

13.1 Introduction

Radio-based passive object sensing is a rapidly developing technology that detects the location and motion of human body parts or associated objects through wireless signals. Compared with conventional vision-based approaches [1], e.g., Leap-Motion [2] and Kinect [3], it is less intrusive and unaffected by ambient light conditions or the sunlight interference. Catalyzed by the proliferation of mobile communication devices, this technology holds potential to enable new ubiquitous user–mobile interfaces and spurs a wide range of applications. The latest passive sensing technology can accurately distinguish a prescribed set of body/limb

[1]University of Wisconsin–Madison, USA
[2]University of California, USA

movement [4] based on Doppler patterns or signal strength variation. Hand gesture sensing [5] is also achievable by training a pattern-matching model.

The 60-GHz mmWave spectrum represents the most promising venue to realize ubiquitous high-precision wireless sensing. It centers at 60-GHz unlicensed mmWave band and spanning 14-GHz spectrum [6]; 60-GHz radios are standardized in IEEE 802.11ad [7], and anticipated to penetrate one-third of wireless links by 2018 [8]. Commercial 802.11ad-capable smartphones have already been demonstrated [9]. Such 60-GHz mmWave devices can be repurposed for wireless sensing, and they possess unique advantages compared with conventional microwave band (2.4 or 5 GHz). A shorter wavelength can create stronger reflections from small objects (e.g., a pen), since wireless signals cannot easily bypass objects larger than wavelength. More importantly, mmWave devices like 802.11ad allow for miniaturized phased array with dozens of antenna elements, which together create highly directional "pencil beams." Such directional beams are electronically steerable, thereby creating a new dimension for object localization/tracking.

We envision two new application scenarios that can be powered by the high-precision mmWave wireless sensing:

- **Fine-grained passive motion tracking using mmWave radios.** Conceptually, empowered by multiple wireless devices, we can create an interactive trackpad on any conventional surface, e.g., whiteboard or desktop, and track a writing object (e.g., a stylus pen or marker) wirelessly. This way, we can deploy large touch screens in a more flexible and economic way than traditional graphic tablets. This vision entails tracking a small writing object with at least sub-centimeter precision. Such precision has proven feasible in certain active radio-frequency (RF) sensing systems like Tagoram [10]. However, they are applicable only to objects instrumented with an radio-frequency identification (RFID) tag or radio receiver.
- **Reconstructing the reflection environment using mmWave radios.** Owing to the high directionality, 60-GHz radios commonly leverage reflections to steer around obstacles. Dependence on directivity and reflection, however, makes the network performance a sensitive function of node placement and environmental characteristics (e.g., geometrical layout and reflectivity of ambient surfaces). If the 60-GHz radio can "sense" (model) the major reflectors in the environment, the sensing information can in turn help optimize access point placement and maximize network capacity and robustness.

In this chapter, we discuss two example designs of the mmWave passive-sensing systems that target the above application scenarios, respectively.

The first system, mTrack, harnesses the unique advantages of mmWave to track the trajectory of a writing object at high precision. It uses one transmitter (Tx) to emit 60-GHz signals and illuminates a handhold object that roams on a trackpad area. Part of the signals are diffused by the object's surface and captured by two directional receivers. mTrack leverages the received signal strength (RSS) and phase (relative to Tx) to localize and track the object. Simply put, by measuring the arriving direction of reflected signals, each receiver (Rx) can estimate the object's relative angle and pinpoint its initial *location* on the trackpad. When the object moves, the reflected signals' path length varies, which alters the phase, providing salient hints for *tracking* the object with sub-wavelength resolution.

Yet in practical environment, these principles entail unique challenges. First, reflected signals from irrelevant background objects can severely distort the target-reflected signal and affect phase-tracking accuracy. To meet this challenge, we take unique advantage of short wavelength of mmWave signals and develop the *phase counting and regeneration* (PCR) algorithm to recover the legitimate phase change induced by target movement only. Second, mTrack leverages steerable 60-GHz Rx antenna to find the matching point and identify the object's relative angle. Although 60-GHz signals are commonly simplified as pseudo-optical [11], they can be widely diffused after hitting a small object. Moreover, practical 60-GHz phased-array antennas can only switch between a discrete set of directions. mTrack overcomes this issue by reconstructing the ideal continuous scanning results from discrete sampling, which can work even if the switching angle is much larger than beamwidth.

The second system, E-Mi, enables reflector learning through mmWave radio's eyes by conducting accurate tracing of mmWave signals' propagation paths. Specifically, E-Mi needs to disentangle all the none-line-of-sight (NLoS) paths for each Rx location and estimates each path's angle of arrival (AoA), angle of departure (AoD) and length. This differs from the vast literature of phased-array localization algorithms that only exploit the line-of-sight (LoS) path [12,13].

The core challenge in E-Mi is how to sense the environment using mmWave radios which can only measure the RSS and phase between each other. E-Mi leverages the well-known sparsity of mmWave channels [14–16]: from the 60-GHz radios' eyes, there are usually only a few *dominating reflectors* in practical environment. E-Mi samples the environment by fixing the Tx radio and moving the Rx to a few locations. At each location, the radio channel comprises one LoS path, and many NLoS ones. By measuring the RSS/phase, E-Mi traces back all NLoS propagation paths, uses a geometrical model to locate where the paths hit reflectors and eventually reverse-engineers the location and reflectivity of dominating reflectors.

The remainder of this chapter is organized as follows. Section 13.2 discusses the overview of mTrack architecture. Sections 13.3 and 13.4 describe how mTrack localizes and tracks a passive object with sub-centimeter accuracy based on mmWave communication radios. Section 13.5 presents the system implementation and evaluation of mTrack. Section 13.6 depicts the key system components of E-Mi. Sections 13.7 and 13.8 present a computational framework that optimizes the 60-GHz network deployment through environment sensing from mmWave radio's eyes. Section 13.9 describes the evaluation of E-Mi. The final section concludes the chapter.

13.2 mTrack: an overview

Figure 13.1 shows a typical mTrack setup. One quasi-omni Tx and two directional Rx antennas are placed to form a right angle, and track an object within a

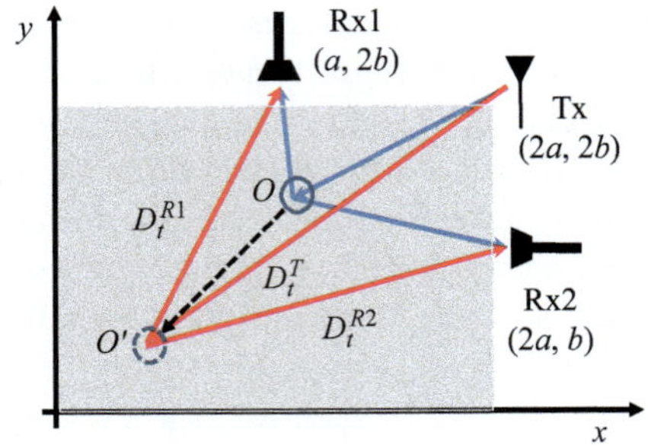

Figure 13.1 mTrack's system setup. Gray area denotes tracking region

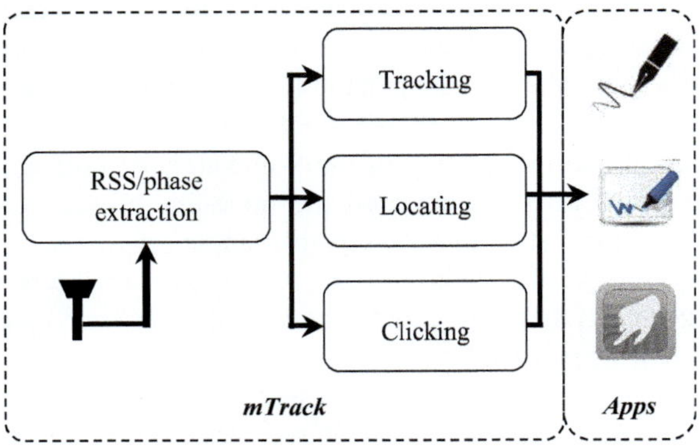

Figure 13.2 mTrack's system components and functionalities

rectangular trackpad region ($0 < x < 2a$, $0 < y < 2b$). a and b are set to 50 cm by default.

Both the quasi-omni and directional beam patterns can be readily generated via 802.11ad devices' phased-array antennas [7,17]. The quasi-omni Tx beam in mTrack is used to illuminate the trackpad area. Typical quasi-omni beamwidth in 60-GHz ranges from 20° to 180°. The transmit signal is able to cover the tracking region, without the need to adjust the Tx's beam direction for scanning target. On the other hand, the Rx adopts highly directional antennas, for two reasons. First, with a high antenna gain, it can substantially boost signal quality, enabling mTrack to capture the weak reflections from target. Second, a directional Rx can help isolate the leakage interference from the Tx, which may otherwise overwhelm the target-reflected signals. Using highly directional antennas reduces the beams' intersecting area. The Rx keeps track of the target by adaptively steering the beam direction. It also knows when the target starts to move and the moving tendency by measuring the trend and amount of phase shifting (detailed in Section 13.3).

Figure 13.2 illustrates mTrack's design components and their interaction. The *phase-tracking* module continuously tracks the target's location by measuring and manipulating phase incremental at each Rx. It resolves the grand challenge from background reflections and achieves millimeter-scale tracking accuracy. The *anchor point acquisition* (APA) module estimates the target's initial position through beam searching. APA has lower location resolution than phase tracking. Hence, it is only opportunistically called on to provide absolute location value for phase tracking and prevent its error accumulation. The *touch-detection* module detects user lifting the writing object or reposition it on the trackpad. Finally, the tracking/locating/touching output can serve a variety of applications, including not only trackpad, but also handwriting transcription, gesture recording, etc.

13.3 Phase-based fine-grained mmWave tracking

13.3.1 Basic successive tracking algorithm

13.3.1.1 Translating phase change into path-length change

We first model the relation between target's moving distance and relative phase change between a pair of Tx and Rx. The model builds the foundation for phase tracking.

Suppose the transmitter sends a single-tone signal of frequency f_c, represented as: $T_b(t) = Ae^{j(2\pi f_c t)}$, where A is the transmit signal magnitude. After target reflection and propagation, the Rx-captured signal becomes

$$R_b(t) = \eta A e^{j[2\pi f_c(t-\tau)+\Phi_1]}$$

where η is the attenuation factor. Signal propagation time $\tau = \dfrac{D_t^T + D_t^R}{c}$, where c is the light velocity, and D_t^T and D_t^R are the distance from target to the Tx and Rx, respectively. Φ_1 denotes the phase change caused by the reflection on the target's surface, which is a constant for the same material. Then the *phase shift*, defined as the received signal's phase change from time $t-1$ to t, and labeled as $\Delta\arg(R_b)_{t-1}^{t}$, is proportional to the change of total path length due to movement:

$$\Delta\arg(R)_{t-1}^{t} = \frac{2\pi\left(D_t^T + D_t^R - D_{t-1}^T - D_{t-1}^R\right)}{\lambda_c} \tag{13.1}$$

where λ_c is the wavelength corresponding to carrier frequency f_c.

13.3.1.2 From distance tracking to 2D tracking

Figure 13.1 illustrates mTrack's basic model setup. When deploying mTrack, the Tx and two Rxs are placed at $(2a, 2b)$, $(a, 2b)$ and $(2a, b)$, respectively. Both a and b are known.

Consider two consecutive signal samples $t-1$ and t, while the target is moving. Based on (13.1), we have

$$\left(D_t^T + D_t^{R1}\right) = \left(D_{t-1}^T + D_{t-1}^{R1}\right) + \frac{\lambda_c \Delta\arg(R1)_{t-1}^{t}}{2\pi}$$

$$\left(D_t^T + D_t^{R2}\right) = \left(D_{t-1}^T + D_{t-1}^{R2}\right) + \frac{\lambda_c \Delta\arg(R2)_{t-1}^{t}}{2\pi}$$

In addition with an inherent geometrical constraint of those signal path lengths (Figure 13.1), we can solve for three unknown path lengths D_t^T, D_t^{R1} and D_t^{R2} with three sets of equations, given that the corresponding path lengths at $t-1$ have already been obtained. This computation runs in a successive manner, with initial values D_0^T, D_0^{R1}, and D_0^{R2} given by mTrack's APA module (to be discussed in Section 13.4). Finally, the target's absolute coordinate at time t is computed using the triangulation method, given the path length to the Tx and both Rxs' locations. Algorithm 13.1 summarizes the successive phase tracking in mTrack.

Algorithm 13.1: *Basic phase-tracking algorithm*

1: *Input: phase shift* $\Delta\arg(R1)_{t-1}^{t}$ *and* $\Delta\arg(R2)_{t-1}^{t}$
2: ***Output:*** *new location* $o'(x_t, y_t)$
3: ***if*** *isempty*$(D\,T_{t-1}, D\,R1_{t-1}, D\,R2_{t-1})$
4: $o'(x_t, y_t) \leftarrow$ APA() */*Locating module*/*
5: $\{D_{t-1}^T, D\,R1_{t-1}, D\,R2_{t-1}\} \leftarrow \{|\vec{o'} - \vec{o}(Tx)|, |\vec{o'} - \vec{o}(Rx1)|$
6: $|\vec{o'} - \vec{o}(Rx2)|\}$ */*Acquire initial path length*/*
7: ***else*** */*Successive tracking*/*
8: $\{D_t^T, D_t^{R1}, D_t^{R2}\} \leftarrow path\big(D_{t-1}^T.DR\,1_{t-1}, D\,R2_{t-1}, \Delta\arg(R1)_{t-1}^{t}$
9: $\Delta\arg(R2)_{t-1}^{t}\big)$ */*Find new path lengths*/*
10: */*Find new location and Update path lengths*/*
11: $o'(x_t, y_t) \leftarrow \left(\dfrac{D\,R1_t^2 - D\,T_t^2 + 3a^2}{2a}, \dfrac{D\,R2_t^2 - D\,T_t^2 + 3b^2}{2b}\right)$
12: $\{D\,T_{t-1}, D\,R1_{t-1}, D\,R2_{t-1}\} \leftarrow \{D\,T_t, D_t^{R1}, D_t^{R2}\}$
13: ***return*** $o'(x_t, y_t)$

This successive tracking is promising for real-time tracking because the solutions of path length and new location can be expressed in closed-form formulas and computed efficiently by single-step operations.

Notably, using highly directional receive antennas may result in a relatively small detection region. For example, with two Rx antennas of beamwidth 4.5° and 0.5 m away from the tracking region, the intersection area of two beams is only 5 cm × 5 cm. mTrack expands the detection region by adaptively steering the beam, whenever it gauges that the target may fall out of the Rx's beam coverage in the next 0.5 s, assuming its speed is consistent within this short period.

13.3.2 Tracking under background reflection

13.3.2.1 Impact of background reflection on phase shift

Though directional antennas can alleviate multipath effect, antenna beamwidth still needs to be broad enough to cover a reasonably large detection region. Thus, background reflection is unavoidable. We model how the background reflection affects mTrack's successive tracking and derive insights that help disentangle target reflection from background reflection. We use vector form on the I-Q plane to

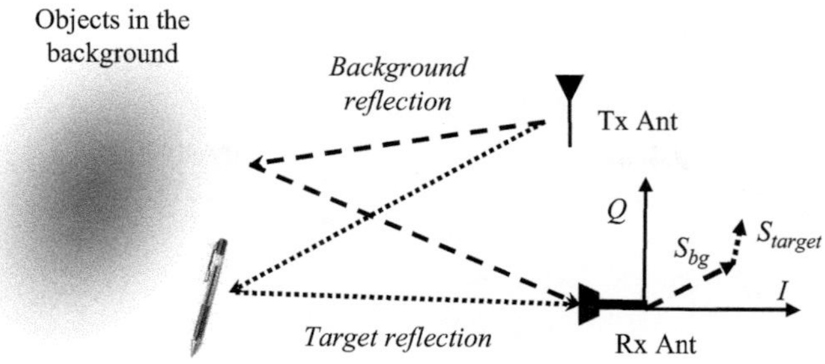

Figure 13.3 Model of the multipath reflection

represent a signal sample (Figure 13.3). Since baseband signal is a single carrier, background reflections can aggregate into a single signal sample $\vec{S}_{bg}$. Similarly, target-reflected signal sample is modeled as $\vec{S}_{trg}$. The demodulated signal at the Rx $\vec{S}_{rec}$ at any time is the sum of two reflected components:

$$\vec{S}_{rec} = \vec{S}_{trg} + \vec{S}_{bg} \tag{13.2}$$

Phase shift of $\vec{S}_{trg}$ carriers information about the moving distance of target, but the Rx can only measure the mixed signal $\vec{S}_{rec}$. So, can this mixed signal's phase still translate into target's moving distance? We answer the question by analyzing how the phase of $\vec{S}_{rec}$ cycles as the target moves over half-wavelength (i.e., the phase of $\vec{S}_{trg}$ cycles from 0 to 2π). The phase cycling pattern in turn depends on the magnitude of target reflection $|\vec{S}_{trg}|$, which we classify into three ranges, as shown in Figure 13.4.

(a) Target-dominated (TD) case: Suppose background reflection is stationary as the target moves across half-wavelength. Accordingly, as $\vec{S}_{trg}$ rotates its phase through one cycle, $\vec{S}_{rec}$ crosses the I-plane once, resulting in phase shift of 2π. Thus, phase shift can still reveal the moving distance when target travels integer multiples of half-wavelength $\lambda_{0.5}$.
(b) Background-dominated (BD) case: As the phase of $\vec{S}_{trg}$ goes through a 2π cycle, the measured phase of $\vec{S}_{rec}$ does not cross the I-plane and phase shift is smaller than 2π. The stronger the background reflection, the smaller it will be. Accordingly, the moving distance estimated from the phase shift of $\vec{S}_{rec}$ is always shorter than actual value.
(c) Intermediate (ITM) case between the above two: As shown in Figure 13.4(c), $\vec{S}_{rec}$ crosses the I-plane twice, causing phase jump at each crossing point. Due to phase ambiguity of measured signal $\vec{S}_{rec}$, the measured phase shift also will not reveal the correct moving distance.

Why is it challenging to deal with?
Multipath reflection from background is a unique challenge for mTrack which aims to track small passive objects at high precision. Unlike radar-like systems that track

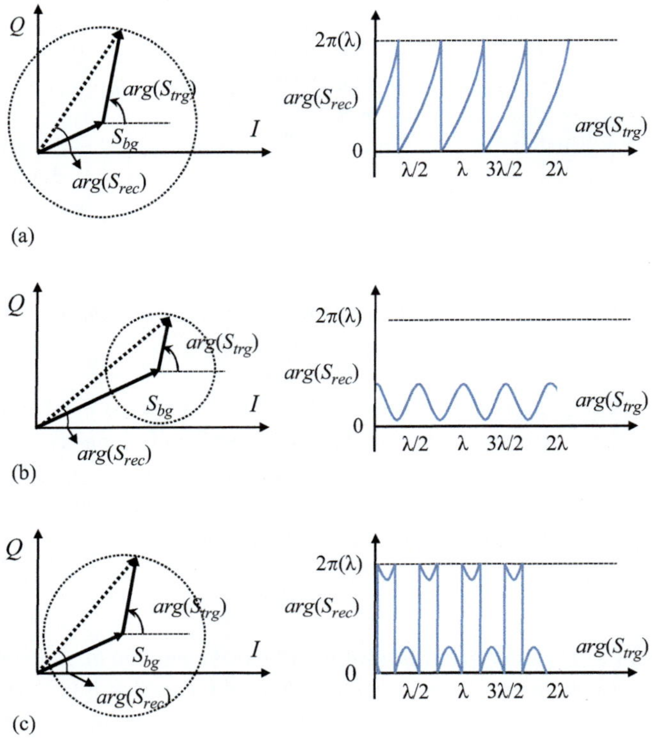

Figure 13.4 Impact of (static) background reflection on phase shift. (a) TD case: Target reflection dominates background. (b) BD case: Target reflection is much weaker than background. (c) ITM case

large targets (e.g., human body [18,19]), reflections from small objects tend to be weaker or comparable to background reflections. Due to the high sensitivity of mmWave, background reflection can disturb the signal phase even as the target moves over a short distance. Radar systems can use ultra-wideband signals to resolve the time of arrival of reflected signal paths, thus separating the foreground and background [20]. In contrast, mTrack relies on a narrow-band (single-tone) continuous wave that mixes background and target reflection.

Unlike active tracking systems [10,21] where reflected paths originate coherently from the same radio Tx, in passive tracking, only parts of the reflected signals contain desired information and need to be extrapolated. Background suppression in active tracking systems is much simpler, since the signal phase changes are caused by baseband modulation, which operates at much higher rate (e.g., multiple kilohertzs in RFID [10]) than background variation. Hence, any background reflection can be considered as relatively static and filtered out by a DC filter [22]. This approach is not applicable to passive tracking systems where phase changes at slow rate due to target movement.

13.3.2.2 Phase counting and regeneration

Since the phase shift directly from the received signal $\vec{S}_{rec}$ is contaminated by background reflection, mTrack employs a PCR algorithm to remove such impact, thus limiting the tracking error. PCR collects a sufficient number of samples to recover the target-reflected signal from noise and background reflection. Under background reflection, although the absolute phase change of measured signal is no longer linear *w.r.t.* target's moving distance, its phase still exhibits a periodic pattern (Section 13.3.2.1). PCR harnesses this phenomenon to *identify/count the phase cycles* of measured signal, based on which it *regenerates* the phase change caused purely by target reflection. Then, it can simply run the basic phase-tracking Algorithm 13.1 to translate phase change into moving distance.

The three different patterns of background contamination (Section 13.3.2.1) impose significant challenges to this approach. First, the three cases may be mixed together, and the ITM case has severe phase-aliasing effects. Second, we need to identify the periodicity based only on a few phase cycles to maintain the timeliness of tracking. Both challenges deter simple periodicity identification algorithms like Fast Fourier Transform. PCR addresses these challenges through three key steps (Figure 13.5 shows an example).

Reducing ITM to BD

We address the first issue by detecting the ITM case, and convert it to the BD case which has no phase aliasing effect. For detection, PCR leverages the unique phase "jumps" in the ITM case caused by aliasing, i.e., neighboring two jumps are in opposite directions (Figure 13.4). Formally, given a time series of N phase samples $x[1], x[2], \ldots, x[N]$, it sequentially searches for a point m_i with phase jump:

$$m_i = \underset{2 \le i \le N}{\arg\min}\, find(|x[i] - x[i-1]| > \pi)$$

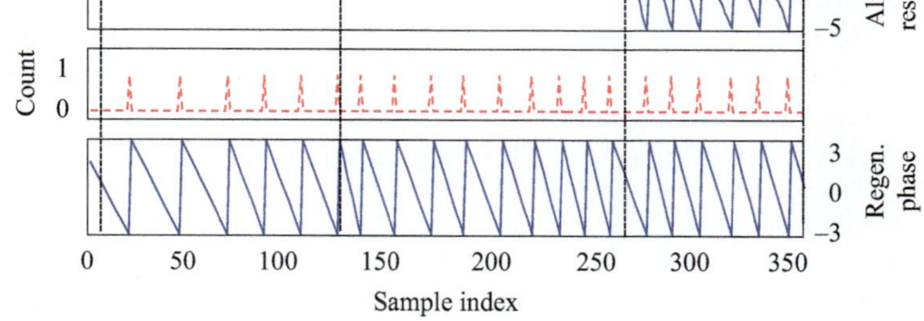

Figure 13.5 An example snapshot of using PCR to remove background reflection

It then checks if there exists a follow-on point that together constitutes the feature in ITM:

$$m_j = \arg\min_{m_i \le j \le N} find(|x[j] - x[j-1]| > \pi) \tag{13.3}$$

$$s.t.\ m_j - m_i < T_h \tag{13.4}$$

$$|x[m_i] - x[m_j]| > \pi \tag{13.5}$$

where T_h is the largest periodicity that PCR can detect. Equation (13.5) ensures the phase jumps at m_i and m_j are opposite.

Once an ITM case is detected, we can reduce it to the BD case by compensating the difference between discontinuous series:

$$x[i] = x[i] - (x[m_i] - x[m_i - 1]), \forall i \in [m_i, m_j) \tag{13.6}$$

Periodicity counting

To reliably detect the periodicity T of the time series of samples after alias removal, we adopt the phase coherence analysis in [23]. Let sequence $\mathbf{L}[i,d] = \{x[i+jd] \mid \forall 0 \le j < \lfloor N/d \rfloor\}$, where $\{\}$ denotes concatenation of samples. Since $\mathbf{L}[i,d]$ are samples that are separated by d points in the given time series, by the periodicity definition, variance of $\mathbf{L}[i,d]$ will be minimized, i.e., 0, when $d = T$. Hence, we find the periodicity T by iterating possible values in range $[T_l, T_h]$ for all sequence i:

$$T = arg\ min_t\ var\left(\left\{\overline{\mathbf{L}[i,t]} \mid \forall 1 \le i \le t\right\}\right), \forall t \in [T_l, T_h]$$

where $\overline{\mathbf{L}[i,t]}$ is $\mathbf{L}[i,t]$ with its mean value subtracted.

Regeneration

PCR will generate the counting sequence (red curve in Figure 13.5) based on the periodicity detected in the time series samples. The correct phase pattern is regenerated by creating 2π phase shift between any two neighboring counts. The trend of phase shift, either increasing or decreasing, is available from the basic successive tracking algorithm.

We emphasize that PCR is uniquely designed for passive tracking using small-wavelength mmWave signals. It assumes that the background reflection does not change as the target moves over $\lambda_{0.5}$, which may not hold when using microwave signals. In addition, its resolution is around $\lambda_{0.5}$, so PCR's precision worsens to multiple centimeters when using microwave signals.

Here are a few aspects that worth further discussion.

Object shape and size

mTrack is designed to track writing objects of small size, like pen or marker, which can be approximated as a point reflector, because their dimension (with diameter ≈ 1 cm) is much smaller than its distance to the antennas (>20 cm). Objects with deformable shape or large size, e.g., human hand and body, do not possess these properties, and hence cannot use mTrack's model-driven tracking algorithm. Similarly, mTrack cannot differentiate phase change due to large hand-shape deformation. Thus, users need to keep a relatively consistent hand shape when writing to avoid confusing the tracking algorithm. They also need to ensure a sizable part of the pen is exposed to the Tx/Rx to reflect the mmWave signals.

Phased-array antennas

Due to hardware limitation, we are only able to evaluate our concept with directional antennas. Beam pattern of a horn antenna differs slightly from that of an 802.11ad phased array in two aspects: *(i)* The beamwidth of a phased array is dependent on steering direction. For example, for a linear antenna array, beamwidth along 0° direction can be twice of that along 90° [17]. However, since beamforming codebook is predefined and known, we can still compensate the gain distortion of discrete sampling. The systematic solution to this problem will be left for our future work. *(ii)* Side lobes of a phased array may be relatively large compared with that of a directional horn antenna. Side lobes may capture signals from undesired directions. However, since they usually have much smaller gain compared to the main lobe, as long as there is an RSS peak when pointing to the target, mTrack's beam-steering-based APA mechanism can still work.

Tracking multiple objects

mTrack can be potentially extended to track multiple targets simultaneously. When the objects are sufficiently separated, and not falling in the same beam direction, the Rx radios can steer the beam toward different objects in a time-interleaving manner and update their tracking trajectories sequentially. Further exploration of this idea is left for our future work.

13.4 RSS-based APA

APA complements phase tracking in two aspects: *(i)* estimating initial location, which serves as an anchor point to bootstrap the successive tracking; *(ii)* opportunistically calibrating the tracking results to prevent error accumulation.

13.4.1 Locating through discrete beam steering

There is a key challenge in locating the target: practical 60-GHz antennas can only be steered at discrete angles. We resolve this challenge by leveraging the roll-off pattern of the antenna gain. The gain of a 60-GHz directional Rx antenna (either horn antenna or phased array [17]) decreases continuously near its beam edge. Accordingly, as a signal source moves from the beam center toward beam edge, RSS falls off smoothly. The fall-off trend is preserved even if Rx steers itself in discrete steps.

Built on this observation, mTrack realizes discrete spatial sampling of the target's direction, thereby relaxing the beam resolution requirement. In analog-to-digital signal conversion, a continuous signal can be preserved by interpolating it across discrete samples, using sinc as the basis function. Similarly for mTrack, the continuous spatial RSS variation can be reconstructed by interpolating the discrete

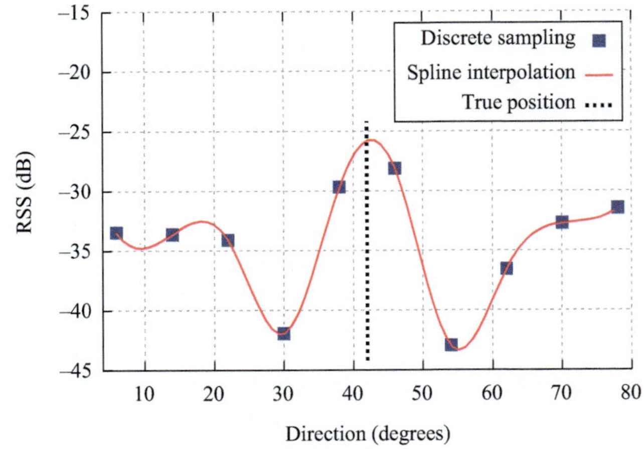

Figure 13.6 Spline interpolation improves granularity of direction estimation from discrete steering

samples, using antenna gain pattern (over angles) as basis function. Yet, since the exact gain pattern is unavailable, we choose to use *spline interpolation* instead. Spline interpolation performs piecewise polynomial approximation to the discrete samples with an objective that minimizes curve bending. It can thus recover the peak RSS position. To perfectly reconstruct continuous variation, angle of beam steering for each step is required to be smaller than $2\times$ the beamwidth.

Figure 13.6 shows one example test result, with discrete RSS samples (blue dots) at 8° beam-steering steps, along with the interpolated curve. We can see that the interpolated maximum RSS position is very close to the ground truth.

Following mTrack's setup (Figure 13.1), the target's angle relative to each Rx antenna can be translated into its absolute location via simple geometrical manipulation. We omit the details for the sake of space.

13.4.2 Background RSS subtraction

Similar to the successive phase tracking, the RSS values captured during APA may be contaminated by background reflection. However, RSS is much less sensitive than phase because the locating result will be affected *only if* the background change alters the peak in the RSS scanning. Thus, mTrack employs a simple iterative background RSS subtraction method in APA. Before use, each Rx collects the RSS of background reflection for each steering angle. When a new APA is invoked, the background RSS at each steering direction is subtracted from the measured RSS. During the writing process, it knows the precise angular location of the target by combining phase tracking and initial APA. Hence, it periodically rescans the background RSS by steering off the target position.

During APA, the tracking module has to be temporarily blocked. However, since mmWave phased-array antenna can instantaneously switch its beam (at *ns*-level latency as specified in 802.11ad [24]), the impact on tracking is negligible.

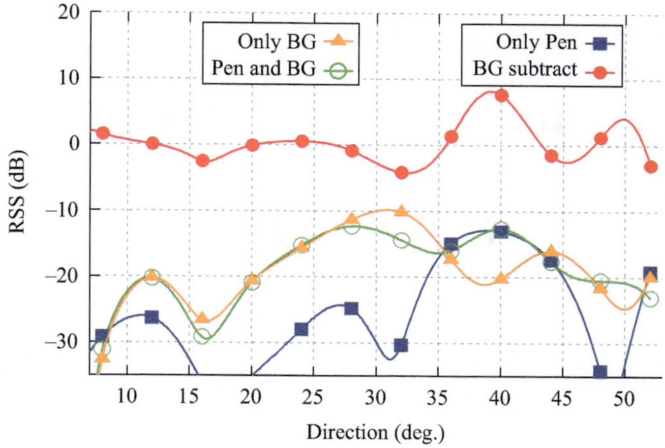

Figure 13.7 APA improves locating accuracy that is disturbed by background reflection

Figure 13.7 illustrates one test experiment, where a metal-surface cup is placed 25 cm behind the detection region and a pen is used as tracking target. When there is only pen (blue curve) or only background (yellow curve) in the environment, mTrack can locate the direction of target accurately (in 40° and 30°, respectively). When the tracking target is mixed with background object, the system is misled to take the later as locating result (in green curve) due to its stronger reflected signal strength. In the red curve, background subtraction cancels the background reflection from cup, and thus the target position can be accurately located.

13.4.3 Opportunistic calibration

To curtail the error accumulation from successive phase tracking, APA is invoked opportunistically to recalibrate the absolute target position.

mTrack adopts a k/θ test to determine when to invoke a calibration. APA module periodically estimates anchor points. If past k anchor points all fall in a sector of angle θ originating from current position, the deviation of tracking trajectory can be considered as sufficiently large. In this case, the calibration will be applied by updating current position to the latest anchor point. In practice, mTrack empirically chooses $k = 3$ and $\theta = 120°$.

13.5 Implementation and evaluation of mTrack

13.5.1 Implementation

We use a custom-built 60-GHz software-radio testbed (Figure 13.8) to evaluate the viability and performance of mTrack. The testbed uses WARP [25] as baseband processing unit, but extends its carrier frequency to 60 GHz. Baseband digital

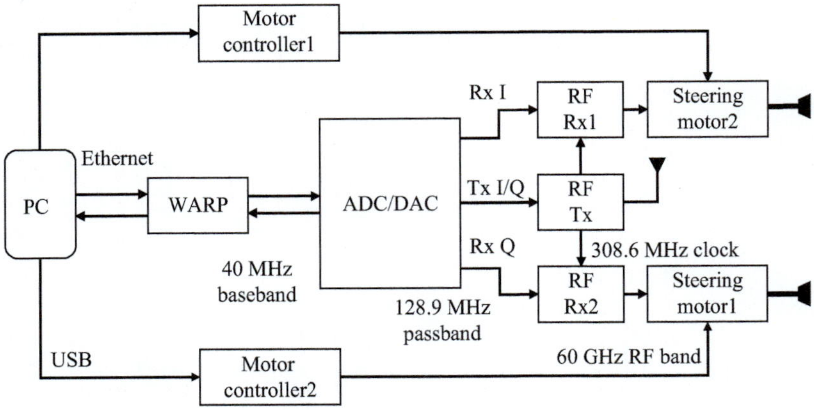

Figure 13.8 Hardware architecture of mTrack

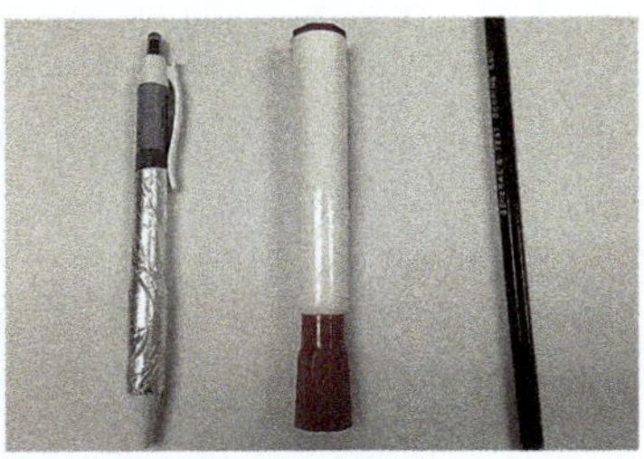

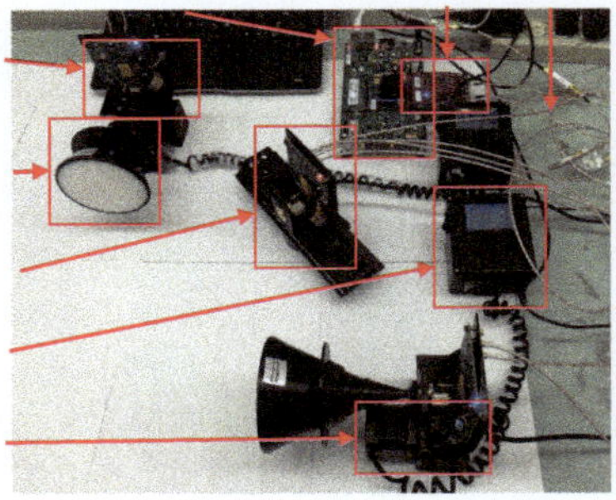

Figure 13.9 mmWave testbed. Testing metal-surfaced pen, plastic marker and wood pencil. Radios are placed 10 cm beside 50 cm × 50 cm writing region

waveforms are generated in a host PC, forwarded to WARP, and converted to analog using a high-speed digital-to-analog converter (DAC). Resulting analog signals are carrier-modulated and transmitted by the Vubiq 60-GHz RF front end [26]. Two Rxs are synchronized by a 308.6-MHz external clock signal sourcing from Tx oscillator. Received signals will be decoded following reverse path, and eventually reach the PC host which runs mTrack's algorithms.

Tx and Rx RF front ends each have a waveguide module as interface to 60-GHz antennas (Figure 13.9). Tx uses its original waveguide module PEM-001 as antenna, which is quasi-omnidirectional (180° beamwidth). Tx sends 5-MHz single-carrier baseband signals modulated by 60-GHz carrier waves. Regarding Rx, we are unaware of any programmable 60-GHz phased-array antenna. Thus, in our prototype, signals are captured by a directional Rx with horn antenna PE9881-34 of 3.4° beamwidth. To emulate the effect of beam steering, the Rx antenna is mounted on a motion control system [27] that is connected to the PC host and allows programmable azimuth rotation at 1° granularity.

We also implement the RSS/phase extraction module on the 60-GHz testbed. It generates and sends single-tone frames (3,000 samples each) through Tx and estimates a pair of RSS and phase values from each received frame. To avoid phase ambiguity, the change between any two phase values cannot exceed π. By default, the extraction module samples RSS/phase at 300 Hz. Thus, the highest tracking speed that our prototype can support is 2.5 mm × 300 Hz/2 = 37.5 cm/s. The speed is sufficiently fast to capture normal handwriting. Note that 300 Hz phase sampling rate is far below our hardware limit. With 40 MHz bandwidth, the phase sampling rate can be up to 40 MHz/3,000 = 13 KHz, equivalently translated to a tracking speed of 26 m/s.

13.5.2 Performance on a trackpad

We now evaluate mTrack's performance in a real trackpad application. The experiments are conducted in an office environment with natural background (drywall, metal cabinet, a user and occasional human walking by). A 50 cm × 50 cm writing region is created on a wood table. To test the precision of APA in locating anchoring points, the user rests the pen tip on 40 random locations, ensuring the bottom part of the pen is exposed to the antennas. mTrack steers the antennas with granularity of 8°. To test phase tracking, the user draws ten circles and ten triangles (with 20 cm perimeter) following printed trajectories in the normal handwriting speed. Since human-hand deformation will affect phase tracking, testers hold the middle portion of pen, while directional antennas point to the bottom portion. Due to lack of time synchronization between user writing trajectory and tracking estimation, we approximate the tracking error as the minimum projection distance from mTrack's location estimation to the trajectory.

13.5.2.1 Localization/tracking error across a large region

Distance between the target and the Rx determines the reflected signal strength and hence may affect mTrack's accuracy. The Tx and Rxs are placed at coordinates (100, 100) cm, (50, 100) cm and (100, 50) cm, respectively. To quantify such location-dependent error, we partition the writing area into 10 cm × 10 cm squares, and repeat the previous precision test on each square. Figures 13.10 and 13.11 plot the APA and phase-tracking error across all squares within a 90 cm × 90 cm region.

When the pen is close to both Rxs (distance <60 cm), mTrack can achieve high accuracy with APA/tracking error of <1.5 cm and <8 mm, respectively. Accuracy starts degrading when the target moves over 70 cm away from Rx, and hence signal-to-nose ratio drops. Nonetheless, the tracking error is still within 1.5 cm even when the target is 90 cm from the Rx. mmWave attenuates to almost noise floor at 100 cm owing to high pathloss of mmWave signals. We expect at least two ways of scaling the writing region: increasing the transmit power, and placing more Rxs along the x- and y-axes. We leave such exploration for future work.

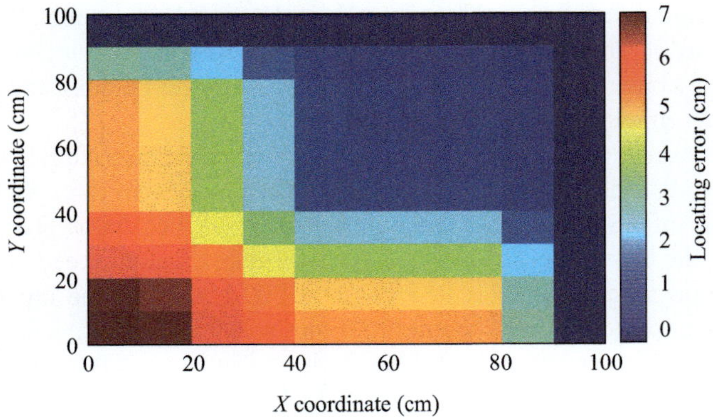

Figure 13.10 Map of APA error

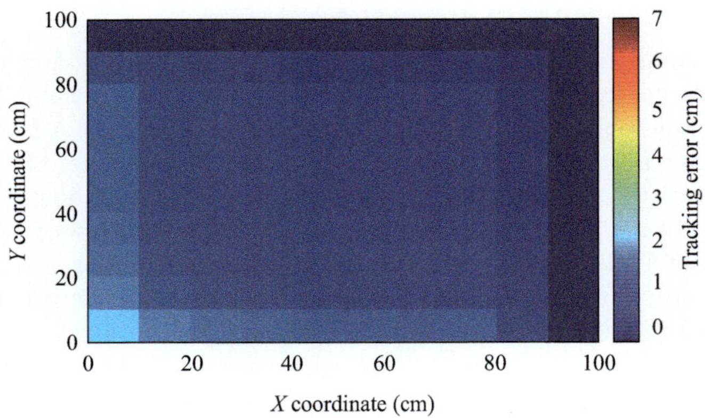

Figure 13.11 Map of phase-tracking error

13.5.3 Application of mTrack

We integrate mTrack with a word recognition application to showcase its potential in computer–human interaction. In specific, we record the writing trace from mTrack word by word, and then export the trace to control the mouse on a PC that runs the MyScript Stylus [28] word recognition software. mTrack's event detector is used to segment the connections between characters in a word. Figure 13.12 shows two examples when we use mTrack to track the character/word trajectory, which exhibits high fidelity even if each character spans only 1 or 2 cm.

We further recruit 7 users to each write 100 random characters and 50 words using metal-surfaced pen on a 15 cm × 15 cm writing region. The words are randomly picked from the standard MacKenzie set [29], which well represents the usage frequency of English words. Figure 13.13 plots the recognition accuracy across users. The accuracy for both character and word ranges from 81% to 89% across users. Users also perform the same test on a real trackpad, and the

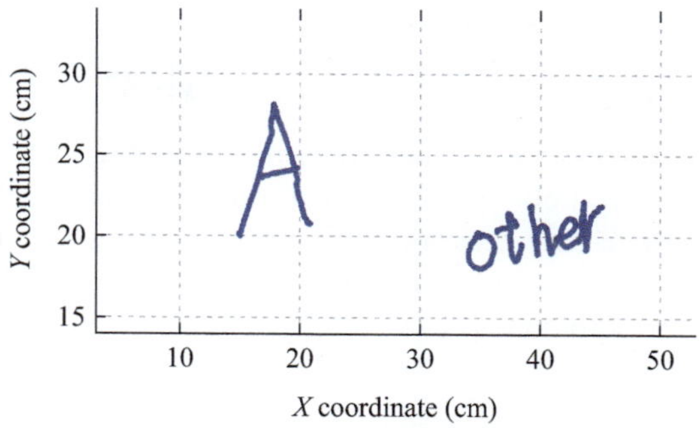

Figure 13.12 Example letter and word by mTrack

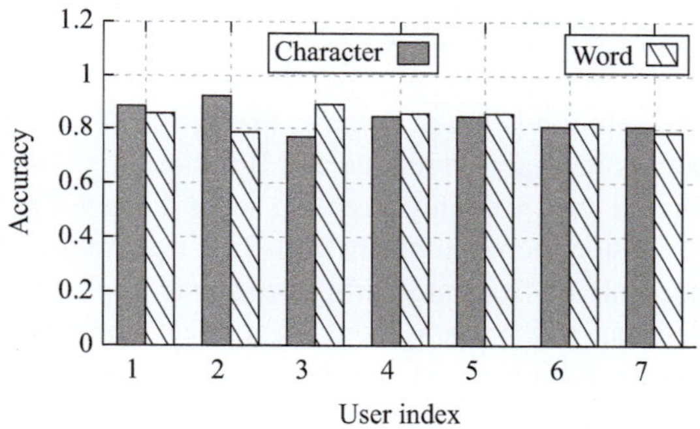

Figure 13.13 Character and word recognition accuracy

recognition accuracy is from 88% to 92%, which shows that mTrack's performance is close to the trackpad. Considering the potential penetration of portable mmWave radios [8,9], we believe mTrack holds potential as a new form of "in-situ" transcription system.

13.6 E-Mi: an overview

E-Mi samples the RSS/phase between a pair of 60-GHz AP and client (also denoted as Tx and Rx), and uses the samples as input to two major modules: *(i) Multipath resolution framework* (Section 13.7), which estimates the geometry, i.e., $\langle AoA, AoD, length \rangle$, of each propagation path and also discriminates their RSS/phase. *(ii) Dominant reflector reconstruction* (Section 13.8), which locates the reflecting points (i.e., spots where the paths hit the reflector), and reconstructs the layout/reflectivity of dominating reflectors, forming a coarse environment map.

E-Mi requires the Tx and Rx to be equipped with phased-array antennas of practical size (default to 16-element, as in typical 802.11ad radios [30]). It does not need a customized PHY layer—it only requires the channel state, which is a portable function on many commodity Wi-Fi devices [31] and expected to be available in the 802.11ad products. Although E-Mi works in a constrained environment that can be illuminated by Tx's signals, the Tx can be moved to different positions to extend its coverage.

When scanning the wireless channel, E-Mi places the Tx and Rx well above the ground, so that they only "see" dominant reflectors like walls and furnitures. They affect the average-case network performance, which is of utmost interest for network planners. In case such reflectors change their locations, we can accommodate the changes by rerunning E-Mi.

13.7 Multipath resolution framework

Multipath resolution framework (MRF) estimates the $\langle AoA, AoD, length\rangle$ of dominating paths between the Tx and Rx. As illustrated in the Figure 13.14, the AoA and AoD are determined by the relative positions of dominant reflector, Tx and Rx, and independent of the beam pattern of phased arrays. To estimate these intrinsic parameters, a naive solution is to use *beam scanning*: the Tx/Rx may steer over all possible combinations of beam directions and find the ones with high RSS. However, a 60-GHz phased array can only steer to a set of discrete directions (e.g., a 16-element one can only steer between beams with 22.5° separation [17]). The discrete beam scanning prevents us from measuring the signal angle precisely. Moreover, unlike horn antennas, phased-array antennas have imperfect directionality—besides the main beam, their antenna pattern bears many side lobes which interfere the AoA/AoD estimation.

E-Mi's MRF introduces three mechanisms (Figure 13.14) to meet the challenges. *(i)* We first estimate the dominating AoD and AoA directions, originating from Tx and ending at Rx, respectively. We adapt a classical signal angle estimation algorithm to 60-GHz phased arrays, which enables super-resolution (i.e., finer resolution than discrete steering by generating a continuous angular spectrum, and unaffected by the imperfect beam shape of phased arrays). *(ii)* We design a *virtual beamforming* (vBeam) scheme that pairs the AoD and AoA directions belonging to

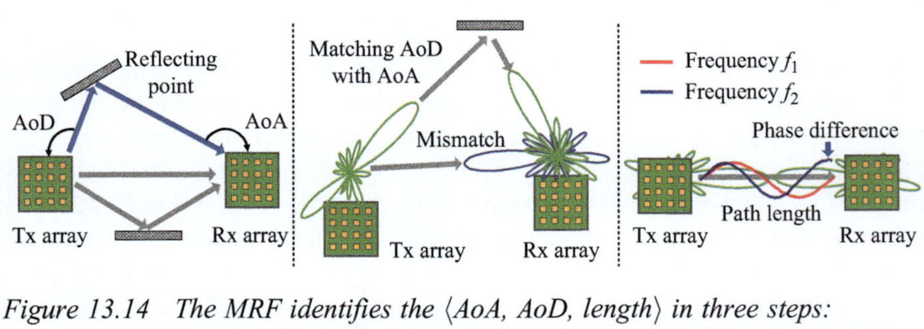

Figure 13.14 The MRF identifies the ⟨AoA, AoD, length⟩ in three steps: (1) identify the dominant AoA and AoD; (2) pair AoA and AoD directions that belong to the same path; (3) estimate the length of each path

the same NLoS path. *(iii)* We employ a *multitone ranging* scheme to estimate the total length of each path.

13.7.1 Estimate path angles using phased arrays

Conventional multi-antenna Rxs can estimate signal AoA using *angular spectrum analysis*, which singles out the arrival angles with strong signal strength [12,13]. However, such analysis needs to *isolate* the signals on each antenna element using digital phased arrays (Figure 13.15(a)). Practical 60-GHz radios use *analog phased arrays* (Figure 13.15(b)), which have a single input/output, comprising a weighted sum of individual antenna's signals that obfuscate each antenna' signals.

To overcome this limitation, a natural way is to vary the weights and obtain a system of equations to solve for individual signals. Suppose there are N elements on the receiving phased array, and $S = [S_1, S_2, \ldots, S_N]^T$ denotes the signals on the N individual antenna elements. When the phased array imposes row vector of weight $\mathbf{w}_1$ on its antenna elements, the received signal becomes $y_1 = \mathbf{w}_1 S$. Suppose the array switches across N different weight vectors to receive the same signals by N times. The weights constitute a two-dimensional (2D) matrix $\mathbf{W}_R = [\mathbf{w}_1; \mathbf{w}_2; \ldots; \mathbf{w}_N]$ with each row being a weight vector. Then, the N output signals form a vector: $\mathbf{Y} = \mathbf{W}R\mathbf{S}$. So, one can simply use $\mathbf{S} = \mathbf{W}R^{-1}\mathbf{Y}$ to recover $\mathbf{S}$, and hence isolate the signals on each antenna element. Figure 13.15(c) illustrates an example where $N = 4$.

In practice, the weight vectors in a 60-GHz phased array are built into hardware and can only be selected from a predefined group, called *codebook*. The key question is, can we find a set of weight vectors to form a matrix $\mathbf{W}R$ that is invertible? The answer is positive: we can find the weight matrix from 60-GHz codebook that is orthogonal (i.e., $\mathbf{W}R\mathbf{W}R^H = I$, where $(\cdot)^H$ denotes the conjugate transpose), and hence invertible. The beamforming codebook ensures orthogonality between weight vectors because it will maximize the isolation across different beam patterns [17,32].

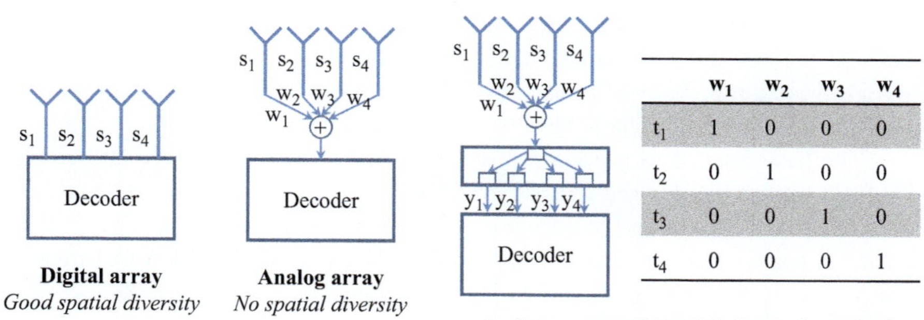

(a) (b) (c) Analog array emulates digital array by collecting signals using different weights over time

Figure 13.15 Isolating individual antenna's signals on an analog phased array, by switching across different weight vectors

To estimate the AoD, a symmetrical operation is needed at the Tx. Suppose the Tx phased array has M antenna elements, then it uses M different sets of weights to transmit the signals by M times, which similarly constitute a transmit matrix $\mathbf{W}T$. We populate $\mathbf{S}$ into an N-by-M matrix. Each element (i,j) of $\mathbf{S}$ represents the signals on ith Rx antenna, when the jth Tx antenna element is triggered. Then, the received signals of analog array becomes $\mathbf{Y} = \mathbf{W}R\mathbf{S}\mathbf{W}^H T$. Each column/row in matrix $\mathbf{Y}$ contains received signals measured using a specific transmitting/receiving weight vector. By way of a similar orthogonality argument as the Rx, we can recover $\mathbf{S}$ as follows: $\mathbf{S} = \mathbf{W}^H R\mathbf{Y}\mathbf{W}T$.

Isolation of individual antenna's signals allows us to apply multiple signal classification (MUSIC) [33], an eigen angle analysis algorithm to jointly estimate the AoA/AoD, in the same way as in digital phased arrays [12]. MUSIC can achieve a scalable resolution with more antenna elements and extricate the discrete beam shape of phased arrays. Specifically, we measure the preamble signals sent/received by standard 60-GHz radios [34], which are sent repeatedly across packets, and across different Tx/Rx beam patterns. We isolate the preamble signals sent/received by different Tx/Rx antenna elements. Then, we run MUSIC to compute the angular spectrum, essentially the likelihood of signals coming from different angles. Finally, we find the peaks in the angular spectrum that are larger than the noise floor and take the corresponding directions as AoAs/AoDs of dominating paths.

A few additional operations are worth noting: *(i)* Since each AoD is pairwise to an AoA *w.r.t.* the same dominant reflector, we remove the excessive AoA/AoD estimations of smaller eigenvalue and make sure the number of AoA and AoD values are equal. *(ii)* To ensure the consistency of reference direction, i.e., 0 degree, in the measurement, the antenna's orientation can be simply kept at a fixed direction, or be compensated by motion sensors in the mobile device. *(iii)* MUSIC is adopted only for AoA and AoD estimation. The RSS estimation and AoA/AoD pairing of each path is accomplished by the virtual beamforming, which will be detailed next.

13.7.2 Virtual beamforming: match path angles

E-Mi's virtual beamforming (vBeam) algorithm serves two purposes. First, the AoAs and AoDs identified above do not have a pairwise mapping. The vBeam can pair up the AoA and AoD values that belong to the same path. Second, the received signals $\mathbf{S}$ are a mix from all propagation paths. To estimate the length of each path (Section 13.7.3), their signals have to be separated from each other.

The basic idea is to process the received signal matrix $\mathbf{S}$ offline and emulate Tx/Rx beamforming toward specific directions. This allows us to generate arbitrary beam patterns, bypassing the codebook constraint of phased arrays. Then, vBeam uses a *beam matching metric* to single out each pair of AoD and AoA directions that belong to the same path.

13.7.2.1 Beam generation

vBeam generates weight vectors of specific beam patterns and applies them to signals from different antenna elements. Whereas the weight vectors can be computed using conventional delay-sum beamformer [35], vBeam applies a beam-nulling technique instead, which beamforms to the desired AoA/AoD directions while nulling signals from other AoA/AoD directions. *This effectively steers the phased-array's side lobes toward directions from which there is no signal coming, and thus helps suppress irrelevant signals.*

Suppose the AoA and AoD identified above are denoted by vectors $\Theta = [\theta_1, \theta_2, \ldots, \theta_K]$ and $\mathbf{\Phi} = [\phi_1, \phi_2, \ldots, \phi_K]$, with K being the number of dominant paths. Denote $\mathbf{a_r}(\theta_i)$ and $\mathbf{a_t}(\phi_i)$ as column weight vectors that beamform toward AoA/AoD angle θ_i and ϕ_i. Take the Rx side as an example, the nulling beam vector $\mathbf{a_r^{null}}(\theta_i)$, which beamforms to θ_j for $j = i$ and nulls other θ_j for $j \neq i$, can be directly derived from $\mathbf{a_r}(\theta_i)$ by [36].

13.7.2.2 Beam matching

Suppose vBeam beamforms toward AoA angle θ_i and AoD angle ϕ_j using the foregoing approach. In order to determine whether θ_i and ϕ_j belong to the same propagation path, we design a *beam matching metric* **F**, which manifests a high value if and only if θ_i and ϕ_j match to the same path. **F** is computed by

$$\mathbf{F}[i,j] = \mathbb{E}\left[|\,\mathbf{a_r^{null}}(\theta_i)\mathbf{S}\mathbf{a_t^{null}}(\phi_j)^H|^2\right], \forall 1 \le i,j \le K$$

where the inner part of the above equation applies the virtual beams to signal matrix **S**, and outer expectation computes the corresponding RSS. Since it is difficult to find an absolute gauge threshold, E-Mi adopts an iterative algorithm (Algorithm 13.2). It starts with the largest metric and takes the corresponding AoA/AoD as a pair. Then it removes values of pairwise AoA/AoD from the row and column of the matching matrix **F** and repeats the above procedure to find the next largest matching metric. This approach works well for paths of different signal strengths.

Algorithm 13.2: *Virtual beamforming*

1: ***procedure*** vBeam(Θ, $\mathbf{\Phi}$, $\mathbf{S}$)
2: ***for*** *i = 1:K, j = 1:K do* ▷ *Loop for Rx and Tx arrays*
3: $F[i,j] = \mathrm{E}\left[|\mathbf{a_r}^{null}(\hat{\theta}_i)\mathbf{S}\mathbf{a_t}^{null}(\hat{\phi}_j)^H|^2\right]$ ▷ *Beamform RSS*
4: ***end for***
5: ***for*** *i = 1:K do* ▷ *Beam matching*
6: *[*I_row, I_col*]* ← max_subscript(**F**) ▷ *Subscript of maximum*
7: $\mathbf{F}[I\_row,:] \leftarrow 0;\ \mathbf{F}[:,I\_col] \leftarrow 0;$
8: $\widehat{\Theta}[i] \leftarrow \Theta[I\_row];\ \widehat{\mathbf{\Phi}}[i] \leftarrow \mathbf{\Phi}[I\_col]$
9: ***end for***
10: ***return*** $\widehat{\Theta}$ *and* $\widehat{\mathbf{\Phi}}$ ▷ *pairwise AoA and AoD*
11: ***end procedure***

Once vBeam identifies all the pairwise AoA/AoD, it can *isolate path i's signal* $\mathbf{S}_{\mathbf{i}}^{\mathrm{path}}$ by projecting the entire signal matrix **S** toward path *i*'s AoA and AoD:

$$\mathbf{S}_{\mathbf{i}}^{\mathrm{path}} = \mathbf{a}_{\mathbf{r}}^{\mathbf{null}}\left(\widehat{\Theta}[i]\right)\mathbf{S}\mathbf{a}_{\mathbf{t}}^{\mathbf{null}}\left(\widehat{\boldsymbol{\Phi}}[i]\right)^{H} \tag{13.7}$$

where $\widehat{\Theta}$ and $\widehat{\boldsymbol{\Phi}}$ are matrices of the pairwise AoA and AoD. E-Mi then further estimates the RSS of signal isolated from each path.

13.7.3 Multitone ranging: estimate path length

E-Mi estimates each path *i*'s length by processing its signals $\mathbf{S}_{\mathbf{i}}^{\mathrm{path}}$, using a *multitone ranging* mechanism. Modern communication systems such as 60-GHz 802.11ad commonly adopt orthogonal frequency division multiplexing (OFDM), which modulates signals across different frequency tones (called *subcarriers*). The phase of each subcarrier can be measured using built-in channel estimators. Suppose a subcarrier has frequency f_1, then its phase increases linearly with propagation path length d, following $2\pi f_1 d/c$. Our multitone ranging leverages the *phase divergence* among OFDM subcarriers, caused by their frequency difference. Given two subcarriers with frequency f_1 and f_2, their phase divergence at distance d equals $\Delta\varphi = 2\pi(f_2 - f_1)d/c$, where c is the speed of light. f_1, f_2 are known and $\Delta\varphi$ can be measured. So we can easily map $\Delta\varphi$ back to d.

To improve resilience to channel noise, E-Mi harnesses diversity from many subcarriers in 802.11ad-like communication systems. Suppose we have isolated the preamble signal $\mathbf{S}_{\mathbf{i}}^{\mathrm{path}}$ along path i (Section 13.7.2). Suppose L subcarriers are located at frequencies $f_1, f_2, \ldots, f_L$, and the Rx-measured phase values are $\varphi_1, \varphi_2, \ldots, \varphi_L$. Then we estimate path i's length via the following optimization framework:

$$d^* = \arg\max_d \left| \sum_{i=1}^{L} e^{j\left(\varphi_i - \frac{2\pi f_i d}{c}\right)} \right| \tag{13.8}$$

The RHS of (13.8) computes the difference between measured phase and theoretical phase over distance in the phaser domain. Essentially, the optimal distance estimation d^* leads to the closest match between these two sets of phase values.

In practice, due to the carrier frequency offset (CFO) between Tx and Rx, the subcarriers will bear unknown phase shifts, which contaminates the phase measurement. We cannot apply the standard CFO compensation technique [37] in this case because it will simultaneously nullify the phase divergence. We address this problem using a *reference calibration* scheme. Specifically, we first separate the Tx and Rx by a known distance d_0 and measure the phase value $\varphi_i(d_0)$ of each subcarrier i. When the Tx/Rx moves to a new (unknown) distance d, the CFO can be canceled by computing their phase difference: $\varphi_i(d) - \varphi_i(d_0) = \dfrac{2\pi f(d-d_0)}{c}$. Substituting φ_i by $\varphi_i - \varphi_i(d_0)$, (13.8) can be reformulated as

$$d^* = \arg\max_d \left| \sum_{i=1}^{L} e^{j\left(\varphi_i - \varphi_i(d_0) - \frac{2\pi f_i (d-d_0)}{c}\right)} \right| \tag{13.9}$$

We note that the phase divergence has an aliasing effect: if the phase difference between two subcarriers exceeds 2π, it will wrap and cause ambiguity. To maximize the unambiguous ranging distance, we should maximize the cycle length of phase divergence, or equivalently minimize the frequency separation between subcarriers (denoted as $f_{\min}$). The unambiguous range is thus determined by $\frac{c}{f_{\min}}$. For 802.11ad, $f_{\min}$ equals the separation between adjacent subcarriers, i.e., 5.156 MHz [34], or equivalent to up to 58.18 m unambiguous ranging distance, which is sufficient for indoor scenarios.

13.8 Dominant reflector reconstruction

The $\langle AoA, AoD, length \rangle$ of all propagation paths form a set of spatial constraints, allowing E-Mi to locate the *reflecting points*, i.e., points where dominant reflectors "bend" the propagation paths. Consequently, E-Mi can geometrically reconstruct reflectors' position, orientation, and reflectivity.

13.8.1 Locating reflecting points in environment

To locate the reflecting points, E-Mi first *pinpoints the* Rx relative to the Tx, based on the $\langle AoA, AoD, length \rangle$ constraints. Figure 13.16 shows an example. Suppose we obtained the $\langle AoA, AoD, length \rangle$ of a single path. Then, any point along *line segment AB* satisfies the same $\langle AoA, AoD, length \rangle$ constraint and is likely to be the Rx position. Therefore, a single path cannot pinpoint the Rx. But we can resolve the

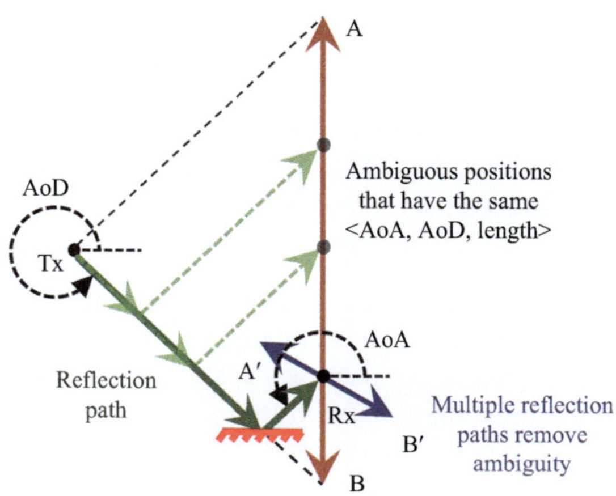

Figure 13.16 Using a single Tx to locate the Rx and reflecting points. Solid green line represents one propagation path. Dotted green lines are alternative paths with the same $\langle AoA, AoD, length \rangle$ constraint

ambiguity by adding another path: the intersection between line segment AB of one path and segment $A'B'$ of another path pinpoints the Rx location.

Practical environment may encounter more than two paths. Denote the path i's $\langle AoA, AoD, length\rangle$ as $\widehat{\theta}_i$, $\widehat{\phi}_i$ and $\widehat{d}_i$. Following Figure 13.16, we can use a simple geometry to represent the intersection formed by all line segments:

$$\mu_i x + \nu_i y = \gamma_i, \tag{13.10}$$

with

$$\begin{aligned}\mu_i &= \sin\left(\widehat{\theta}_i\right) + \sin\left(\widehat{\phi}_i\right), \nu_i = -\left(\cos\left(\widehat{\theta}_i\right) + \cos\left(\widehat{\phi}_i\right)\right)\\ \gamma_i &= x_{tx}\left(\sin\left(\widehat{\theta}_i\right) + \sin\left(\widehat{\phi}_i\right)\right) - y_{tx}\left(\cos\left(\widehat{\theta}_i\right) + \cos\left(\widehat{\phi}_i\right)\right)\\ &\quad + \widehat{d}_i\left(\sin\left(\widehat{\phi}_i\right)\cos\left(\widehat{\theta}_i\right) - \cos\left(\widehat{\phi}_i\right)\sin\left(\widehat{\theta}_i\right)\right)\end{aligned}$$

where (x_{tx}, y_{tx}) and (x_{rx}, y_{rx}) are the Tx and Rx positions, respectively. Assuming the Rx position is intersected by N line segments, (13.10) can be rewritten in a matrix format:

$$\Gamma X = P \tag{13.11}$$

where $X = [x_{rx}, y_{rx}]^T$, $\Gamma = [\mu_1, \ldots, \mu_K; \nu_1, \ldots, \nu_K]^T$ and $P = [\gamma_1, \ldots, \gamma_K]^T$. In practice, due to residual error of measurement, line segments may not intersect on a single point. We thus reformulate (13.11) as a least-square optimization problem:

$$X^* = \arg\min_X \|P - \Gamma X\|^2 \tag{13.12}$$

where X^* estimates the Rx position with minimum error.

Unfortunately, our initial experimental tests found the optimization alone works poorly due two practical factors: *(i)* The $\langle AoA, AoD, length\rangle$ estimation (Section 13.7) contains residual errors, especially for long-range and weak-RSS paths. Such errors may cause intersecting line segments to be close to parallel, which significantly amplifies the Rx location error. *(ii)* The MRF (Section 13.7) may capture high-order reflections that do not follow the model in Figure 13.16. Such mismatch may deviate the estimation arbitrarily away from the real position.

E-Mi introduces two mechanisms to overcome above challenges.

(1) **Weighting the residual error.** We first reformulate the optimization problem in (13.12) to account for MRF's residual errors. The inner term of (13.12) calculates the difference between matrices ΓX and P. Since different paths' geometries have different residual errors, we weight the paths according to the *confidence level* in MRF's estimation, based on the following observation: those paths of shorter length (thus higher RSS) tend to have less error. Thus, we can use the inverse of path length as the weight. Besides, we need to minimize the sum of distance from Rx position to all line segments, which requires normalization by a coefficient $\sqrt{\mu_i^2 + \nu_i^2}$ [38]. The final weight value for path i is

$$w_i = \frac{1}{d_i\sqrt{\mu_i^2 + v_i^2}}$$

which consists of the inverse of path length and the normalization factor. Stacking the weights into a vector $W = diag(w_1, \ldots, w_K)$, the optimization problem (13.12) can be rewritten as $X^* = argmin_X \| W(P - \Gamma X)\|^2$, which can be solved by standard least-square algorithms.

(2) **Filtering higher-order reflection.** To constrain the problem within the geometrical model of first-order reflection (Figure 13.16), we should exclude any higher-order reflection paths in the optimization. Our key observation is that line segment intersections of higher-order reflections tend to randomly distribute and exhibit a larger deviation since they do not fit into the geometry model for the first-order reflection. Therefore, we apply a K-means clustering algorithm to filter out the $p\%$ most significantly deviated line segments that most likely belong to the higher-order reflection. The choice of p value depends on the amount of higher-order reflections. We prefer a larger p value in a highly reflective environment, and otherwise a smaller p. Yet, we find E-Mi is not sensitive to it because most indoor environments have comparable number of dominant reflectors, and an empirical value (e.g., $p = 20$) would suffice.

After determining the client position, the *locations of reflecting points* $\left(x_{ref}^i, y_{ref}^i\right)$ can be estimated by

$$\frac{x_{ref}^i - x_r}{y_{ref}^i - y_r} = \tan\left(\hat{\theta}_i\right), \frac{x_{ref}^i - x_t}{y_{ref}^i - y_t} = \tan\left(\hat{\varphi}_i\right) \tag{13.13}$$

which solves a set of equations following simple geometry in Figure 13.16.

13.8.2 Reconstructing dominant reflector layout and reflectivity

E-Mi reconstructs the dominant reflector geometry (orientation/location/length) and reflectivity, by sampling the wireless channel across a *sparse* set of Rx locations, and locating the corresponding reflecting points following the above steps. It creates a 2D cross section of the environment corresponding to the horizontal plane of Tx/Rx.

13.8.2.1 Reconstructing dominant reflectors' geometry

Ideally, we can move the Rx to many positions, each helping to locate multiple reflecting points. A sufficient number of reflecting points can form a *pixel cloud* that outlines the reflector geometry. However, due to sparsity of the propagation paths [14–16], collecting a dense pixel cloud requires hundreds of Rx positions even for a small office.

To avoid such war-driving, we design a sparse reconstruction method which only samples at a few positions. We abstract the reflectors into two categories: (1) *Specular reflectors:* a large continuous surface, e.g., wall and cabinet, upon which mmWave signals experience specular reflection [39]; (2) *Diffusive reflectors:* small-size objects, e.g., pillar and computer screen, which scatter signals toward a wide range of angles.

(i) *Locating specular reflectors.* We model a specular reflector as a continuum of planar segments, each with different reflectivity or orientation. According to the law of reflection, signals that are reflected by a specular reflector look like emitting from a VS (Figure 13.17). Given the $\langle AoA, AoD, length\rangle$ and reflecting point location of one propagation path, we can use simple geometry to pinpoint the VS position relative to the real TX position, following the law of reflection.

Ideally, each specular segment should have a single VS even as the Rx moves and creates different reflection paths. Yet due to the MRF's residual errors, the VSs estimated at different Rx positions do not exactly overlap, but fortunately they tend to form a cluster (Figure 13.17). We apply the clustering algorithm [40] to isolate the clusters and use each cluster center to represent one VS. Once a reflector segment's VS is identified, its position and orientation is readily determined via a mirror-partitioning between the real source and VS. Further, we extend the reflector segment, and take the size of geometrical shape from pixel cloud as constraint to determine the length of each reflector segment. An example experiment will be provided in Section 13.9.2 to elucidate the procedure.

(ii) *Locating diffusive reflectors.* Observing that a diffusive reflector corresponds to densely concentrated reflecting points (Figure 13.17), we can also apply the clustering algorithm on the reflecting points and use the centers of resulting clusters to represent the diffusive reflectors. Yet this will be interfered by dispersive reflecting points created by specular reflectors. Fortunately, based on the previous step, we can identify and exclude such interfering points. Specifically, we identify them based on their Euclidean distance to the specular reflectors. Threshold is set to 3× the variance of reflector position error, which can isolate a majority of specular reflecting points.

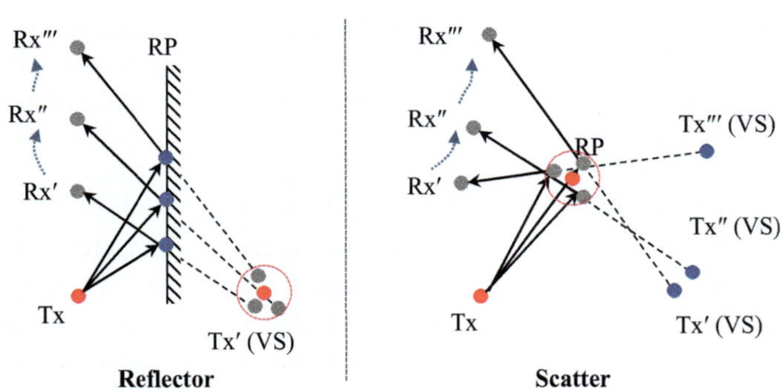

Figure 13.17 Reflecting point (VP) and corresponding virtual source (VS), used to recover the geometry and location of specular/diffusive reflectors

13.8.2.2 Estimating reflection loss

We now describe how E-Mi models reflection loss, the major distorting factor when signals hit the reflector. Other factors such as diffraction may also vary the signal strength but the effect is minor [41]. E-Mi separately models the reflection loss of each path it has identified. Three factors contribute to the propagation loss: free-space pathloss, oxygen absorption O_l, and reflection loss R_l, i.e.,

$$RSS = P_t + G_t + G_r - (20\log_{10}(d) + O_l + R_l)$$

where d is the path length. P_t and G_t, G_r represent the Tx power and Tx/Rx antenna gain, respectively. The O_l almost remains a constant for distance of tens of meters [32]. For each path, d and RSS are known from MRF (Section 13.7). Thus, to obtain R_l, we need to obtain the constant parameters P_t, G_t, G_r, which may not be available in practice. In addition, the constant value O_l is unknown either.

We address this issue by using the LoS path as reference calibration to cancel those unknown factors. First, we can isolate the LoS path's signals from NLoS path's signals by metrics such as shortest path length and strongest RSS. We then estimate the reflection loss of each NLoS reflecting path via a simple subtraction:

$$R_l = RSS_{\mathrm{LOS}} - RSS_{\mathrm{ref}} - 20\log_{10}(d_{\mathrm{ref}}/d_{\mathrm{LOS}}) \tag{13.14}$$

where RSS_{LOS}, RSS_{ref} and d_{LOS}, d_{ref} are RSS and path length for LoS and reflected path. Since each reflecting segment/point may have multiple estimations corresponding to multiple reflection paths, we take the average as its final reflection loss.

13.9 Implementation and evaluation of E-Mi

13.9.1 Implementation

We prototype E-Mi on a custom-built 60-GHz software-radio testbed, which uses WARP [25] to generate and process baseband waveforms under the control of a host PC. The digital signals are converted to/from analog through a high-speed DAC/analog-to-digital converter (ADC), and carrier-modulated by the Pasternack PEM-003 60-GHz RF front end [42].

The physical-layer metrics (e.g., channel state information) in commercial 60-GHz radios, though internally available to the radio vendors, are still not opened to the public yet. The recently developed phased-array antennas in 60-GHz software radios [43] can only support a small number of elements. Therefore, we reproduce the effects of a 60-GHz circular phased array using the time-lapse synthesis method, which follows the way that a phased array modulates, weights and transmits wireless signals. We mount a 60-GHz omnidirectional antenna MI-WAVE 267V (with 360° azimuth and 40° elevation beamwidth) [44] onto Axis360 [27], a programmable motion control system (Figure 13.18). The Axis360 rotates the antenna to discrete positions along a circle, each position corresponding to one antenna element on a real circular array. The array dimension follows empirical

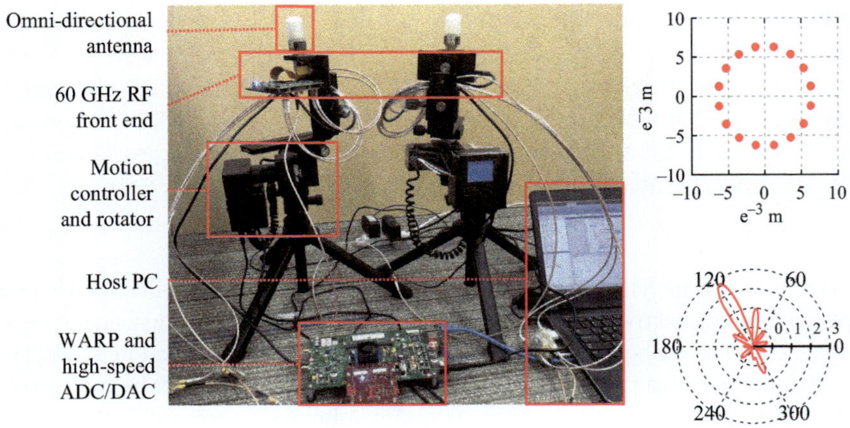

Figure 13.18 Left: Custom-built 60-GHz software-radio platform. Top-right: Antenna placement of a 16-element phased array. Bottom-right: Example beam pattern

recommendations in antenna design [36]. The radius of a 16-element circular array is 6.4 mm, with 2.5 mm (half-wavelength) between adjacent elements. This time-lapse synthesis approach has been adopted and verified by previous works [45,46].

To synthesize a pair of Tx/Rx phased arrays, we apply beamforming weights (based on a standard 802.11ad codebook [17]) and then combine the measurement from all elements within the Rx circular array. Since E-Mi runs in static environment, this time-lapse approach can realistically reproduce a real phased array where all elements are excited concurrently. Besides, each element of a phased-array antenna is expected to have a close to omnidirectional coverage in horizon plane [47]. Figure 13.18 depicts an example Tx phased-array gain pattern generated by this time-lapse approach, and measured using a 3.4° horn Rx. Even though the measurement speed of our platform is slow currently owing to the mechanical antenna rotation, a full-fledged 60-GHz device, that has the electronic phased-array antenna, can transmit a wireless packet at tens of microseconds. The overall sensing time of each location will be at millisecond level.

We implement E-Mi's major modules (Section 13.6) within the software-radio's host PC. Due to limited bandwidth, our platform cannot send 802.11ad-compatible preambles for channel estimation. Instead, the Tx sends five orthogonal tones from 3 to 15 MHz as baseband signals, modulated by 60.48 GHz carrier frequency. This does not obstruct our validation because the narrow band implementation can be considered as only utilizing a few subcarriers in the 2 GHz wide band. When an 802.11ad-compatible device is available, E-Mi can be easily extended to conduct MRF and dominant reflector reconstruction across orthogonal subcarriers over a wideband.

For evaluation purpose, we also use the beam-scanning method to acquire the ground-truth $\langle AoA, AoD, length \rangle$, similar to Rappaport *et al.* [14]. We use a pair of

Tx/Rx radios equipped with directional horn antennas of 3.4° beamwidth [48]. With Axis360, the Tx antenna sweeps the horizontal plane at a step of 3°. Meanwhile, the Rx measures the wireless channel and steers to next step after the Tx completes a full scanning.

13.9.2 Effectiveness of dominant reflector reconstruction

We set up a pair of Tx and Rx, each synthesizing a 16-element phased array. We conduct experiments in a 90 m^2 office environment, which represents a typical indoor environment. The dominating reflectors involve two drywalls, two concrete walls and one pillar. We fix the Tx and randomly move Rx over multiple locations. The result is compared against the ground-truth AoA, AoD (measured using the oracle beam scanning, Section 13.9) and path length (measured using a laser ranger). The ground-truth measurement reveals each link has three to five dominating propagation paths. The estimated AoA, AoD and path length are supplied to the algorithm in Section 13.8.1 to locate the reflecting points for each given Rx position. We move the Rx to 15 uniform positions to reconstruct the dominating reflectors.

13.9.2.1 Accuracy in localizing reflecting points

Since the reflecting point location has a linear, deterministic relation with the Rx position (Section 13.8.1), we mainly focus on evaluating the latter, whose ground truth is obtained via a laser range finder BOSCH DLE40.

Figure 13.19(a) plots the cumulative distribution function (CDF) of localization error. E-Mi can locate the Rx position with mean/90th error of 0.38 and 0.6 m, which is even smaller than the path length estimation, because we apply the minimum least square method that leverages the redundant information of multiple reflected paths to reduce the estimation error. Indeed, when we intentionally eliminate the LoS paths, the performance ("NLoS only") drops due to lower path diversity. The results verify that E-Mi's *reflecting point localization algorithm indeed achieves high precision based on the MRF. More paths provide more diversity and hence higher accuracy.*

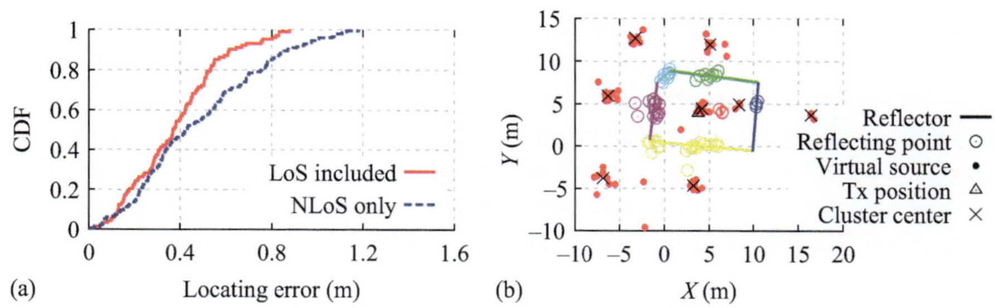

Figure 13.19 Environment reconstruction accuracy: (a) error CDF in the reflecting point localization algorithm; (b) breakdown of reconstruction output for an office environment

Table 13.1 Estimated reflection loss on different reflectors

Reflector color	Reflection loss	Reflector color	Reflection loss
—— drywall	15.87 dB	—— concrete wall	8.60 dB
—— drywall	13.83 dB	—— drywall	20.56 dB
—— concrete wall	5.50 dB	—— pillar/monitor	1.24 dB

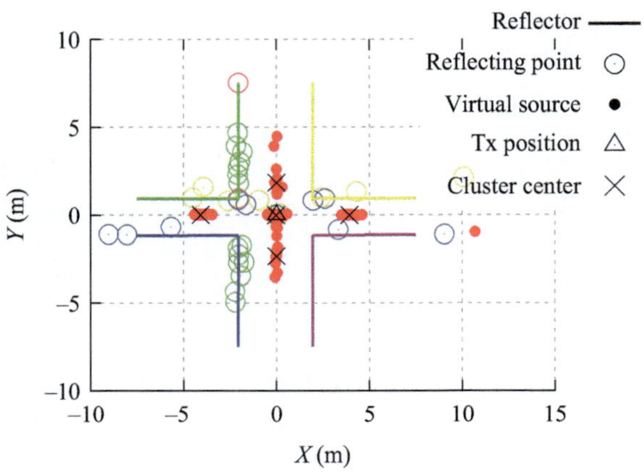

Figure 13.20 Breakdown of reconstruction output for the corridor environment

13.9.2.2 Performance of dominant reflector reconstruction

Recall that, given an estimation of the Rx's and reflecting points' positions, the dominant reflector reconstruction locates the virtual source, specular reflector, and diffusive reflector, respectively. Figure 13.19(b) puts together the output from each step and shows the final reconstruction output based on 15 Rx sampling positions. We observe that the output closely matches the ground truth: four walls (specular reflectors) and one pillar/monitor (diffusive reflector). In effect, even the geometrical size of the reflectors matches the ground truth well with less than 0.3 m error. Table 13.1 lists the estimated reflection loss. The two concrete walls have 10 + dB lower loss than the two drywalls. And the metal pillar/monitor shows even lower loss. Intuitively, this matches with the reflectivity of materials at 60 GHz [49]. We have also evaluated E-Mi in the cross of a corridor (Figure 13.20) and observed a similar level of accuracy in positioning the dominant reflectors.

In summary, *E-Mi can effectively identify the geometry and locations, and distinguish the reflectivity of major reflectors from the 60-GHz radio's eyes.*

13.9.2.3 Accuracy of link performance prediction

We now feed the reconstructed reflector geometry/reflectivity into the parametric ray-tracing engine, and predict the spatial channel for another set of 15 randomly located Tx/Rx pairs. Following the ground-truth measurement (Section 13.9),

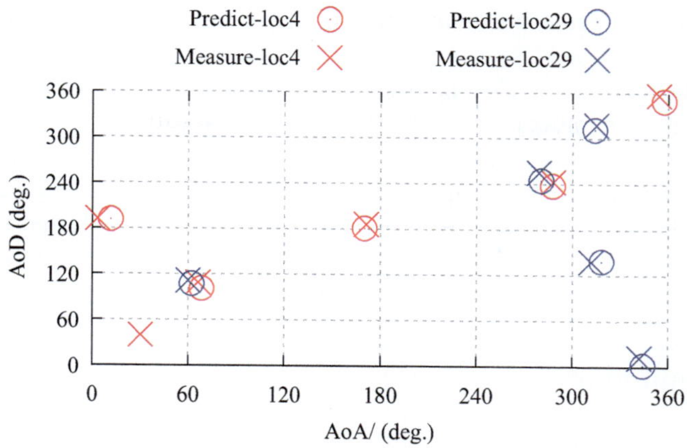

Figure 13.21 Predicted versus measured AoA/AoD patterns for two locations

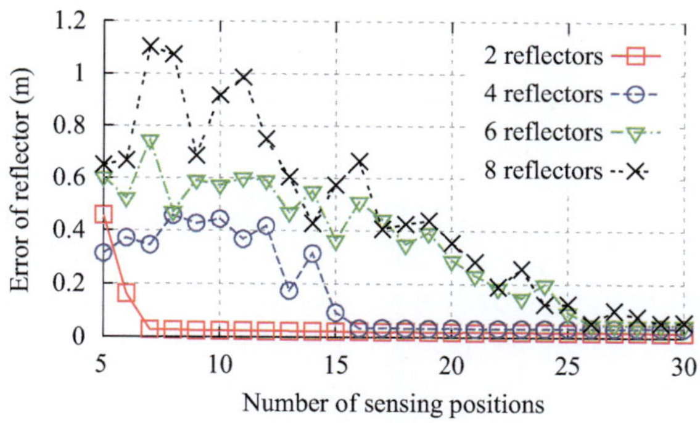

Figure 13.22 Reflector location error versus number of Rx sampling positions

we found these links have 66 paths in total. Figure 13.21 showcases example results from two randomly selected links. We observe that the predicted AoA/AoD patterns are highly consistent with the ground truth. The missing spot (e.g., AoA 30°/AoD 40°) is caused by reflection that is not captured by the Tx/Rx during MRF. We found such spots correspond to signal paths with negligible RSS and hence little impact on network performance. Also, adding more Rx position samples can incrementally reduce the probability of prediction loss.

13.9.2.4 Scalability in complicated environment

We investigate how many sampling locations are needed to reconstruct the dominant reflectors. We simulate an 8 m × 10 m room environment with four concrete walls. The Tx is placed at the center, and Rx randomly moves over 5–30 locations in each test. Across tests, different number of planar reflectors are placed randomly around the Tx/Rx locations. We use the ray tracing to obtain the ground-truth $\langle AoA, AoD, length\rangle$. Figure 13.22 depicts the average error between E-Mi's reconstructed reflector positions and ground truth. Generally, *more dominant reflectors requires more sampling positions*. Yet, even for an 8-reflector environment, E-Mi only needs 20 sampling locations to ensure an accuracy of around 0.2 m. Since the number of dominating reflectors tends to be sparse, the amount of on-site sampling needed in E-Mi is still substantially lower than that of a site-survey. Besides, the reflector accuracy depends on the radio's sensing position, and it tends to be more accurate if the radio can see stronger reflected paths from a reflector. Fortunately, this location sensitivity will be averaged out when the number of sensing positions is sufficiently large.

13.10 Summary

This chapter presents two example system designs that enable passive sensing at super-resolution using commodity 60-GHz mmWave radios. The first system, mTrack, is a high-precision passive-object tracking system that uses 60-GHz signals as sensing medium. mTrack takes advantage of the short wavelength and steerable directional beams of 802.11ad-like 60-GHz radio devices, and uses subtle RSS/phase variation to track passive writing objects on a trackpad-like area. It can track a pen at sub-centimeter level accuracy, which even outperforms existing systems that use radio-instrumented objects. Considering the growing popularity of 60-GHz devices, we believe mTrack can potentially open up a wide range of mobile sensing applications. The second system, E-Mi, is a sensing-assisted paradigm to facilitate 60-GHz networks, whose performance is highly sensitive to reflectors. E-Mi senses the environment from 60-GHz radios' eyes. It "reverse-engineers" the geometry/reflectivity of dominant reflectors, by tracing back the LoS/NLoS paths between a pair of 60-GHz nodes. Such environment information can be to augment a broad range of network planning and protocol reconfigurations.

References

[1] Wachs JP, Kölsch M, Stern H, *et al.* Vision-based Hand-gesture Applications. Communications of the ACM. 2011;54(2):60–71.

[2] DannBerg, Leap Motion Controller is a great toy, but not yet the future of interaction. https://www.theverge.com/2013/7/22/4544706/leap-motion-controller-toy-not-future-of-interaction.

[3] RoboRealm: Microsoft Kinect.; 2013. http://www.roborealm.com/help/Microsoft_Kinect.php.

[4] Pu Q, Gupta S, Gollakota S, *et al.* Whole-home Gesture Recognition using Wireless Signals. In: ACM MobiCom; 2013.

[5] Melgarejo P, Zhang X, Ramanathan P, *et al.* Leveraging Directional Antenna Capabilities for Fine-grained Gesture Recognition. In: ACM UbiComp; 2014.

[6] Vubiq Networks, FCC Adopts New Rules for Unlicensed V-Band, Extending the Coverage to 57–71 GHz. 2016. https://www.vubiqnetworks.com/news-and-events/press-releases/fcc-adopts-new-rules-for-unlicensed-v-band-extending-the-coverage-to-57-71-ghz/

[7] IEEE Standards Association. IEEE 802.11ad PHY Specifications: Enhancements for Very High Throughput in the 60 GHz Band. IEEE Standards; 2012.

[8] Dailywireless. 60 GHz Backhaul for Small Cells; 2013. Available from: http://www.dailywireless.org/2014/01/03/60ghz-backhaul-for-small-cells/.

[9] PC Magazine. Wilocity Unveils Blazing Fast 802.11ad Smartphone Wi-Fi Chip; 2014. Available from: http://www.pcmag.com/article2/0,2817,2454187,00.asp.

[10] Yang L, Chen Y, Li XY, *et al.* Tagoram: Real-time Tracking of Mobile RFID Tags to High Precision using COTS Devices. In: ACM MobiCom; 2014.

[11] Maltsev A, Maslennikov R, Sevastyanov A, Lomayev A, and Khoryaev A. Statistical Channel Model for 60 GHz WLAN Systems in Conference Room Environment. In: Proc. of European Conference on Antennas and Propagation; 2010.

[12] Xiong J, and Jamieson K. ArrayTrack: A Fine-Grained Indoor Location System. In: Presented as part of the 10th USENIX Symposium on Networked Systems Design and Implementation (NSDI 13); 2013.

[13] Joshi K, Hong S, and Katti S. PinPoint: Localizing Interfering Radios. In: Proc. of USENIX NSDI; 2013.

[14] Xu H, Kukshya V, and Rappaport TS. Spatial and Temporal Characteristics of 60-GHz Indoor Channels. IEEE Journal on Selected Areas in Communications. 2002;20(3):620–630.

[15] Smulders P. Exploiting the 60 GHz Band for Local Wireless Multimedia Access: Prospects and Future Directions. IEEE Communications Magazine. 2002;40(1):140–147.

[16] Sur S, Venkateswaran V, Zhang X, *et al.* 60 GHz Indoor Networking through Flexible Beams: A Link-Level Profiling. In: Proc. of ACM SIGMETRICS; 2015.

[17] Wang J, Lan Z, Sum CS, *et al.* Beamforming Codebook Design and Performance Evaluation for 60 GHz Wideband WPANs. In: IEEE Trans. on Vehicular Technology; 2009.

[18] Xu C, Firner B, Moore RS, *et al.* SCPL: Indoor Device-free Multi-subject Counting and Localization Using Radio Signal Strength. In: Proc. of ACM/IEEE IPSN; 2013.

[19] Adib F, Kabelac Z, Katabi D, *et al.* 3D Tracking via Body Radio Reflections. In: Proc. of USENIX NSDI; 2014.

[20] Nanzer J. Microwave and Millimeter-Wave Remote Sensing for Security Applications. Artech House; Norwood, MA, USA, 2012.

[21] Wang J, Vasisht D, and Katabi D. RF-IDraw: Virtual Touch Screen in the Air using RF Signals. In: ACM SIGCOMM; 2014.

[22] Nikitin PV, Martinez R, Ramamurthy S, *et al.* Phase Based Spatial Identification of UHF RFID Tags. In: RFID, 2010 IEEE International Conference on; 2010.

[23] Lindström J, Kokko H, and Ranta E. Detecting Periodicity in Short and Noisy Time Series Data. Oikos. 1997;78(2):406–410.

[24] Valdes-Garcia A, Reynolds S, Natarajan A, *et al.* Single-element and Phased-array Transceiver Chipsets for 60-GHz Gb/s Communications. IEEE Communications Magazine; 2011;49(4):120–131.

[25] Rice University. Wireless Open-Access Research Platform; 2013. http://warp.rice.edu/trac/wiki.

[26] Vubiq 60GHz System; 2014. http://vubiq.com/v60wgd03/.

[27] Axis360 Motion Control System. http://cinetics.com/two-axis360/.

[28] MyScript technology; 2015. http://myscript.com/technology/.

[29] MacKenzie IS, and Soukoreff RW. Phrase Sets for Evaluating Text Entry Techniques. In: ACM CHI Extended Abstracts; 2003.

[30] Wilocity 802.11ad Multi-Gigabit Wireless Chipset; 2013. http://wilocity.com.

[31] Halperin D, Hu W, Sheth A, *et al.* Tool Release: Gathering 802.11n Traces with Channel State Information. ACM SIGCOMM CCR. 2011.

[32] Rappaport TS, Jr RWH, Daniels RC, *et al.* Millimeter Wave Wireless Communications. Prentice Hall; Upper Saddle River, NJ, USA, 2014.

[33] Hayes MH. Statistical Digital Signal Processing and Modeling. John Wiley & Sons; New York, USA, 2009.

[34] IEEE Standards Association. IEEE Standards 802.11ad-2012: Enhancements for Very High Throughput in the 60 GHz Band. IEEE Standards; 2012.

[35] Van Veen BD, and Buckley KM. Beamforming: A Versatile Approach to Spatial Filtering. IEEE ASSP Magazine. 1988;5(2):4–24.

[36] Stevanovic I, Skrivervik A, and Mosig JR. Smart Antenna Systems for Mobile Communications. Ecole Polytechnique Fédérale de Lausanne. 2003.

[37] Tan K, Fang J, Zhang Y, *et al.* Fine-Grained Channel Access in Wireless LAN. ACM SIGCOMM. 2011.

[38] Ballantine JP, and Jerbert AR. Distance from a Line, or Plane, to a Point. The American Mathematical Monthly; 1952;59(4):242–243. doi:10.2307/2306514.

[39] Zhu Y, Zhu Y, Zhao BY, *et al.* Reusing 60GHz Radios for Mobile Radar Imaging. In: Proc. of ACM MobiCom; 2015.

[40] Arthur D, and Vassilvitskii S. K-means++: The Advantages of Careful Seeding. In: ACM-SIAM Symposium on Discrete Algorithms; 2007.

[41] Maltsev A, Maslennikov R, Sevastyanov A, *et al.* Statistical Channel Model for 60 GHz WLAN Systems in Conference Room Environment. In: Antennas and Propagation (EuCAP), 2010 Proceedings of the Fourth European Conference on. IEEE; 2010.

[42] Pasternack Inc. 60 GHz Transmit/Receive (Tx/Rx) Development System; 2015. http://www.pasternack.com/60-ghz-test-development-system-pem003-kit-p.aspx.

[43] Zhang J, Zhang X, Kulkarni P, *et al.* OpenMili: A 60 GHz Software Radio Platform with a Reconfigurable Phased-array Antenna. In: Proc. of ACM MobiCom; 2016.

[44] MI-WAVE Inc. Omni-directional Antenna Series 267; 2014. http://www.miwv.com/millimeter-wave-products/antenna-products/.

[45] Adib F, Kumar S, Aryan O, *et al.* Interference Alignment by Motion. In: ACM MobiCom; 2013.
[46] Kumar S, Gil S, Katabi D, *et al.* Accurate Indoor Localization with Zero Start-up Cost. In: ACM MobiCom; 2014.
[47] Gulbrandsen F. Design and Analysis of an X-band Phased Array Patch Antenna. Norwegian University of Science and Technology; 2013.
[48] Pasternack Inc. WR-15 Waveguide Horn Antenna Operating from 56 GHz to 66 GHz; 2015. http://www.pasternack.com/horn-antenna-50-75-ghz-nominal-34-dbi-gain-wr-15-pe9881-34-p.aspx.
[49] Langen B, Lober G, and Herzig W. Reflection and Transmission Behaviour of Building Materials at 60 GHz. In: Proc. of IEEE PIMRC; 1994.

Chapter 14
Conclusion

Changzhan Gu[1] and Jaime Lien[2]

Radio frequency (RF) sensing based on radar has been a well-established and mature field of engineering and applied sciences since the first radar systems were developed as early as the 1930s [1]. The mainstream radar research and development have been focused on "long range" applications such as the detection and tracking of large moving subjects at long distances in the order of hundreds of meters or even kilometers, including but not limited to air-defense systems [2], atmospheric radars [3], marine radars [4], geophysical monitoring systems [5], and so forth. The "short range" radar often refers to short distances of a few meters or even less than a meter. It was not until 1970s that the short-range radar technology was first introduced for healthcare applications of detecting the small movement of human vital signs [6]. The short-range radar is relatively new but promising across a very broad range of emerging applications such as wireless healthcare [7–9], search and rescue [10–12], medical [13–15], gesture sensing [16], security and surveillance [17–20], and Internet of things (IoT) [21–24], among many others. Due to the short distance nature and modern consumer applications, the well-developed hardware systems and computational methods for long-range radars may not be directly applicable for short-range applications, in which radar sensors must be small enough to fit into compact consumer hardware, run on limited battery power and computational resources, and track dynamic objects such as human hands or chest wall at sub-millimeter accuracy. Using the radar technology for consumer applications is new and would not be possible without the recent technical advancements in RF/microwave, semiconductor, signal processing, machine learning, and computing. Due to the small form factor, low-power consumption, and low cost, we can envision the ubiquitous applications of the short-range radar technology in the coming IoT era, including but not limited to mobile and wearables, smart home hardware, augmented reality/virtual reality (AR/VR), gaming console, healthcare, and many other exciting applications and use cases.

[1]Department of Electronic Engineering, Shanghai Jiao Tong University, China
[2]Google LLC, USA

References

[1] R. Watson-Watt, "Radar in war and in peace," Nature, vol. 156, pp. 319–324, 1945.

[2] O. Adrian, "Future surface radar technology: from air defence to air and missile defence," Proc. IEEE Radar Conf., pp. 49–54, April 2007.

[3] B. Isometal, "The Atmospheric Imaging Radar (AIR) for high-resolution observations of severe weather," Proc. IEEE Radar Conf., pp. 627–632, May 2011.

[4] P.D.L. Williams, "Civil marine radar—a review and a way ahead," The Journal of Navigation, vol. 51, pp. 394–403, 1998.

[5] N.J.S. Stacy, M.P. Burgess, M.R. Muller, and R. Smith, "Ingara: an integrated airborne imaging radar system," Geoscience and Remote Sensing Symposium 1996. IGARSS'96. Remote Sensing for a Sustainable Future. International, vol. 3, pp. 1618–1620, 1996.

[6] J.C. Lin, "Noninvasive microwave measurement of respiration," Proc. IEEE, vol. 63, no. 10, pp. 1530–1530, 1975.

[7] M. Mercuri, P. J. Soh, G. Pandey, *et al.*, "Analysis of an indoor biomedical radar-based system for health monitoring," IEEE Trans. Microw. Theory Techn., vol. 61, no. 5, pp. 2061–2068, 2013.

[8] F. Lin, Y. Zhuang, C. Song, *et al.*, "SleepSense: a noncontact and cost-effective sleep monitoring system," IEEE Transactions on Biomedical Circuits and Systems, vol. 11, no. 9, pp. 189–202, 2017.

[9] M.C. Huang, J.J. Liu, W. Xu, C. Gu, C. Li, and M. Sarrafzadeh, "A self-calibrating radar sensor system design for measuring vital signs," IEEE Transactions on Biomedical Circuits and Systems, vol. 10, no. 2, pp. 352–363, 2016.

[10] K.-M. Chen, D. Misra, H. Wang, H.-R. Chuang, and E. Postow, "An X-band microwave life-detection system," IEEE Trans. Biomed. Eng., vol. BME-33, no. 7, pp. 697–701, 1986.

[11] K.-M. Chen, Y. Huang, J. Zhang, and A. Norman, "Microwave life-detection systems for searching human subjects under earthquake rubble or behind barrier," IEEE Trans. Biomed. Eng., vol. 27, no. 1, pp. 105–114, 2000.

[12] V.M. Lubecke, O. Boric-Lubecke, A. Host-Madsen, and A.E. Fathy, "Through-the-wall radar life detection and monitoring," in Proc. IEEE/MTT-S Int. Microwave Symp., June 2007, pp. 769–772.

[13] C. Gu, R. Li, C. Li, and S.B. Jiang, "A multi-radar wireless system for respiratory gating and accurate tumor tracking in lung cancer radiotherapy," in Proc. 33rd Annu. Int. Conf. IEEE Engineering Medicine Biology Society, Boston, MA, August 2011, pp. 417–420.

[14] C. Gu, R. Li, H. Zhang, *et al.*, "Accurate respiration measurement using dc-coupled continuous wave radar sensor for motion-adaptive cancer radiotherapy," IEEE Trans. Biomed. Eng., vol. 59, no. 11, pp. 3117–3123, 2012.

[15] P. Bernardi, R. Cicchetti, S. Pisa, E. Pittella, E. Piuzzi, and O. Testa, "Design, realization, and test of a UWB radar sensor for breath activity monitoring," IEEE Sensors J., vol. 14, no. 2, pp. 584–596, 2014.

[16] Z. Peng, C. Li, J.-M. Muñoz-Ferreras, and R. Gómez-García, "An FMCW radar sensor for human gesture recognition in the presence of multiple targets," Microwave Bio Conference (IMBIOC) 2017 First IEEE MTT-S International, pp. 1–3, 2017.

[17] E. Yavari, H. Jou, V. Lubecke, and O. Boric-Lubecke, "Doppler radar sensor for occupancy monitoring," Proc. 2013 IEEE Radio Wireless Symp., pp. 316–318.

[18] P. Nuti, E. Yavari, and O. Boric-Lubecke, "Doppler radar occupancy sensor for small-range motion detection," Microwave Conference (APMC) 2017 IEEE Asia Pacific, pp. 192–195, 2017.

[19] P. Weber, A. Premji, T. Nohara, and C. Krasnor, "Low-cost radar surveillance of inland waterways for homeland security applications," in Proc. IEEE Radar Conf. Philadelphia, PA, USA, 26–29 April, 2004.

[20] E. Yavari, C. Song, V. Lubecke, and O. Boric-Lubecke, "System-on-chip based Doppler radar occupancy sensor," in Proc. 2011 Annu. Int. Conf. IEEE Engineering Medicine Biology Society, pp. 1913–1916.

[21] G. Wang, C. Gu, T. Inoue, and C. Li, "A hybrid FMCW-interferometry radar for indoor precise positioning and versatile life activity monitoring," IEEE Transactions on Microwave Theory and Techniques, vol. 62, no. 11, pp. 2812–2822, 2014.

[22] Z. Peng, J. Muñoz-Ferreras, Y. Tang, *et al.*, "A portable FMCW-interferometry radar with programmable low-IF architecture for localization, ISAR imaging and vital-sign tracking," IEEE Transactions on Microwave Theory and Techniques, vol. 65, no. 4, pp. 1334–1344, 2017.

[23] Z. Peng, L. Ran, and C. Li, "A K-band portable FMCW radar with beamforming array for short-range localization and vital-Doppler targets discrimination," IEEE Transactions on Microwave Theory and Techniques, vol. 65, no. 9, pp. 3443–3452, 2017.

[24] J.G.D. Hester, J. Kimionis, R. Bahr, W. Su, B. Tehrani, and M.M. Tentzeris, "Radar & additive manufacturing technologies: the future of Internet of things (IoT)," 2018 IEEE Radar Conference (RadarConf18).

Index